Computational Models in Surveying and Photogrammetry

B.D.F. METHLEY

Lecturer in Land Surveying and Photogrammetry
University of Glasgow

Blackie

Glasgow and London

Blackie & Son Limited,
Bishopbriggs, Glasgow G64 2NZ
7 Leicester Place, London WC2H 7BP

First published 1986

British Library Cataloguing in Publication Data

Methley, B.D.F.
Computational models in surveying and
photogrammetry.
1. Surveying—Mathematics
I. Title
526.9'01'51 TA549

ISBN 0-216-91996-7

Photosetting by Thomson Press (India) Limited, New Delhi.
Printed in Great Britain by Bell & Bain (Glasgow) Ltd.

Preface

This book introduces the subject of computation in surveying and photogrammetry. It develops the mathematical models and algorithms which form the basis of the practical applications of the subject, and it will therefore meet the needs of both the college or university student considering the topics for the first time, and the practitioner requiring an aid to computer software development.

It is assumed that the reader has a basic understanding of surveying and photogrammetry, and mathematics to first year university level. The latter part of the book assumes also that the reader is familiar with differential calculus and matrix algebra. The book will serve as an introductory text which deals solely with the more common computational models, and which includes both surveying and photogrammetry; a text which describes traditional methods, and bridges the gap between these and the more rigorous methods available today; and as a text which takes account of the changes in observational techniques and the variation in computing facilities that are available.

The book is divided broadly into three parts. Chapters 2-4 are concerned with surveying, chapters 5-7 with photogrammetry, and chapters 8-10 with both these topics. The first two parts deal with the subject matter at an elementary level, and include the development of equations to be used with a minimum number of observations. These chapters also include models for the reduction or refinement of observations, and later sections are devoted to 'approximate' methods when redundant observations are included in the data set. The third part is concerned with the more rigorous treatment of observations by least squares, and includes a chapter outlining some elementary statistical concepts. Many of the computational models developed in the book are common to applications of both surveying and photogrammetry, and the book aims to promote awareness of this fact. However, it is possible for a reader with a limited interest in any one of the two topics to follow that topic in a logical sequence.

Appendices are restricted to the problem of linearization (a recurring requirement), and to useful mathematical formulae. Matrix algebra and the solution of simultaneous equations are excluded, since suitable texts covering these topics are readily available. The book includes many computational examples, starting with simple problems which have solutions set out in considerable detail, and progressing to more complex problems where the results are given at intermediate stages. Most of the data are original data collected by staff and students; however, the data used in Examples 7.1-7.3 and 10.9-10.11 were obtained from the Ordnance Survey, and are included with the permission of the Controller of Her Majesty's Stationery Office.

I wish to acknowledge the co-operation and encouragement of colleagues at the University of Glasgow, and I am grateful to the many students who, directly and indirectly, have persuaded me to write this book. Finally, I wish to thank my family for their patience and understanding.

B.D.F.M.

Contents

1 Introduction

Surveying is an activity which has been closely linked to position-fixing and mapping for many centuries, while photogrammetry has become important in the same fields during this century. In recent times instruments and methods have improved enormously as technology has become more sophisticated, with notable advances having taken place in the last few decades. The introduction of aerial photogrammetry, electronic distance measurement, remote sensing, satellite doppler position-fixing, inertial surveying, and sophisticated off-shore surveying systems, represent significant milestones. When these are allied with developments in the processing of digital imagery and automated cartography it becomes clear that the spectrum is very wide.

Against such a background the more traditional role of the surveyor and photogrammetrist seems fairly narrow. However, it is surprising how often principles and methods developed over centuries are simply modified to meet the requirements of new applications. This is particularly evident in the computational processes, where the change in emphasis has been from analogue to digital techniques. Mathematical models which have been used in surveying for many years find application first in photogrammetry and then in digital cartography. Data is held in the form of coordinates, from which further information such as direction, angle, height, area, etc. is derived. Transformations of a data set are applied mathematically, and the results are plotted automatically in order to obtain a map image with a different appearance. Thus the conversion from the analogue to the digital process is evident across a much wider range of the map-making spectrum than in the past.

The change from analogue to digital techniques has been the result of dramatic progress in the design of electronic components, and in the speed and power of electronic computers. In particular, the average practitioner is today capable of purchasing computing facilities which were quite beyond his reach a few decades ago. Therefore computations which at one time were carried out in a central and extremely remote computer bureau, are now possible on-site, and often on-line with the data acquisition instrument. Results may be presented to the observer more or less instantaneously, and decisions regarding further action may be made immediately. The level of sophistication varies—from the total station in surveying which presents reduced horizontal distances and height differences to the surveyor in the field, to the highly sophisticated on-line analytical photogrammetric/graphics workstation. The operation of such systems is often extremely impressive, largely due to the speed with which computations may be carried out today. The algorithms involved, however, are much the same as in the past—the differences being that they are used more often, with more data, at greater speed, and in a different context. Furthermore, the trend has been to use smaller and more powerful microcomputers to do what at one time was only possible on large and expensive mainframe computers.

The chapters that follow describe some basic computational models in surveying and photogrammetry. These include the reduction of observations to the particular reference surfaces being considered, and the assessment of the accuracy of the derived quantities. Attention is focused on the use of cartesian coordinates in a two or three dimensional rectangular system. The methods used for the calculation of these coordinates are explained by considering the problem from two standpoints—firstly, when the minimum number of observations is available, and secondly when more than the minimum is to be used to determine the final result. The latter is subdivided further, a distinction being made between what are termed 'approximate' methods and least-squares methods. It is appreciated, however, that such a distinction is questionable, and it is made purely as a matter of convenience, to distinguish the method rather than the accuracy of the result. Examples are given whenever they are considered necessary. For obvious reasons these are limited in number and scope in order to remain within the constraints of publication, while at the same time representing test material which may be used in developing microcomputer applications. In many cases it is a simple matter to extend the example and the concept to that which would be appropriate for larger problems. The emphasis throughout is to describe techniques which are useful given the computational facilities which are available. Although many other physical quantities are currently measured, attention is concentrated on the measurements of angles and distances, as these are still the most commonly measured quantities.

At times the terminology used may be different to that which is familiar to many practitioners. In particular, and in keeping with the modern view, the term 'adjustment' is avoided as far as possible, and the word 'estimation' used instead. Also 'standard deviation', 'standard error' and 'variance' are used to express dispersion, rather than 'probable error' and 'root mean square error'. Some readers may be surprised at the attention given to semigraphic methods of position-fixing. It is felt that such a section is necessary for two reasons. Firstly, it represents a method which will frequently give acceptable results, is flexible in application, and does not rely on the use of complicated equations or sophisticated computing equipment. Secondly, the processes involved are educational, and lead naturally to an appreciation of what is involved when a wholly analytical solution is implemented by least squares. In the sections on least squares N^{-1} is used repeatedly to express the solution of normal equations, and generally the examples given are trivial enough to justify the calculation of the inverse. In practice this is less likely, and the solution of simultaneous equations—particularly those with symmetric coefficient matrices—is achieved more efficiently by other methods. The significance of N^{-1} in the text therefore is purely symbolic, and does not imply a specific method of solution.

The text is divided into three parts—surveying, photogrammetry, and least-squares methods applied to both. The fundamental mathematical models for surveying and photogrammetry are developed in Chapters 2 and 5 respectively. The reduction of observations in both surveying (Chapter 3) and photogrammetry (Chapter 6) is presented after the chapters on basic mathematical models. Although these chapters might serve as useful reference material, they are not essential reading for those interested in the general problem of deriving coordinates from redundant observations. Some approximate methods of dealing with situations involving redundant observations are outlined in Chapters 4 and 7, and the last three chapters (8, 9, 10) are concerned with the statistical analysis of observations and least-squares estimation. Chapter 10 starts with a summary of applicable formulae in least-squares estimation by observation equations, and this is followed by a presentation of the more common applications in both surveying and photogrammetry. It should be possible for the reader to select those sections which are of interest, and still follow the text systematically.

It is assumed that the reader has a certain level of background knowledge of surveying and photogrammetry. However, the level expected is not high and would be consistent with that obtained from practical experience, or from a first course in the subject. The level of understanding of mathematics is probably higher, and certainly a basic understanding of differential calculus and matrix algebra is required for the last few chapters. Otherwise, the text is considered to be introductory in nature, and designed to whet the appetite of those interested in taking some first steps in the subject. For more advanced aspects the reader is referred elsewhere.

2 Mathematical models in surveying

2.1 Introduction

This chapter is concerned with the derivation of equations for the more commonly used calculations in plane surveying. The assumption is that only the minimum number of observations have been made to solve specific problems, and the equations are used to derive the necessary coordinates. Conversely, if coordinates are known, the equations are used to derive other associated elements of the geometrical framework. In the closing sections the calculation of heights is considered, followed by the two-dimensional transformation of plane coordinates, and calculations associated with the determination of area.

Although the equations have been traditionally associated with plane surveying, many find application in other related activities, such as photogrammetry and digital mapping, where their use could indeed be considered more frequent. What is probably more important is that the equations are often used as a preliminary step to solve computational problems by least squares, when redundant observations are present. In such a case they provide approximate values for unknown parameters which are fairly close to the final least-squares estimates. Thus, in the modern context of automatic data acquisition and processing, the equations become particularly useful.

2.2 Coordinate systems

Measurements or 'observations' in surveying are made in order to establish the relative positions of points or objects. The relationships between the multitude of points surveyed can only be established if the observations are used to establish coordinates of the points on one homogeneous coordinate system. Many different coordinate systems are used in practice, depending on the circumstances, and there are usually good reasons for the choice of a certain system for a particular survey.

A convenient three-dimensional global system would be one with axes at right angles and origin at the centre of the earth. Such a system, referred to as a *geocentric cartesian* system, is shown in Figure 2.1*a*. The XY-plane is the equatorial plane, the XZ-plane contains the Greenwich meridian, and the Z-axis coincides with the axis of rotation of the earth. The origin O would be located near the earth's centre of mass, and a point P on the topographic surface would have cartesian coordinates XYZ.

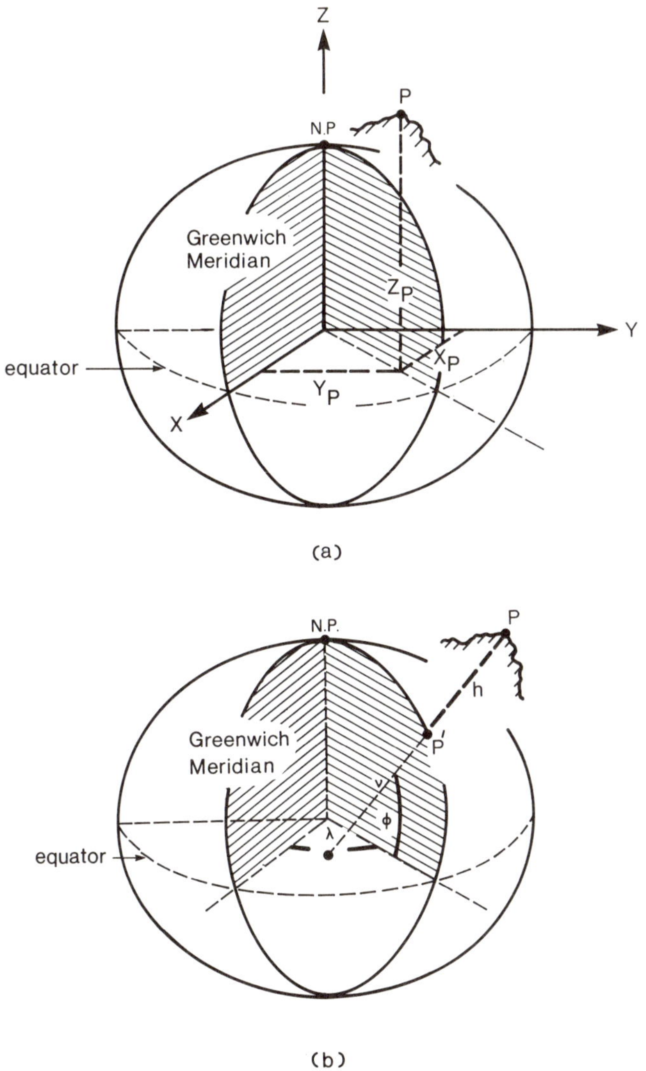

Figure 2.1 Three-dimensional coordinate system: *a*, geocentric cartesian; *b*, spheroidal.

Alternatively, a *spheroid of reference* could be defined whose surface approximates closely to the earth's surface. Such a spheroid is defined by specifying the parameters (size, shape and position) of an ellipse, which on rotation about its minor axis traces out the spheroidal surface. The surface is specified in fairly simple terms mathematically and this, together with the fact that it lies close to the earth's surface (in particular the mean sea-level surface), makes it attractive as a reference surface onto which points are projected. The point P could then be located by a set of curvilinear coordinates—latitude (φ)

and longitude (λ)—and distance ($\nu + h$): as shown in Figure 2.1*b*. The quantity ν is one of two principal radii of curvature of the spheroid surface at P', below P, and h is the height of P above the surface. Because ν is a function of the latitude of the point and the constants defining the ellipse, it is easily calculated, and it is normal practice to specify only h as the third coordinate.

The curved surface of the spheroid is often referred to as the *horizontal datum*, being the reference surface which is approximately perpendicular to the plumbline at all points. It is the surface to which measured angles and distances are referred or reduced. The *vertical* or *height datum* is the surface to which vertical distances are referred or reduced. This is normally either the geoid (or mean sea-level) surface or the spheroid surface.

Surveys involving the measurement of length and direction of very long lines, and the corresponding reduction to geocentric coordinates or spheroid surface, are termed *geodetic surveys*. They require a high order of precision both in the measurement phase and in the reduction of the observations. Thus the mathematical models are necessarily more complete. *Plane surveys*, on the other hand, involve observations over shorter distances and cover areas of such limited extent that many simplifying assumptions are made. The spheroid surface for the area of interest is often assumed to be a plane surface; or the curvature is accounted for by applying corrections to observations according to specific map projection formulae. Either way, the result is a set of coordinates for points which do not correspond exactly with the positions on the curved spheroid surface. This is the well-known problem of projecting a curved surface onto a plane, and the amount of deformation is a function of the extent of the survey.

The plane surface, which is used as a reference, is tangential to the spheroid at a local point of origin (O), as shown in Figure 2.2. Alternatively, in the case of a map projection, it is the surface onto which the positional information

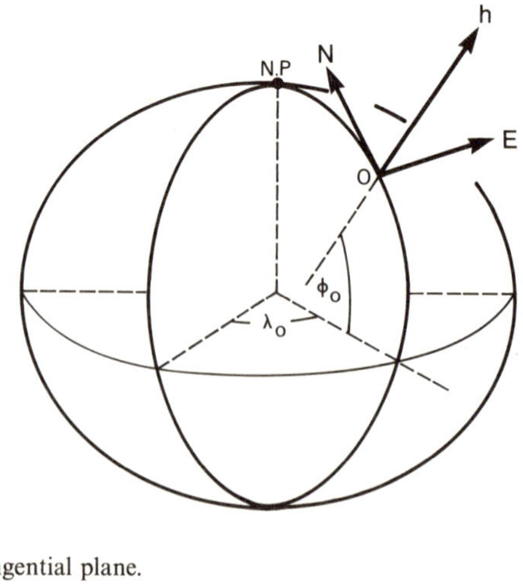

Figure 2.2 Local tangential plane.

is projected, using map projection formulae. The E, N-axes define a rectangular cartesian system with the N-axis generally directed towards north. Because surveying observations are made in or close to the horizontal plane the computations for planimetric position (E, N) and height (h) generally take place independently.

The greatest amount of activity in surveying takes place at the level of plane surveying rather than geodetic surveying, and the mathematical models developed in the following sections refer largely to this category.

2.3 Angles, directions, bearings

Angles in surveying are normally considered as clockwise rotations. Horizontal angles are measured about the plumbline, or about the normal to the spheroid if the two are considered to be coincident. Figure 2.3*a* shows the angle between two points B and C as measured when a theodolite is set up at point A. The value of the angle could be obtained by setting the horizontal circle to read 0° when the instrument is pointing to B, and then reading the value when C is sighted. This is not a very practical approach, however, and it is usual to record both readings with an arbitrary setting of the horizontal circle, and then derive the value of the angle by subtracting one observation from the other.

For example, if the horizontal circle readings to B and C are

$$\begin{array}{lr} B & 29°15'46'' \\ C & 134°32'08'' \end{array}$$

the clockwise angle BAC would be

$$134°32'08'' - 29°15'46'' = 105°16'22''$$

and the clockwise angle CAB would be

$$29°15'46'' - 134°32'08'' = -105°16'22''$$

which may be converted to a positive value by adding 360°, i.e.

$$-105°16'22'' + 360° = 254°43'38''$$

In practice it is more convenient to add 360° to the observation in the first place. This is justified since a direction will not change if a multiple of 360° is added to it. The clockwise angle CAB may then be found from

$$389°15'46'' - 134°32'08'' = 254°43'38''$$

Thus the recorded values are angles with respect to a zero reference which points in some arbitrary direction, and it is the *difference* between two observations that establishes the value of the angle between the two targets sighted. Figure 2.3*b* illustrates a round of observations from point A to B,

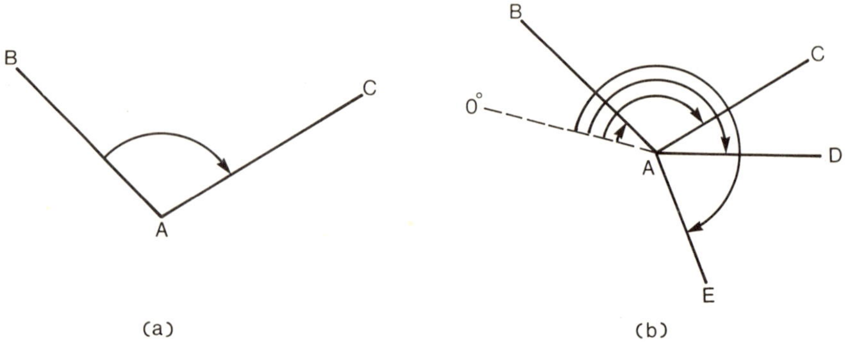

Figure 2.3 Horizontal angle observations: *a*, horizontal angle *BAC*; *b*, horizontal directions.

C, *D*, *E*, with 0° pointing in an unspecified direction. Because 0° is pointing in an arbitrary direction, the observations are conveniently referred to as observed *directions*.

Let these directions be d_B, d_C, d_D, d_E. If the 0° reference direction was pointing north the observations would be *bearings*. Thus a bearing is a clockwise angle with respect to north. Bearings may be required before observations are used to derive coordinates of points, and provided the bearing from *A* to one of the points is known then the bearings for the others may be derived. If α_B is the known bearing for the line *AB*, as shown in Figure 2.4, then an *orientation correction* (θ) may be derived from

$$\theta = \alpha_B - d_B \tag{2.1}$$

which may then be added to all the other directions in order to obtain bearings. Thus

$$\left.\begin{aligned} \alpha_C &= d_C + \theta \\ \alpha_D &= d_D + \theta \\ &\vdots \\ &\text{etc.} \end{aligned}\right\} \tag{2.2}$$

Notice that the orientation correction θ is also the bearing of the zero reference line.

A numerical example is given in Table 2.1. The orientation correction above was derived from a single known bearing. A better value may be calculated if a number of points with known bearings are included in the round of observations, and a mean orientation correction derived. The orientation correction may be regarded as a means of compensating for systematic error in the observations. Later, when the more rigorous processing of observations is considered, it will be seen to be an additional unknown which must be solved.

A known bearing may be a value established from known coordinates of the endpoints of a line, or from some other field method such as azimuth observations in field astronomy, or from observations with a gyrotheodolite, etc. The most common, however, is the method using rectangular coordinates.

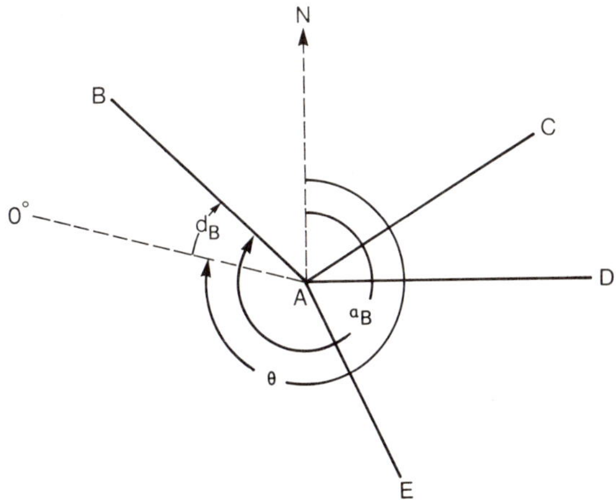

Figure 2.4 Orientation correction θ.

Table 2.1 Orientation of observations

Point sighted	Observed direction ° ′ ″	Orientation correction ° ′ ″	Bearing ° ′ ″
B	29 15 46	−4 01 44	25 14 02
C	134 32 08	−4 01 44	130 30 14
D	254 18 21	−4 01 44	250 16 37
E	3 06 13	−4 01 44	359 04 29

2.4 Rectangular coordinates

A rectangular coordinate system, such as that shown in Figure 2.5, is necessary in plane surveying in order that the correct relative positions of points or objects may be established. The system is defined by specifying the position of the origin *O* and the direction of the ordinate *N*: it follows therefore that the other axis *OE* is at right angles to *ON*. The origin may be chosen to be a point near the centre of the area, thus reducing the magnitude of the coordinates involved; or it may be a convenient point outside the area, thus making all coordinate values positive. The origin may also be established by assigning coordinate values to any point included in the survey.

Traditionally the ordinate *ON* is directed towards north, but variations are common, including systems directed towards south, and others with quite arbitrary directions. The latter are often used temporarily because of the lack of orientation information: when the information becomes available the coordinates are normally transformed into a system oriented towards north. Once the direction of *ON* is established it is referred to as *grid north*, and the

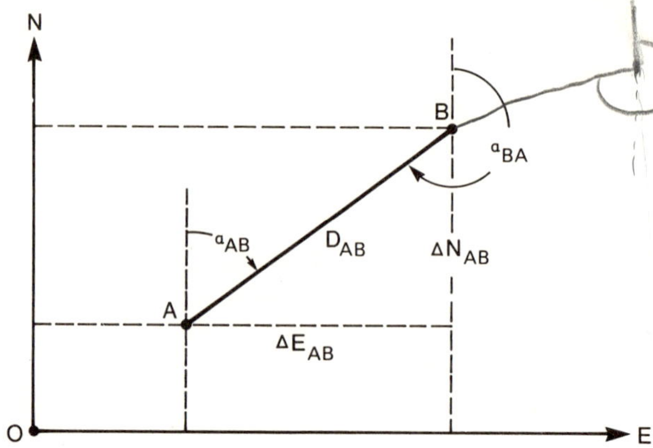

Figure 2.5 Plane rectangular coordinate system.

axis as the northings axis. *E*, *N*-coordinates therefore correspond to *x*, *y*-coordinates in plane coordinate geometry. An important difference however, is that whereas in coordinate geometry angles are reckoned anticlockwise from the *x*-axis, in surveying they are clockwise from the *N*-axis.

A line joining two points *A* and *B*, when projected orthogonally onto the *E* and *N* axes, will produce the coordinate differences ΔE and ΔN. The signs for these differences are defined as follows:

$$\begin{aligned} \Delta E_{AB} &= E_B - E_A \\ \Delta N_{AB} &= N_B - N_A \end{aligned} \tag{2.3}$$

where the subscript *AB* signifies 'for the direction *A* to *B*'. Thus

$$\begin{aligned} \Delta E_{BA} &= -\Delta E_{AB} \\ \Delta N_{BA} &= -\Delta N_{AB} \end{aligned}$$

2.5 The 'join' and 'polar' calculations

The bearing α_{AB} for the line is found from

$$\tan \alpha_{AB} = \frac{\Delta E_{AB}}{\Delta N_{AB}}, \qquad \text{or} \qquad \cot \alpha_{AB} = \frac{\Delta N_{AB}}{\Delta E_{AB}} \tag{2.4}$$

and the *reverse* or *back bearing* α_{BA} from

$$\alpha_{BA} = \alpha_{AB} \pm 180^\circ \tag{2.5}$$

which, when the limiting value of 360° is taken into account, effectively means

$$\left.\begin{aligned} \alpha_{BA} &= \alpha_{AB} + 180^\circ \qquad \text{if } \alpha_{AB} \leqslant 180^\circ \\ \alpha_{BA} &= \alpha_{AB} - 180^\circ \qquad \text{if } \alpha_{AB} \geqslant 180^\circ \end{aligned}\right\} \tag{2.6}$$

This should be evident from an inspection of Figure 2.5.

The distance D_{AB} may be obtained from any of the following equations:

$$\left.\begin{aligned} D_{AB}^2 &= \Delta E_{AB}^2 + \Delta N_{AB}^2 \\ D_{AB} &= \frac{\Delta E_{AB}}{\sin \alpha_{AB}} \\ D_{AB} &= \frac{\Delta N_{AB}}{\cos \alpha_{AB}} \end{aligned}\right\} \quad (2.7)$$

The choice between the formulae depends on the circumstances—the value of α_{AB}, the number of significant digits which can be held by the calculator, etc. In eqns (2.4) the functions tan and cot become sensitive to small changes as the value of α approaches 0° (or 180°) and 90° (or 270°) respectively, and the more reliable formula should be used on each occasion. A useful rule of thumb is to select the formula which has the largest absolute difference as denominator. In eqns (2.7) sin and cos are sensitive for the same two directions. Again, a useful rule is to select the formula containing the larger coordinate difference—larger, that is, in absolute terms. With modern powerful calculators, which hold many significant digits, it is probably wiser to use the first of the three formulae (which always contains the largest difference), and rely on the other two as a means of checking the result.

Equations (2.4) are applicable as they stand for bearings between 0° and 180°, provided the correct sign conventions are used in establishing ΔE and ΔN. It may be necessary to add 180° or 360° to obtain the correct absolute values; and

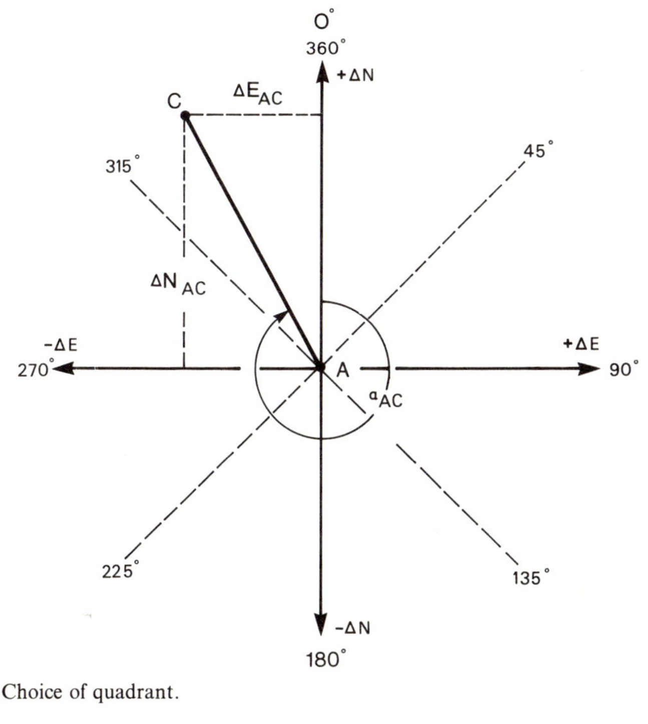

Figure 2.6 Choice of quadrant.

the matter of establishing the correct quadrant is settled by inspection of the signs and magnitudes of ΔE and ΔN. Thus in Figure 2.6, if A is regarded as a local origin, and the axes define coordinate differences, the bearing AC has negative ΔE and positive ΔN, placing it in the fourth quadrant. This, together with the fact that the absolute value of ΔN is greater than ΔE, establishes that the bearing lies between 315° (i.e. 270° + 45°) and 360°.

The calculation of bearing and distance from rectangular coordinates, which is conveniently referred to as a '*join*' calculation, is probably the most fundamental and important calculation in surveying. It is used frequently either to check existing data or to prepare data for further more complex operations—such as coordinate system transformations or least-squares estimation. It is essential therefore to verify that the results of a join calculation are correct. The simplest method is to carry out the reverse calculation, conveniently referred to as the '*polar*' calculation. If the coordinates of a point A are known, and the bearing α_{AB} and distance D_{AB} are known, then the coordinates of B may be found from

$$\begin{aligned} E_B &= E_A + D_{AB} \sin \alpha_{AB} \\ N_B &= N_A + D_{AB} \cos \alpha_{AB} \end{aligned} \tag{2.8}$$

which are simply a rearrangement of eqns (2.7).

The formulae are completely general for bearings from 0° to 360°, provided the sign conventions indicated in Figure 2.6 are adhered to. The distance D is always positive.

If the angle entered as argument for a particular trigonometric function must be a value between 0° and 90° then the following procedure is useful.

(i) Establish the *sign* of the function from Figure 2.6.
(ii) Identify the *multiple of* 90° (0°, 90°, 180°, or 270°) that marks the beginning of the quadrant containing the bearing.

Table 2.2 Functions of angles between 0° and 90° for bearings between 0° and 360°

Required function of a bearing α	Required function of the corresponding angle between 0° and 90°: Quadrant 1	2	3	4
sin	sin	cos	sin	cos
cos	cos	sin	cos	sin
tan	tan	cot	tan	cot
cot	cot	tan	cot	tan
sec	sec	cosec	sec	cosec
cosec	cosec	sec	cosec	sec

(iii) Subtract the multiple from the bearing in order to establish the *argument*.
(iv) If the multiple found in (ii) is 90° or 270° then use the complementary function—i.e. cos instead of sin. etc.

Table 2.2 shows the full range of functions to be used if this procedure is followed. Verification of the results may be established, as with the join, by carrying out the reverse calculation.

The following numerical example serves to illustrate some of the points mentioned so far.

Example 2.1

Points *A*, *B*, *C*, and *D* have the following rectangular coordinates:

	E (m)	*N* (m)
A	8246.18	4102.96
B	7781.63	2593.70
C	5997.73	3912.56
D	7279.56	4643.10

(a) Calculate the 'joins' *AC*, *BC*, *CD*.
(b) Calculate the coordinates of points *P* and *Q*, based on the following observations made at point *C*:

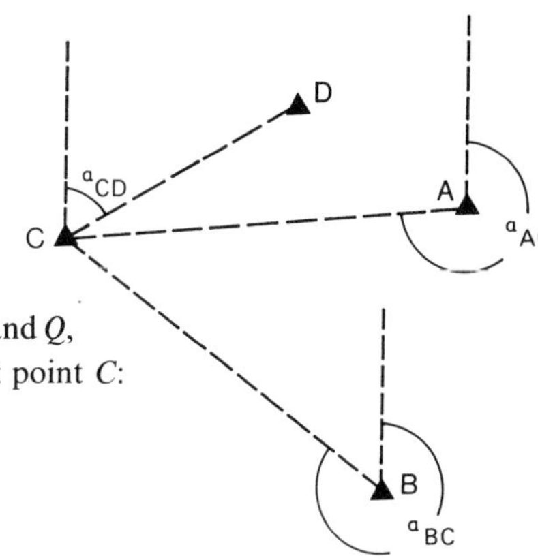

Figure 2.7

At C

To	Observed direction °	′	″	Observed distance (m)
A	65	04	29	—
P	78	41	22	3224.51
B	106	23	21	—
Q	322	11	43	2796.32
D	40	13	59	—

Solution

(a) *For AC*:
$\Delta E_{AC} = E_C - E_A = -2248.45$
$\Delta N_{AC} = N_C - N_A = -\ \ 190.40$
$\cot \alpha_{AC} = -190.40/-2248.45 = 0.084681$; therefore $\alpha_{AC} = 85°09'35''$ or $265°09'35''$; and since both ΔE and ΔN are negative, the bearing is in the third quadrant.
Therefore $\alpha_{AC} = 265°\ 09'\ 35''$.
$D_{AC} = ((-2248.45)^2 + (-190.40)^2)^{1/2} = 2256.497$ m
or $D_{AC} = \Delta E_{AC}/\sin \alpha_{AC} = -2248.45/-0.996434 = 2256.497$m

For BC:
$\Delta E_{BC} = E_C - E_B = -1783.90$
$\Delta N_{BC} = N_C - N_B = 1318.86$
$\cot \alpha_{BC} = 1318.86/-1783.90 = -0.739313$; therefore $\alpha_{BC} = -53°31'26''$ (i.e. $306°28'34''$) or $126°28'34''$ and since ΔE is negative and ΔN positive, the bearing is in the fourth quadrant;

Therefore $\alpha_{BC} = 306°28'34''$.

$D_{BC} = ((-1783.90)^2 + (1318.86)^2)^{1/2} = 2218.488\text{m}$

or $D_{BC} = \Delta E_{BC}/\sin \alpha_{BC} = -1783.90/-0.804106 = 2218.488\text{m}$

For CD: $\Delta E_{CD} = E_D - E_C = 1281.83$

$\Delta N_{CD} = N_D - N_C = 730.54$

$\cot \alpha_{CD} = 730.54/1281.83 = 0.569920$; therefore $\alpha_{CD} = 60°19'13''$ or $240°19'13''$; and since both ΔE and ΔN are positive, the bearing is in the first quadrant.

Therefore $\alpha_{CD} = 60°19'13''$.

$D_{CD} = ((1281.83)^2 + (730.54)^2)^{1/2} = 1475.390\text{m}$

or $D_{CD} = \Delta E_{CD}/\sin \alpha_{CD} = 1281.83/0.868807 = 1475.390\text{m}$

(b) The observations at *C* must be oriented to establish bearings to *P* and *Q*, and the observations to *A*, *B* and *D*, together with the computed bearings *AC*, *BC*, *CD* above, may be used to derive the orientation correction, as follows:

At C

To	Computed bearing ° ′ ″	Observed direction ° ′ ″	Orientation correction ° ′ ″
A	85 09 35	65 04 29	+20 05 06
B	126 28 34	106 23 21	+20 05 13
D	60 19 13	40 13 59	+20 05 14
		Mean =	+20 05 11

The mean orientation correction of $+20° 05' 11''$ is applied to the observations to *P* and *Q*, to give the following oriented observations:

At C

To	Bearing ° ′ ″
P	98 46 33
Q	342 16 54

The 'polars' *CP* and *CQ* may now be calculated:

$$E_P = E_C + D_{CP} \sin \alpha_{CP} = 5997.73 + 3186.760 = 9198.490\,\text{m}$$

$$N_P = N_C + D_{CP} \cos \alpha_{CP} = 3912.56 + (-491.960) = 3420.600\,\text{m}$$

and

$$E_Q = E_C + D_{CQ} \sin \alpha_{CQ} = 5997.73 + (-851.026) = 5146.704\,\text{m}$$

$$N_Q = N_C + D_{CQ} \cos \alpha_{CQ} = 3912.56 + 2663.674 = 6576.234\,\text{m}$$

If the angles and distances between a number of consecutive points have been observed, as indicated in Figure 2.8, then the bearings may be derived, and a series of polars may be calculated in order to fix the positions of the points.

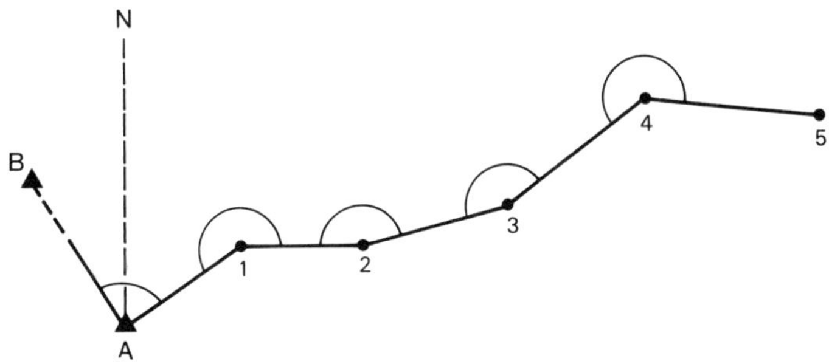

Figure 2.8 Open traverse.

Such a configuration constitutes a *traverse*, and in the situation shown it is considered an open traverse—i.e. an unchecked traverse. The adoption of coordinates based on the computation of an open traverse is not recommended, because of the likelihood of an error going undetected. Therefore additional observations are essential, and these should be such that they expose any error in observation or computation. Traverse computation is considered again in sections 4.3, 4.4, and 10.2.

2.6 Triangulation intersection

If a certain minimum number of elements in a triangle is known then the other elements may be calculated. Viewed differently, if the coordinates of two

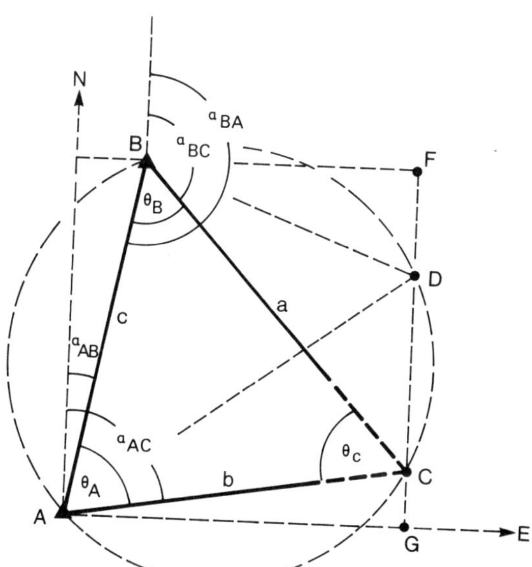

Figure 2.9 Triangulation intersection.

of the three points are known and certain observations made, then the coordinates of the third point may be calculated. With the coordinates of all three points established any of the elements of the triangle may be calculated. Figure 2.9 shows points A and B, which have known coordinates. Angles θ_A and θ_B have been observed at A and B in order to establish the coordinates of point C.

The calculations described in the previous section may be used in sequence to derive the coordinates of C, and the steps would be:

(i) Calculate the *join AB* to obtain values for the distance c and the bearing α_{AB}.

(ii) Derive a value for the *angle* θ_C:

$$\theta_C = 180 - (\theta_A + \theta_B)$$

(iii) Use the sine formula to derive *distances* a and b:

$$\frac{a}{\sin\theta_A} = \frac{c}{\sin\theta_C}, \qquad \frac{b}{\sin\theta_B} = \frac{c}{\sin\theta_C}$$

(iv) Derive the *bearings* α_{AC} and α_{BC}:

$$\alpha_{AC} = \alpha_{AB} + \theta_A$$

$$\alpha_{BC} = \alpha_{BA} - \theta_B$$

(v) Calculate *polars* AC and BC to obtain two sets of coordinates for C, which should be identical within rounding errors and the limits of the functions used.

The calculation of two sets of coordinates serves as a check against an error in calculation; the same purpose would be served by calculating two joins AC and BC once the coordinates of C are known. However, they do *not* verify that the correct data or observations were used in the first place.

Other variations on the data supplied for triangulation arise from time to time. For example, the coordinates of A and the bearing and distance AB might have been the initial data. It is clear, however, that a polar calculation for AB would soon establish the coordinates of B, and the steps outlined above would then apply.

A solution which provides the coordinates of the unknown point directly, rather than as a sequence of steps, is more relevant in the age of the powerful calculator or microcomputer. Two sets of formulae are derived below; the first is based on *angles* at A and B, and the second on *bearings* at A and B.

Solution based on angles

In order to avoid an ambiguity as to which side of the line AB the point C is located, a clockwise convention will be adopted—i.e. the points are designated alphabetically in a clockwise order, as shown in Figure 2.9.

An inspection of Figure 2.9 reveals the following identities:

$$\left.\begin{aligned} FD &= BF\cot\theta_A = (E_C - E_B)\cot\theta_A \\ DG &= AG\cot\theta_B = (E_C - E_A)\cot\theta_B \end{aligned}\right\} \qquad (2.9)$$

and

$$\begin{aligned} N_B - N_A &= FD + DG \\ &= E_C\cot\theta_A - E_B\cot\theta_A + E_C\cot\theta_B - E_A\cot\theta_B \\ &= E_C(\cot\theta_A + \cot\theta_B) - E_A\cot\theta_B - E_B\cot\theta_A \end{aligned}$$

Therefore

$$E_C = \frac{E_A\cot\theta_B + E_B\cot\theta_A - N_A + N_B}{\cot\theta_A + \cot\theta_B} \qquad (2.10a)$$

and similarly

$$N_C = \frac{N_A\cot\theta_B + N_B\cot\theta_A + E_A - E_B}{\cot\theta_A + \cot\theta_B} \qquad (2.10b)$$

Solution based on bearings

Functions based on bearings may be derived as follows. From eqn (2.4),

$$\Delta N_{AC} = \Delta E_{AC}\cot\alpha_{AC}$$

i.e.

$$(N_C - N_A) = (E_C - E_A)\cot\alpha_{AC} \qquad (2.11)$$

therefore

$$N_C - E_C\cot\alpha_{AC} + E_A\cot\alpha_{AC} - N_A = 0 \qquad (2.12)$$

Also

$$(N_C - N_B) = (E_C - E_B)\cot\alpha_{BC}$$

therefore

$$N_C - E_C\cot\alpha_{BC} + E_B\cot\alpha_{BC} - N_B = 0 \qquad (2.13)$$

Eliminating N_C, by subtracting eqn (2.11) from eqn (2.12),

$$E_C(\cot\alpha_{AC} - \cot\alpha_{BC}) - E_A\cot\alpha_{AC} + E_B\cot\alpha_{BC} + N_A - N_B = 0$$

Therefore

$$E_C = \frac{E_A\cot\alpha_{AC} - E_B\cot\alpha_{BC} - N_A + N_B}{\cot\alpha_{AC} - \cot\alpha_{BC}} \qquad (2.14a)$$

and by substitution in eqn (2.11)

or

$$\left.\begin{aligned} N_C &= N_A + (E_C - E_A)\cot\alpha_{AC} \\ N_C &= N_B + (E_C - E_B)\cot\alpha_{BC} \end{aligned}\right\} \qquad (2.14b)$$

In a similar manner, but starting with the first of eqns (2.4),

$$\left.\begin{aligned} \Delta E_{AC} &= \Delta N_{AC} \tan \alpha_{AC} \\ \text{or} \qquad (E_C - E_A) &= (N_C - N_A) \tan \alpha_{AC} \end{aligned}\right\} \qquad (2.15)$$

Therefore the following relationships may be derived:

$$\left.\begin{aligned} N_C &= \frac{N_A \tan \alpha_{AC} - N_B \tan \alpha_{BC} - E_A + E_B}{\tan \alpha_{AC} - \tan \alpha_{BC}} \\ \text{and} \\ E_C &= E_A + (N_C - N_A) \tan \alpha_{AC} \\ \text{or} \\ E_C &= E_B + (N_C - N_A) \tan \alpha_{BC} \end{aligned}\right\} \qquad (2.16)$$

Either eqns (2.14) or eqns (2.16) may be used. However, to avoid the instability caused by bearings which are nearly parallel to either axis, it is wise to be selective in the choice. A simple rule of thumb is that the absolute value of the denominator should be as close to 1 as possible. To achieve this, identify the bearing which is more nearly parallel to either coordinate axis; if that axis is 90° or 270° use eqns (2.14), if 0° or 180° use eqns (2.16).

It is worth noting that the equations representing the solution for E_C, N_C may be obtained using matrix algebra. Equation (2.11) is rearranged, and the appropriate equation for each of the two bearings α_{AC} and α_{BC} is established. These are

$$\left.\begin{aligned} N_A - E_A \cot \alpha_{AC} &= -E_C \cot \alpha_{AC} + N_C \\ N_B - E_B \cot \alpha_{BC} &= -E_C \cot \alpha_{BC} + N_C \end{aligned}\right\} \qquad (2.17)$$

which may be written in the form of matrices as

$$\begin{pmatrix} N_A - \cot \alpha_{AC} \cdot E_A \\ N_B - \cot \alpha_{BC} \cdot E_B \end{pmatrix} = \begin{pmatrix} -\cot \alpha_{AC} & 1 \\ -\cot \alpha_{BC} & 1 \end{pmatrix} \cdot \begin{pmatrix} E_C \\ N_C \end{pmatrix}$$

or

$$\boldsymbol{y} = \boldsymbol{A} \cdot \boldsymbol{x}$$

where $\boldsymbol{x}$ is the vector of unknown coordinates of point C. The solution is found from

$$\boldsymbol{x} = \boldsymbol{A}^{-1} \boldsymbol{y} \qquad (2.18)$$

The inverse of $\boldsymbol{A}$ (using Cramer's rule) is

$$\boldsymbol{A}^{-1} = \frac{1}{\cot \alpha_{BC} - \cot \alpha_{AC}} \begin{pmatrix} 1 & -1 \\ \cot \alpha_{BC} & -\cot \alpha_{AC} \end{pmatrix}$$

and therefore, substituting in eqn (2.18),

$$\begin{pmatrix} E_C \\ N_C \end{pmatrix} = \frac{1}{\cot \alpha_{BC} - \cot \alpha_{AC}} \begin{pmatrix} 1 & -1 \\ \cot \alpha_{BC} & -\cot \alpha_{AC} \end{pmatrix} \begin{pmatrix} N_A - \cot \alpha_{AC} \cdot E_A \\ N_B - \cot \alpha_{BC} \cdot E_B \end{pmatrix}$$

which becomes

$$\begin{pmatrix} E_c \\ N_c \end{pmatrix} = \frac{1}{\cot\alpha_{BC} - \cot\alpha_{AC}} \begin{pmatrix} N_A - \cot\alpha_{AC}\cdot E_A - N_B + \cot\alpha_{BC}\cdot E_B \\ N_A\cot\alpha_{BC} - \cot\alpha_{AC}\cot\alpha_{BC}\cdot E_A - N_B\cot\alpha_{AC} \\ + \cot\alpha_{AC}\cot\alpha_{BC}\cdot E_B \end{pmatrix}$$

Therefore

$$E_C = \frac{E_A\cot\alpha_{AC} - E_B\cot\alpha_{BC} - N_A + N_B}{\cot\alpha_{AC} - \cot\alpha_{BC}} \tag{2.19a}$$

N_C may be simplified by dividing the numerator and denominator by $\cot\alpha_{AC}\cot\alpha_{BC}$, giving

$$N_C = \frac{N_A\tan\alpha_{AC} - N_B\tan\alpha_{BC} - E_A + E_B}{\tan\alpha_{AC} - \tan\alpha_{BC}} \tag{2.19b}$$

These results are identical to those in eqns (2.14) and (2.16).

Example 2.2

P is an unknown point which is east of the line AB. Use the data given below to derive the coordinates of P by the following three methods:
(a) the method involving a sequence of steps;
(b) the direct solution involving observed angles, and given by eqns (2.10); and
(c) the direct solution involving observed bearings, and given by eqns (2.19).

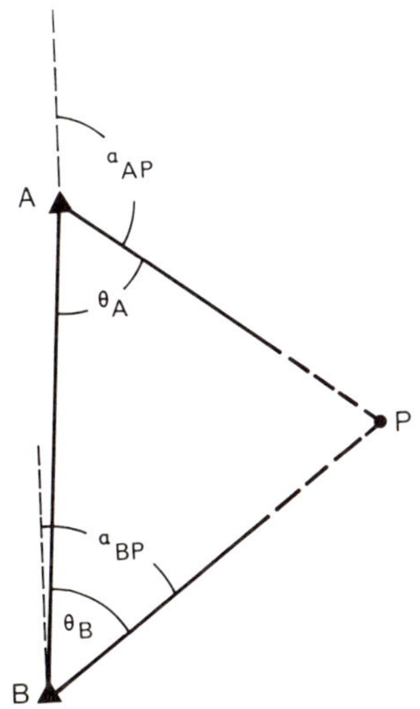

Figure 2.10

Point	Coordinates E (m)	N	Angle	Observed value ° ′ ″
A	6091.13	3631.61	$BAP(\theta_A)$	57 06 01
B	5973.88	1561.00	$ABP(\theta_B)$	50 08 10

Solution

(a) (i) The calculation of a 'join' for the line AB gives the values $\Delta E_{AB} = -117.25$ and $\Delta N_{AB} = -2070.61$, therefore $\alpha_{AB} = 183°14'27''$, and $D_{AB} = 2073.927$ m.

(ii) The angle θ_P is derived as follows:

$$\theta_P = 180° - (57°06'01'' + 50°08'10'') = 72°45'49''$$

(iii) The distances D_{AP} and D_{BP} are calculated as follows:

$$D_{AP} = \frac{D_{AB}\sin\theta_B}{\sin\theta_p} = 1666.735; \qquad D_{BP} = \frac{D_{AB}\sin\theta_A}{\sin\theta_P} = 1823.195.$$

(iv) The bearings α_{AP} and α_{BP} are derived as follows:

$$\alpha_{AP} = \alpha_{AB} - \theta_A = 183°\,14'27'' - 57°\,06'01'' = 126°\,08'26''$$
$$\alpha_{BP} = \alpha_{BA} + \theta_B = \quad 3°\,14'27'' + 50°\,08'10'' = \quad 53°\,22'37''$$

(v) The 'polars' AP and BP are calculated as follows:

$$E_P = E_A + D_{AP}\sin\alpha_{BP} = 6091.13 + 1346.010 = 7437.140\text{m}$$
$$N_P = N_A + D_{AP}\cos\alpha_{AP} = 3631.61 + (-982.987) = 2648.623\text{m}$$

and

$$E_P = E_B + D_{BP}\sin\alpha_{BP} = 5973.88 + 1463.255 = 7437.135\text{ m}$$
$$N_P = N_B + D_{BP}\cos{}_{BP} = 1561.00 + 1087.623 = 2648.623\text{ m}$$

The two sets of coordinates agree to the nearest centimetre, hence the calculation is checked.

(b) The second method derives the coordinates directly, using the observed angles and eqns (2.10). Note that the points have been designated in an anti-clockwise order in Figure 2.10, therefore the subscripts A and B in eqns (2.10) are interchanged, and the result is that the last two terms in the numerator change sign. Thus,

$$E_P = \frac{E_A\cot\theta_B + E_B\cot\theta_A + N_A - N_B}{\cot\theta_A + \cot\theta_B}$$

$$= \frac{6091.13 \times 0.835059 + 5973.88 \times 0.646922 + 3631.61 - 1561.00}{1.481982}$$

$$= 7437.137\text{ m}$$

$$N_P = \frac{N_A\cot\theta_B + N_B\cot\theta_A - E_A + E_B}{\cot\theta_A + \cot\theta_B}$$

$$= \frac{3631.61 \times 0.835059 + 1561.00 \times 0.646922 - 6091.13 + 5973.88}{1.481982}$$

$$= 2648.620\,\text{m}$$

(c) The third method derives the coordinates directly, using the bearings and eqns (2.14) or (2.16). In this case the bearing α_{AP} is more nearly parallel to the eastings axis, therefore eqns (2.14) will be used:

$$E_P = \frac{E_A \cot \alpha_{AP} - E_B \cot \alpha_{BP} - N_A + N_B}{\cot \alpha_{AP} - \cot \alpha_{BP}}$$

$$= \frac{6091.13 \times (-0.730297) - 5973.88 \times 0.743290 - 3631.61 + 1561.00}{-1.473587}$$

$$= 7437.137\,\text{m}$$

$$N_P = N_A + (E_P - E_A) \cot \alpha_{AP}$$

$$= 3631.61 + (1346.007 \times (-0.730297)) = 2648.625\,\text{m}$$

Comment

(i) All the results agree to within the limits set by the given data.
(ii) Although a clockwise convention is necessary when the method of angles is used in (b), it is not required when the method of bearings is used.

2.7 Trilateration intersection

The trilateration problem arises when distances rather than angles have been measured from points with known coordinates. Figure 2.11 shows a triangle in

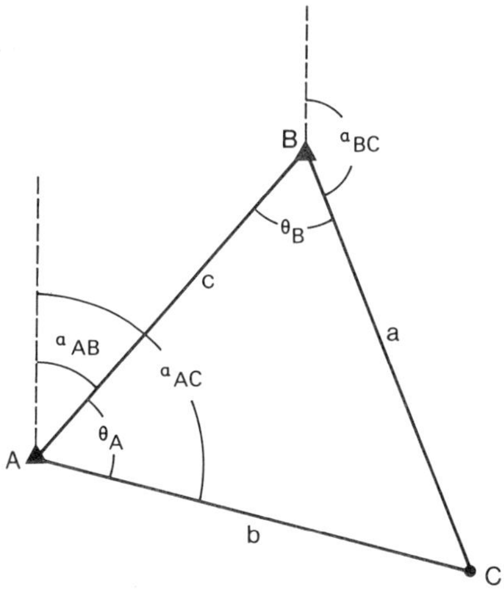

Figure 2.11 Trilateration intersection.

which A and B are known points and the distances AC and BC have been measured in order to fix the position of C.

Two methods of calculation are described: the first is a series of calculations carried out in sequence, and the second involves formulae which give the coordinates of C directly.

The solution as a sequence of calculations

(i) Calculate the *join AB* to obtain the bearing α_{AB} and distance c.
(ii) Calculate the *angle* θ_A from the cosine formula:

$$\cos\theta_A = \frac{b^2 + c^2 - a^2}{2bc}$$

(iii) Derive the *bearing* α_{AC}:

$$\alpha_{AC} = \alpha_{AB} + \theta_A$$

(iv) Calculate the *polar AC* to obtain the coordinates of C.
(v) Repeat the steps (ii)–(iv) for the angle θ_B and the bearing α_{BC}, in order to *check* the calculations.

The direct solution

If C is the unknown point and ABC are the points noted in a *clockwise* order, then the derivation is based on eqns (2.10):

$$\left.\begin{aligned} E_C &= \frac{E_A\cot\theta_B + E_B\cot\theta_A - N_A + N_B}{\cot\theta_A + \cot\theta_B} \\ N_C &= \frac{N_A\cot\theta_B + N_B\cot\theta_A + E_A - E_B}{\cot\theta_A + \cot\theta_B} \end{aligned}\right\} \tag{2.20}$$

As only sides are observed, $\cot\theta_A$ and $\cot\theta_B$ must be converted into functions of a, b and c. If

$$2s = a + b + c$$

then the area (Δ) of the triangle is given by

$$\Delta = (s(s-a)(s-b)(s-c))^{1/2}$$

Area is also found from

$$\Delta = \tfrac{1}{2}\cdot bc\sin\theta_A = \tfrac{1}{2}\cdot ac\sin\theta_B$$

Therefore

$$\sin\theta_A = \frac{2\Delta}{bc}, \qquad \sin\theta_B = \frac{2\Delta}{ac}$$

and since

$$\cos\theta_A = \frac{b^2+c^2-a^2}{2bc}, \qquad \cos\theta_B = \frac{a^2+c^2-b^2}{2ac}$$

therefore

$$\cot\theta_A = \frac{b^2+c^2-a^2}{4\Delta}, \qquad \cot\theta_B = \frac{a^2+c^2-b^2}{4\Delta}$$

Substitution in eqns (2.20) gives

$$\left.\begin{aligned} E_c &= \frac{E_A+E_B}{2} + \frac{(E_A-E_B)(a^2-b^2)}{2c^2} - \frac{2\Delta(N_A-N_B)}{c^2} \\ N_c &= \frac{N_A+N_B}{2} + \frac{(N_A-N_B)(a^2-b^2)}{2c^2} + \frac{2\Delta(E_A-E_B)}{c^2} \end{aligned}\right\} \quad (2.21)$$

Example 2.3

P is an unknown point which is south-east of the line *AB*. Use the data given below to derive the coordinates of *P* by the following two methods:

(a) the method involving a sequence of steps; and
(b) the direct solution based on eqns (2.21).

Point	Coordinates E (m)	Coordinates N (m)	Line	Observed distance (m)
A	5997.73	3912.56	*AP*	1120.76
B	7279.56	4643.10	*BP*	574.53

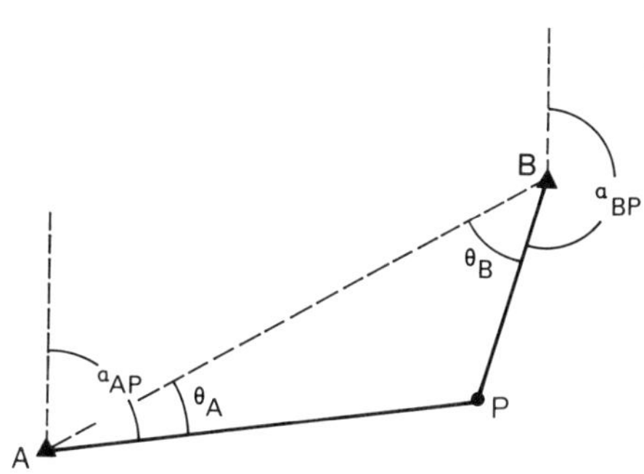

Figure 2.12

Solution

(a) (i) The calculation of the 'join' *AB* gives the values $\Delta E_{AB} = 1281.83$ and $\Delta N_{AB} = 730.54$, therefore $\alpha_{AB} = 60°19'13''$, and $D_{AB} = 1475.390$ m.

(ii) The angle θ_A is derived using the cosine formula:

$$\cos\theta_A = \frac{D_{AP}^2 + D_{AB}^2 - D_{BP}^2}{2D_{AP}D_{AB}} = \frac{1120.76^2 + 1475.39^2 - 574.53^2}{2 \times 1120.76 \times 1475.390} = 0.938217;$$

therefore, $\theta_A = 20°14'44''$.

(iii) The bearing α_{AP} is derived as follows:

$$\alpha_{AP} = \alpha_{AB} + \theta_A = 60°19'13'' + 20°14'44'' = 80°33'57''$$

(iv) The 'polar' AP is calculated as follows:

$$E_P = E_A + D_{AP}\sin\alpha_{AP} = 5997.73 + 1105.602 = 7103.332\,\text{m}$$
$$N_P = N_A + D_{AP}\cos\alpha_{AP} = 3912.56 + 183.705 = 4096.265\,\text{m}$$

(v) In a similar manner,

$$\cos\theta_B = \frac{D_{BP}^2 + D_{AB}^2 - D_{AP}^2}{2D_{BP}D_{AB}} = \frac{574.53^2 + 1475.392^2 - 1120.76^2}{2 \times 574.53 \times 1475.390} = 0.737775;$$

therefore $\theta_B = 42°27'28''$.

$$\alpha_{BP} = \alpha_{BA} - \theta_B = 240°19'13'' - 42°27'28'' = 197°51'45''$$
$$E_P = E_B + D_{BP}\sin\alpha_{BP} = 7279.56 + (-176.228) = 7103.332\,\text{m}$$
$$N_P = N_B + D_{BP}\cos\alpha_{BP} = 4643.10 + (-546.835) = 4096.265\,\text{m}$$

(b) The second method derives the coordinates directly, using eqns (2.21). Points A and B are in a clockwise order from P, therefore eqns (2.21) are used as they stand. Thus,

$$s = \frac{D_{AB} + D_{AP} + D_{BP}}{2} = 1585.340$$

$$\Delta = (s(s - D_{AB})(s - D_{AP})(s - D_{BP}))^{1/2} = 286103.7911$$

$$E_P = \frac{E_A + E_B}{2} + \frac{(E_A - E_B)(D_{BP}^2 - D_{AP}^2)}{2D_{AB}^2} - \frac{2\Delta(N_A - N_B)}{D_{AB}^2}$$

$$= 6638.645 + 272.650 - (-192.036) = 7103.332\,\text{m}$$

$$N_P = \frac{N_A + N_B}{2} + \frac{(N_A - N_B)(D_{BP}^2 - D_{AP}^2)}{2D_{AB}^2} + \frac{2\Delta(E_A - E_B)}{D_{AB}^2}$$

$$= 4277.830 + 155.389 + (-336.954) = 4096.265\,\text{m}$$

2.8 Resection

A resection calculation is necessary when angle observations have been made from only the unknown point. In field operations there may be distinct advantages in avoiding observations from known points, when observations from the unknown point alone would suffice. The problem that arises in such a situation is that the observations lack orientation.

Figure 2.13*a* shows an unknown point P from which angle observations were made to known points A and B. The angle θ_P remains the same regardless

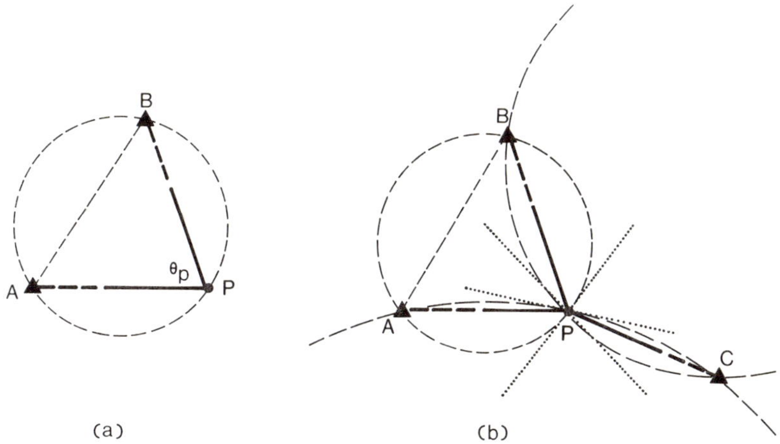

Figure 2.13 Resection observations.

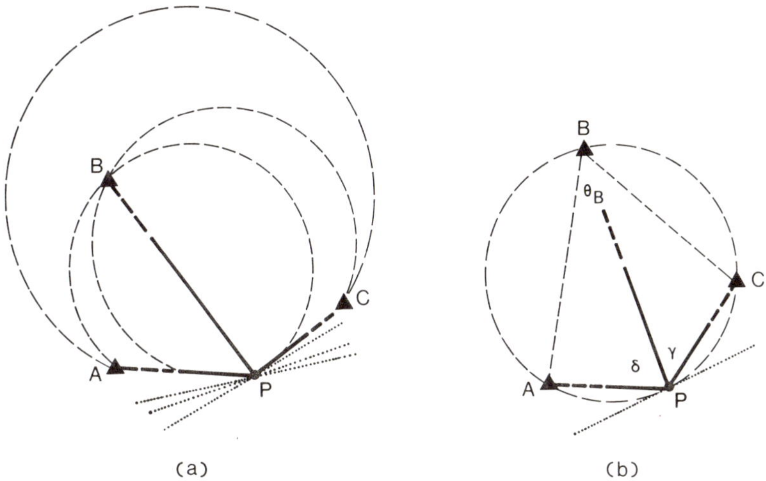

Figure 2.14: *a*, Weak resection; *b*, no resection: the 'danger' circle.

of *P*'s position on the segment of the circle passing through the three points. Thus an additional item of information is required before the coordinates of *P* may be established. The additional information could be the bearing of either line *PA* or *PB*—say established from observations with a gyrotheodolite. Alternatively, and more usually, it would be an observation to a third known point *C*, as shown in Figure 2.13*b*.

The observation to the third point allows the construction of two more circles, and it is the intersection of the three circles—i.e. the loci of the apexes of the individual angles—that determines the position of *P*. The size of the angles between the *tangents* to the circles at *P* determines the strength of the fix. These tangents are shown dotted in Figure 2.13*b*, and it is clear that they intersect with reasonably large angles—60° being the ideal. Thus a strong fix for P would be obtained in this case. Figure 2.14 illustrates what happens when the configuration of points is such that the position of P becomes unstable.

In Figure 2.14*a* the angles between the tangents are small, indicating a weak fix which should be avoided. In Figure 2.14*b* the tangents are coincident, indicating no fix at all. This latter case is the well-known *danger circle* situation, which is to be avoided at all costs. It is clear from the diagram that this condition is present when

$$\theta_B + \gamma + \delta = 180° \tag{2.22}$$

θ_B may be established from known coordinates of A, B, C, and γ and δ have been observed; therefore the situation may be assessed without the need to draw diagrams. If the sum of the angles is within 30° of 180° the determination of the position of P would be weak, and it would be wise to select another group of three known points. If a reconnaissance map does not exist it is good practice to plot the points to scale so that a realistic picture of the pattern of points may be obtained, and the most suitable group of three known points selected.

There are several different ways of carrying out a resection calculation. Two methods are described below: the first involves a sequence of basic calculations which have been described in previous sections, and the second involves formulae which give the coordinates directly.

The solution as a sequence of calculations

Figures 2.15*a* and 2.15*b* illustrate two possible cases; the first shows P, the unknown point, to be inside the triangle formed by the known points; and the second shows P outside the triangle. The solution, known as the *Collin's Point* method, involves calculating the coordinates of an auxiliary point X in the first instance, and using these values as a means of orienting the observations at P. The angles α, β, γ, δ, λ at P may be derived from the observations. Angles equal to these may be identified at points A and B, based on the theorem that angles in the same segment of a circle are equal. The sequence of calculations would then be as follows:

(i) Calculate the *join AB*, and derive the *bearings* α_{AX} and α_{BX}.
(ii) Calculate the *coordinates* of X by triangulation intersection.
(iii) Calculate the *join XC* and so obtain the *bearing* α_{XC}.
(iv) Derive the *orientation correction* θ from

$$\theta = \alpha_{XC} - d_C$$

where d_C is the observed direction to C.

(v) Apply the orientation correction to all the observed directions from P to obtain *oriented bearings*.

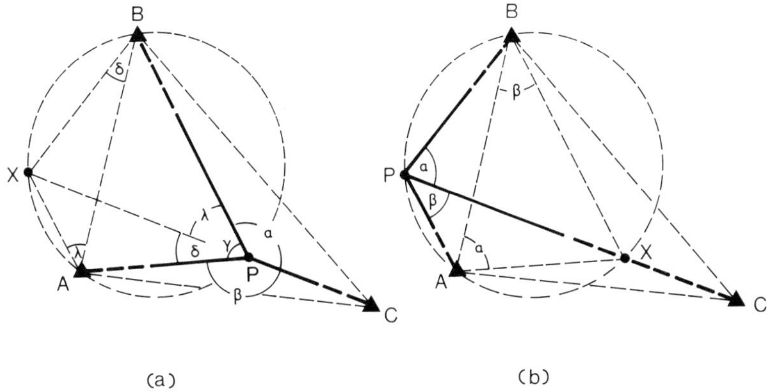

Figure 2.15 Collin's Point resection.

(vi) Find the reverse bearings for two of these oriented bearings, and calculate the *coordinates* of *P* by triangulation intersection.

(vii) Use the third bearing to *check* the calculation. This may be achieved in one of two ways; either by calculating another triangulation intersection and verifying that the coordinates are identical with those found in (vi), or by using the coordinates derived in (vi) to calculate a join *PC* and verifying that the bearing agrees with that derived in (v).

The Collin's Point method has the advantage that it is based on a series of fundamental survey calculations in common use, which are not likely to be forgotten, and which may be carried out on the simplest of calculators.

The example below illustrates the method.

Example 2.4

Theodolite observations have been made from a point *P* to three control points *A*, *B* and *C*. The observations and the coordinates of the control points are given below. Calculate the coordinates of *P*.

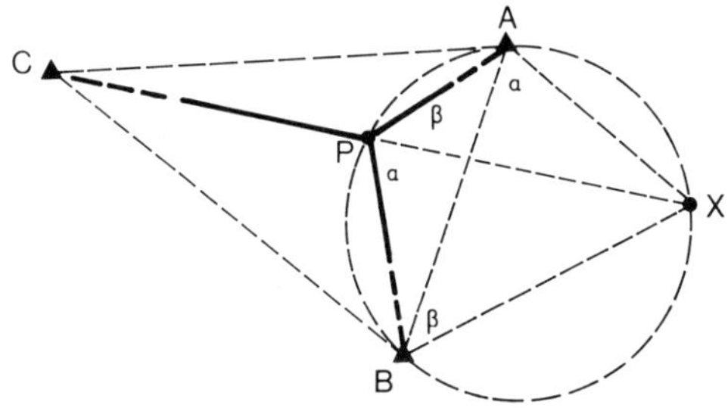

Figure 2.16

Coordinates

Point	E (m)	N (m)
A	8246.18	4102.96
B	7781.63	2593.70
C	5997.73	3912.56

Observations at P:

To	Direction ° ′ ″
A	57 45 00
B	171 13 11
C	280 08 12

Solution

(i) The 'join' calculation for AB gives

$$\alpha_{AB} = 197°\ 06'30'' \text{ and } D_{AB} = 1579.137\,\text{m}$$

If the observed directions are d_A, d_B, and d_C, then the angles α and β are found as follows:

$$\alpha = d_B - (d_C - 180°) = 171°13'11'' - 100°08'12'' = 71°04'59''$$

$$\beta = (d_C - 180°) - d_A = 100°08'12'' - 57°45'00'' = 42°23'12''$$

Therefore, the bearings α_{AX} and α_{BX} are

$$\alpha_{AX} = \alpha_{AB} - \alpha = 197°\ 06'30'' - 71°\ 04'59'' = 126°\ 01'31''$$

$$\alpha_{BX} = \alpha_{BA} + \beta = 17°06'30'' + 42°23'12'' = 59°29'42''$$

(ii) The bearings α_{AX} and α_{BX}, and the known coordinates of A and B, are used to calculate the coordinates of X by triangulation intersection, and these are found to be

	E	N
X	9184.788 m	3420.388 m

(iii) A 'join' XC is calculated:

$$\alpha_{XC} = 278°46'43''; \quad D_{XC} = 3224.837\,\text{m}$$

(iv) The orientation correction θ is found:

$$\theta = \alpha_{XC} - d_C = 278°46'43'' - 280°08'12'' = -1°21'29''$$

(v) The orientation correction is applied to the observations to give oriented bearings:

$$\alpha_{PA} = d_A + \theta = 57°45'00'' + (-1°21'29'') = 56°23'31''$$

$$\alpha_{PB} = d_B + \theta = 171°13'11'' + (-1°21'29'') = 169°51'42''$$

$$\alpha_{PC} = d_C + \theta = 280°08'12'' + (-1°21'29'') = 278°46'43''$$

(vi) The reverse bearings α_{AP} and α_{BP}, and the known coordinates of A and B, are used to calculate the coordinates of P by triangulation intersection and these are found to be

	E	N
P	7589.759 m	3666.702 m

(vii) The coordinates of P, derived in (vi), and the coordinates of C, are used to calculate a 'join' PC:

$\alpha_{PC} = 278°46'44''$; $D_{PC} = 1610.901$ m

The value of α_{PC} agrees with the value derived in (v), hence the resection calculation is verified.

The direct solution

The equations for the direct solution (referred to as Tienstra's Resection) are derived from two sets of formulae. The first are standard trigonometric formulae which, when related to Figure 2.17, are

$$\left.\begin{aligned}(m+n)\cot\theta &= n\cot\theta_A - m\cot\theta_B\\ (m+n)\cot\theta &= m\cot\sigma - n\cot\phi\end{aligned}\right\} \qquad (2.23)$$

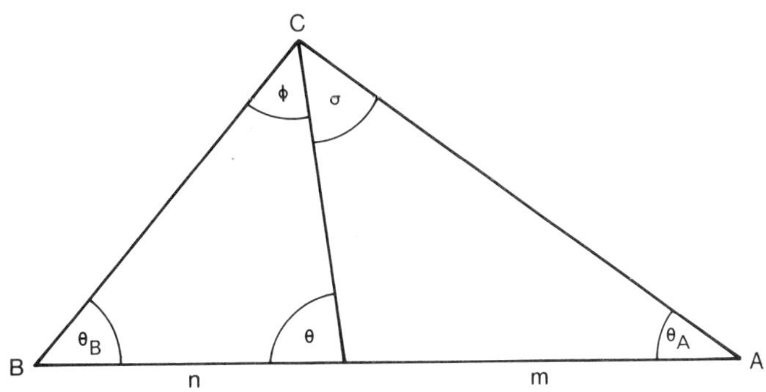

Figure 2.17

Equations (2.23) may be applied to the situation shown in Figure 2.18, which is similar to that shown in Figure 2.15*a*, but with more detail included.

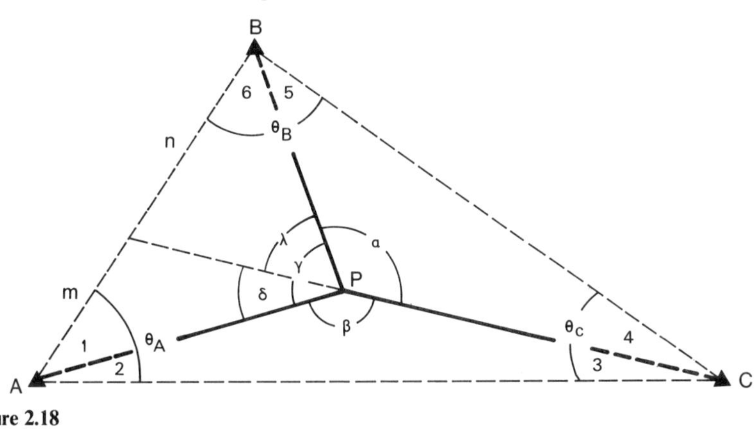

Figure 2.18

The following equations are obtained when the triangles ABP and ABC are considered.
From the triangle ABP:

$$(m+n)\cot\theta = n\cot 1 - m\cot 6 \tag{2.24}$$

$$(m+n)\cot\theta = m\cot\delta - n\cot\lambda \tag{2.25}$$

and since

$$\delta = 180 - \beta, \lambda = 180 - \alpha$$

therefore

$$(m+n)\cot\theta = -m\cot\beta + n\cot\alpha \tag{2.26}$$

From the triangle ABC:

$$(m+n)\cot\theta = n\cot\theta_A - m\cot\theta_B \tag{2.27}$$

$$(m+n)\cot\theta = m\cot 3 - n\cot 4 \tag{2.28}$$

Equating the right-hand sides of eqns (2.24) and (2.28),

$$m(\cot 3 + \cot 6) = n(\cot 1 + \cot 4) \tag{2.29}$$

and for eqns (2.26) and (2.27),

$$m(\cot\theta_B - \cot\beta) = n(\cot\theta_A - \cot\alpha) \tag{2.30}$$

From eqns (2.29) and (2.30),

$$\frac{n}{m} = \frac{\cot 3 + \cot 6}{\cot 1 + \cot 4} = \frac{\cot\theta_B - \cot\beta}{\cot\theta_A - \cot\alpha} \tag{2.31}$$

Let

$$\left.\begin{aligned} n_A &= 1/(\cot\theta_A - \cot\alpha) \\ n_B &= 1/(\cot\theta_B - \cot\beta) \\ n_C &= 1/(\cot\theta_C - \cot\gamma) \end{aligned}\right\} \tag{2.32}$$

then

$$\frac{n_A}{n_B} = \frac{\cot 3 + \cot 6}{\cot 1 + \cot 4} = \frac{1/(\cot\theta_A - \cot\alpha)}{1/(\cot\theta_B - \cot\beta)} \tag{2.33}$$

and similarly

$$\frac{n_B}{n_C} = \frac{\cot 1 + \cot 4}{\cot 2 + \cot 5} = \frac{1/(\cot\theta_B - \cot\beta)}{1/(\cot\theta_C - \cot\gamma)} \tag{2.34}$$

Therefore

$$(\cot 3 + \cot 6):(\cot 1 + \cot 4):(\cot 2 + \cot 5) = n_A:n_B:n_C$$

and

$$\left.\begin{aligned}\frac{n_A}{n_A+n_B+n_C}&=\frac{\cot 3+\cot 6}{\cot 1+\cot 2+\cot 3+\cot 4+\cot 5+\cot 6}\\ \frac{n_B}{n_A+n_B+n_C}&=\frac{\cot 1+\cot 4}{\cot 1+\cot 2+\cot 3+\cot 4+\cot 5+\cot 6}\\ \frac{n_C}{n_A+n_B+n_C}&=\frac{\cot 2+\cot 5}{\cot 1+\cot 2+\cot 3+\cot 4+\cot 5+\cot 6}\end{aligned}\right\}\quad(2.35)$$

The second set of formulae are eqns (2.10). When these are related to Figure 2.18 the following equations are obtained:

$$\left.\begin{aligned}E_P&=\frac{E_A\cot 6+E_B\cot 1-N_A+N_B}{\cot 1+\cot 6}\\ N_P&=\frac{N_A\cot 6+N_B\cot 1+E_A-E_B}{\cot 1+\cot 6}\end{aligned}\right\}\quad(2.36)$$

Therefore

$$\left.\begin{aligned}E_P(\cot 1+\cot 6)&=E_A\cot 6+E_B\cot 1-N_A+N_B\\ N_P(\cot 1+\cot 6)&=N_A\cot 6+N_B\cot 1+E_A-E_B\end{aligned}\right\}\quad(2.37)$$

Similarly

$$\left.\begin{aligned}E_P(\cot 3+\cot 2)&=E_C\cot 2+E_A\cot 3-N_C+N_A\\ N_P(\cot 3+\cot 6)&=N_C\cot 2+N_A\cot 3+E_C-E_A\end{aligned}\right\}\quad(2.38)$$

and

$$\left.\begin{aligned}E_P(\cot 5+\cot 4)&=E_B\cot 4+E_C\cot 5-N_B+N_C\\ N_P(\cot 5+\cot 4)&=N_B\cot 4+N_C\cot 5+E_B-E_C\end{aligned}\right\}\quad(2.39)$$

Adding the first equations of the pairs (2.37), (2.38) and (2.39) gives

$$\begin{aligned}E_P(\cot 1+\cot 2+\cot 3+\cot 4+\cot 5+\cot 6)\\ =E_A(\cot 3+\cot 6)+E_B(\cot 1+\cot 4)\\ +E_C(\cot 2+\cot 5)\end{aligned}$$

Therefore, considering also eqns (2.35),

$$E_P=E_A\cdot\frac{n_A}{n_A+n_B+n_C}+E_B\cdot\frac{n_B}{n_A+n_B+n_C}+E_C\cdot\frac{n_C}{n_A+n_B+n_C}$$

Therefore

$$\left.\begin{aligned} E_P &= \frac{n_A E_A + n_B E_B + n_C E_C}{n_A + n_B + n_C} \\ \text{and similarly} \qquad & \\ N_P &= \frac{n_A N_A + n_B N_B + n_C N_C}{n_A + n_B + n_C} \end{aligned}\right\} \tag{2.40}$$

Once the angles at A, B, C have been found from the known coordinates, and the angles α, β, γ have been derived from the observations at P, the values of n_A, n_B, n_C are calculated using eqns (2.32). Checks on the calculations are provided by ensuring that the following equations are satisfied:

$$\left.\begin{aligned} \theta_A + \theta_B + \theta_C &= 180^\circ \\ \alpha + \beta + \gamma &= 360^\circ \\ \cot\theta_A \cot\theta_B + \cot\theta_B \cot\theta_C + \cot\theta_C \cot\theta_A &= 1 \\ \cot\alpha\cot\beta + \cot\beta\cot\gamma + \cot\gamma\cot\alpha &= 1 \end{aligned}\right\} \tag{2.41}$$

Equations (2.40) are then used to obtain the coordinates. The advantages in using the equations are

(i) The formulae are easily remembered. Eastings are in terms of eastings only, and northings in terms of northings. Furthermore, an arbitrary local origin may be chosen for computational purposes; even to the extent of making the easting and/or northing coordinate of one known point 0.
(ii) The values of $(\cot\theta_A - \cot\alpha)$, etc. indicate the strength of the resection. If these values are 0 the point is on the danger circle, and if they are small a weak resection is the result.

The one disadvantage is that the equations break down if the three known points are collinear.

Example 2.5 illustrates a convenient way to set out the computation if the calculator storage capacity is limited. The data is the same as that used in the previous example for Collin's Point resection.

Example 2.5

The known coordinates and the observations are

Coordinates Point	E (m)	N (m)	*Observations at P* To	Direction ° ′ ″
A	8246.18	4102.96	A	57 45 43
B	7781.63	2593.70	B	171 13 11
C	5997.73	3912.56	C	280 08 12

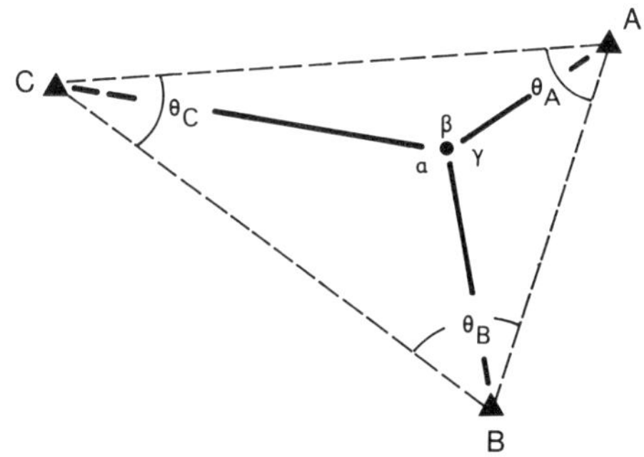

Figure 2.19

Solution

The calculation of 'joins' between the three known points produces bearings, from which the corresponding angles may be derived. These angles plus those derived from the observations are given below.

Line	Bearing		Angle		Angle
	° ′ ″		° ′ ″		° ′ ″
AB	197 06 30	θ_A	68 03 05	α	108 55 01
BC	306 28 34	θ_B	70 37 56	β	137 37 31
CA	85 09 35	θ_C	41 18 59	γ	113 27 28

Table 2.3 Example on Tienstra's Resection

Point		Angle ° ′ ″	cot	*n*	*E* (m)	*N* (m)
A	θ_A	68 03 05	0.402983	1.341039	8246.18	4102.96
	α	108 55 01	−0.342707			
B	θ_B	70 37 56	0.351524	0.690782	7781.63	2593.70
	β	137 37 31	−1.096110			
C	θ_C	41 18 59	1.137620	0.636312	5997.73	3912.56
	γ	113 27 28	−0.433936			
P			$n_A + n_B + n_C =$	2.668134	7589.684	3666.805

Clearly such a table is not required if the calculator or computer storage facility is such that it becomes unnecessary to record intermediate values.

2.9 Intersections in general

Sections 2.6 to 2.8 deal with the computations more frequency encountered in surveying. However, other intersection problems arise from time to time;

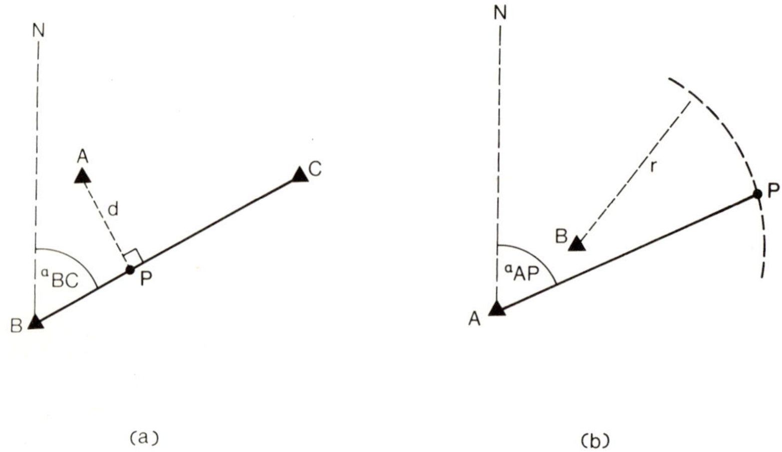

Figure 2.20

one problem is the determination of the coordinates of the foot of a perpendicular from a known point to a straight line. Figure 2.20*a* shows points *A*, *B*, *C* which have known coordinates, and the coordinates of *P* at the foot of the perpendicular are required.

The problem could be solved by a sequence of calculations, namely

(i) establish the *bearing* α_{BC} from the known coordinates of *B* and *C*;
(ii) add 90° to α_{BC} to obtain the bearing α_{AP}; and
(iii) calculate the coordinates of *P* by *triangulation* using the bearings α_{BC} and α_{AP}, and the known coordinates of *A* and *B*.

Example 2.6

The coordinates of points *A*, *B*, *C*, and the data derived from these values are:

Point	*E* (m)	*N* (m)	Line	Bearing ° ′ ″	Distance (m)
A	5997.73	3912.57	*AB*	60 19 13	1475.390
B	7279.56	4643.10			
C	7103.26	4096.30			

Calculate the coordinates of *P*, the foot of the perpendicular from *C* to *AB*.

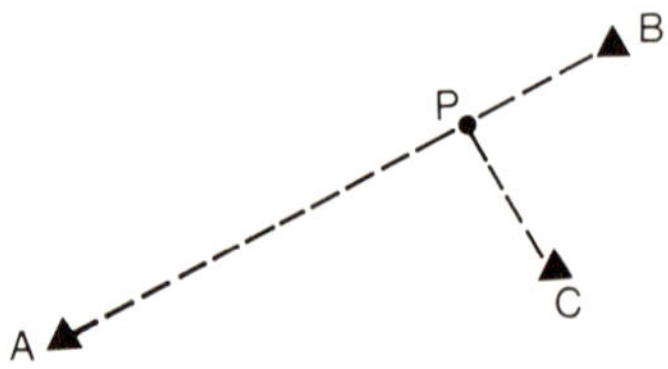

Figure 2.21

Solution

The bearing α_{CP} is derived:

$$\alpha_{CP} = \alpha_{AP} - 90° = 60°19'13'' - 90° = 330°19'13''$$

and the point P is calculated by triangulation intersection from the bearings α_{AP} and α_{CP}:

	E	N
P	6911.251 m	4433.205 m

In Figure 2.20*a*, d is the shortest distance from A to the line BC, and once coordinates of P have been established a 'join' calculation for the line AP will determine this distance.

Alternatively d may be found directly, as follows. The *two-point form* of the equation of the line BC is

$$\frac{E - E_B}{E_C - E_B} = \frac{N - N_B}{N_C - N_B} \tag{2.42}$$

which may be rearranged to give

$$\Delta N_{BC} E - \Delta E_{BC} N + (E_C N_B - E_B N_C) = 0 \tag{2.43}$$

The *normal form* (in the sense of perpendicular) of the equation of the line BC is eqn (2.43) divided by $+(\Delta E_{BC}^2 + \Delta N_{BC}^2)^{1/2}$.

Therefore, if $\quad D_{BC} = +(\Delta E_{BC}^2 + \Delta N_{BC}^2)^{1/2}$,

the normal form is

$$\frac{\Delta N_{BC}}{D_{BC}} \cdot E - \frac{\Delta E_{BC}}{D_{BC}} \cdot N + \frac{(E_C N_B - E_B N_C)}{D_{BC}} = 0 \tag{2.44}$$

If A is a point with known coordinates which is off the line BC, then the shortest distance (d) from it to the line is the amount by which the equation is not satisfied when the coordinates of A are substituted. Thus,

$$\left.\begin{aligned} d &= \frac{\Delta N_{BC}}{D_{BC}} \cdot E_A - \frac{\Delta E_{BC}}{D_{BC}} \cdot N_A + \frac{(E_C N_B - E_B N_C)}{D_{BC}} \\ \text{or} \quad d &= E_A \cos \alpha_{BC} - N_A \sin \alpha_{BC} + \frac{(E_C N_B - E_B N_C)}{D_{BC}} \end{aligned}\right\} \tag{2.45}$$

Thus, with d and the bearing α_{AP} known, the coordinates of P may be calculated if necessary from the 'polar' AP.

Example 2.7

Use the same data as Example 2.6 and eqn (2.45) to derive the perpendicular distance d from C to the line AB.

Solution

Equation (2.45) applied to this problem becomes

$$d = E_C \cos\alpha_{AB} - N_C \sin\alpha_{AB} + \frac{(E_B N_A - E_A N_B)}{D_{AB}}$$

$$= 3517.188 - 3558.893 + 429.532$$

$$= 387.827\,\text{m}$$

Another possible intersection problem is that shown in Figure 2.20*b*. The coordinates of the point P are required, being the intersection of a circle and a straight line. The circle has its centre at B, which has known coordinates, and the radius (r) of the circle is also known. The straight line AP is defined by the known coordinates of A and the bearing α_{AP}.

The equation for the line AP is

$$\tan\alpha_{AP} = \frac{\Delta E_{AP}}{\Delta N_{AP}} = \frac{E_P - E_A}{N_P - N_A} \tag{2.46}$$

and for the circle

$$r^2 = \Delta E_{BP}^2 + \Delta N_{BP}^2 = (E_P - E_B)^2 + (N_P - N_B)^2 \tag{2.47}$$

If eqn (2.46) is rearranged it may be written

$$E_P = (N_P - N_A)\tan\alpha_{AP} + E_A$$

which may be substituted in eqn (2.47) to give

$$r^2 = ((N_P - N_A)\tan\alpha_{AP} - E_B)^2 + (N_P - N_B)^2 \tag{2.48}$$

N_P is the only unknown in eqn (2.48), and when values for the other quantities are substituted the equation becomes a quadratic equation in N_P. The roots are calculated and the choice of either root will be evident from an inspection of the diagram.

Example 2.8

Points A and B have known coordinates which are given below. P is a point which must be fixed, using an observed bearing from A and an observed distance from B.

Coordinates

Point	E (m)	N (m)
A	7589.69	3666.80
B	8246.18	4102.96

Observations

$\alpha_{AP} = 60°31'20''$

$D_{BP} = 852.61\,\text{m}$

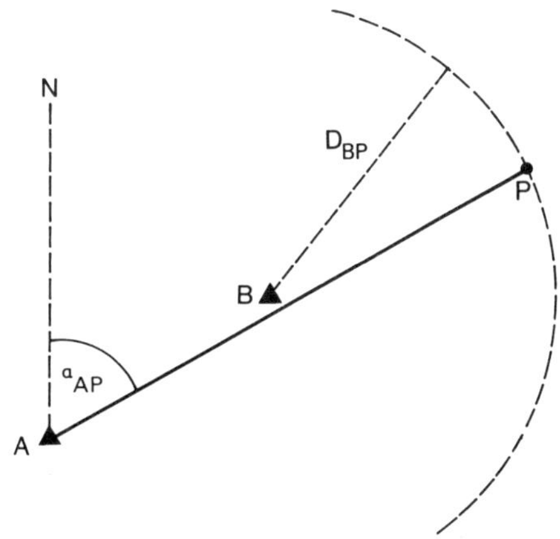

Figure 2.22

Solution

Equation (2.48) applied to this problem becomes

$$D_{BP}^2 = ((N_P - N_A)\tan\alpha_{AP} - E_B)^2 + (N_P - N_B)^2$$

For convenience, let point A have coordinates 0, 0. Therefore the new coordinates are

	E'	N'
	(m)	
A	0	0
B	656.49	436.16

Therefore

$$D_{BP}^2 = (N'_P \tan\alpha_{AP} - E'_B)^2 + (N'_P - N'_B)^2$$

and when values are substituted.

$$852.61^2 = (1.769095\,N'_P - 656.49)^2 + (N'_P - 436.16)^2$$

Therefore

$$4.12969\,N'^2_P + (-3195.10)\,N'_P + (-10729) = 0$$

and

$$x = \frac{-b \pm (b^2 - 4ac)^{1/2}}{2a}$$

where $x = N'_P$, $a = 4.12969$, $b = -3195.10$, $c = -105729$.

Therefore, the two roots are

$$N'_P = 805.476, \quad \text{or} \quad -31.785$$

and the corresponding E'_P values are

$$E'_P = E'_A + (N'_P - N'_A)\tan\alpha_{AP} = 1424.963 \quad \text{or} \quad -56.231$$

Thus the E, N values are

$$E_P = E_A + E'_P = 9014.653 \quad \text{or} \quad 7533.459$$
$$N_P = N_A + N'_P = 4472.276 \quad \text{or} \quad 3635.015$$

and an inspection of the diagram establishes that the former set are the required pair of coordinates; thus

	E	N
P	9014.653	4472.276

The 'joins' AP and BP verify this result:

$$\alpha_{AP} = 60°31'20'', \quad \text{and} \quad D_{BP} = 852.611\,\text{m}$$

Sections 2.6 and 2.7 dealt with direct methods of obtaining the coordinates of intersected points when a minimum number of observations was used. Intersection problems may also be solved by using the principle of linearization. Equations which express a non-linear relationship between observations and the coordinates of the point to be fixed may be made linear in the vicinity of the point, and then solved simultaneously to provide a solution to the problem.

The method relies on assigning initial approximate values for the coordinates of the intersected point, and the solution of two linear equations establishes corrections to these values in order to obtain improved values for the coordinates. The improved values become new approximate values, and the process is repeated until the corrections are negligible. Figure 2.23 shows situations involving trilateration and triangulation respectively. A and B are

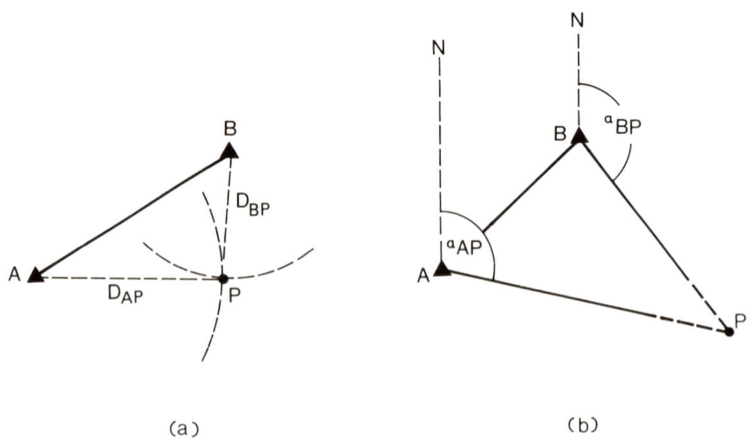

Figure 2.23: *a*, trilateration; *b*, triangulation.

points with known coordinates, and observations have been made from these points to fix the position of P.

In trilateration the equation for the locus of a distance measurement (D) is

$$D^2 = \Delta E^2 + \Delta N^2 \tag{2.49}$$

and when it is linearized using Taylor's theorem (see Appendix A) it becomes

$$2D \cdot \mathrm{d}D = 2\Delta E \cdot \mathrm{d}(\Delta E) + 2\Delta N \cdot \mathrm{d}(\Delta N)$$

or

$$\mathrm{d}D = \left(\frac{\Delta E}{D}\right)_0 \cdot \mathrm{d}(\Delta E) + \left(\frac{\Delta N}{D}\right)_0 \cdot \mathrm{d}(\Delta N) \tag{2.50}$$

where

(i) subscript 0 implies that the expression is evaluated using approximate (provisional) values for the coordinates of P;
(ii) $\mathrm{d}D$ is the observed value minus the computed value for D; the computed value being derived from the provisional coordinates of P; and
(iii) $\mathrm{d}(\Delta E)$ and $\mathrm{d}(\Delta N)$ are small unknown corrections to ΔE and ΔN respectively.

In order to simplify the notation, eqn (2.50) may be written

$$\mathrm{d}D = r \cdot \mathrm{d}(\Delta E) + s \cdot \mathrm{d}(\Delta N) \tag{2.51}$$

where

$$r = (\Delta E/D)_0 \quad \text{and} \quad s(\Delta N/D)_0.$$

Now

$$\Delta E_{AP} = E_P - E_A$$
$$\Delta N_{AP} = N_P - N_A$$

therefore

$$\left.\begin{aligned} \mathrm{d}(\Delta E)_{AP} &= \mathrm{d}E_P - \mathrm{d}E_A \\ \mathrm{d}(\Delta N)_{AP} &= \mathrm{d}N_P - \mathrm{d}N_A \end{aligned}\right\} \tag{2.52}$$

So, for the observation AP, eqn (2.50) becomes

$$\mathrm{d}D_{AP} = r_{AP}(\mathrm{d}E_P - \mathrm{d}E_A) + s_{AP}(\mathrm{d}N_P - \mathrm{d}N_A) \tag{2.53a}$$

and similarly, for the observation D_{BP},

$$\mathrm{d}D_{BP} = r_{BP}(\mathrm{d}E_P - \mathrm{d}E_B) + s_{BP}(\mathrm{d}N_P - \mathrm{d}N_B) \tag{2.53b}$$

Thus linear equations have been established relating small changes in the measured distances to corresponding changes in the coordinates of the end points of the line. Normally the coordinates of the control point will be held

fixed, therefore dE_A, dN_A, dE_B, dN_B will be equal to 0, and the terms will not appear in the equations.
Therefore

$$\left.\begin{aligned} dD_{AP} &= r_{AP}dE_P + s_{AP}dN_P \\ dD_{BP} &= r_{BP}dE_P + s_{BP}dN_P \end{aligned}\right\} \tag{2.54}$$

where

dE_P, dN_P are the unknowns—corrections to provisional coordinates of P;

dD_{AP}, dD_{BP} are known discrepancies—the observed minus the computed values; they must be derived precisely, to an order of precision equivalent to the observations, and this implies that the computed distances must also be to the same order.

$\left.\begin{matrix} r_{AP}, s_{AP} \\ r_{BP}, s_{BP} \end{matrix}\right\}$ are known coefficients, derived from the known coordinates of A and B and the provisional coordinates of P; the number of digits used should be consistent with the size of the discrepancies; note that these coefficients are also $\sin\alpha_{AP}$, $\cos\alpha_{AP}$, $\sin\alpha_{BP}$, $\cos\alpha_{BP}$, where α_{AP} and α_{BP} are bearings for the lines.

The values for the unknowns dE_P, dN_P, are derived from a solution of the two simultaneous equations. These values are added to the provisional values of P to provide improved values for the point. The computation is repeated using the new provisional values, and the iterations continue until the corrections to the coordinates become negligible. The number of iterations depends on the accuracy of initial approximate values; however, in practice it is a simple matter to obtain values reliable enough that only one computational cycle is necessary.

Example 2.9

Point P is to be fixed using distance measurements made from two control points A and B. The data given in Example 2.3 is used again for this problem.

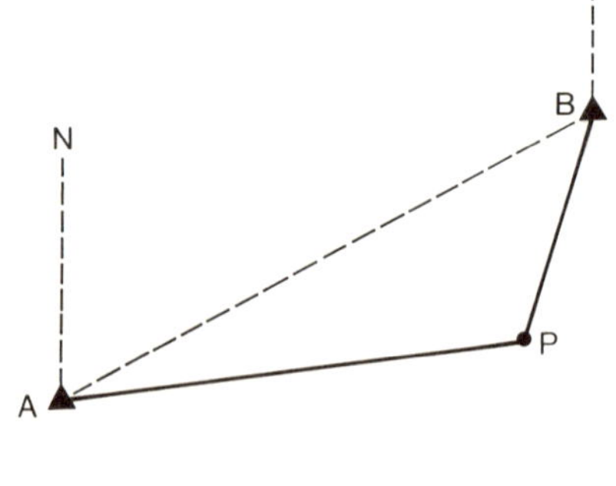

Figure 2.24

Coordinates: Point	E (m)	N (m)	*Observations:* Line	Distance (m)
A	5997.73	3912.56	AP	1120.76
B	7279.56	4643.10	BP	574.53

Solution

Let the approximate coordinates of P be

E	N
7103m	4096m

and when these are used to calculate the 'joins' AP and BP the 'computed' values for bearing and distance are obtained. These values, together with the observed values, allow the determination of the required elements of eqns (2.54).

Line	Computed bearing °	′	″	Computed distance (m)	Observed distance (m)	Obs. minus comp. dist. (dD)	$\sin\alpha$ (r)	$\cos\alpha$ (s)
AP	80	34	36	1120.389	1120.76	0.371	0.9865	0.1637
BP	197	53	09	574.884	574.53	−0.354	−0.3071	−0.9517

Therefore, eqns (2.54) applied to this problem become

$$0.371 = 0.9865\mathrm{d}E_P + 0.1637\mathrm{d}N_P$$
$$-0.354 = -0.3071\mathrm{d}E_P - 0.9517_d N_P$$

and the simultaneous solution provides values for $\mathrm{d}E_P$ and $\mathrm{d}N_P$. The equations may be expressed in matrix notation as

$$\boldsymbol{Ax} = \boldsymbol{b}$$

where

$$\boldsymbol{A} = \begin{pmatrix} 0.9865 & 0.1637 \\ -0.3071 & -0.9517 \end{pmatrix}, \quad \boldsymbol{x} = \begin{pmatrix} \mathrm{d}E_P \\ \mathrm{d}N_P \end{pmatrix}, \quad \boldsymbol{b} = \begin{pmatrix} 0.371 \\ -0.354 \end{pmatrix}$$

and the solution is found from

$$\boldsymbol{x} = \boldsymbol{A}^{-1}\boldsymbol{b} = \frac{1}{-0.888580}\begin{pmatrix} -0.9517 & -0.1637 \\ 0.3071 & 0.9865 \end{pmatrix}\begin{pmatrix} 0.371 \\ -0.354 \end{pmatrix} = \begin{pmatrix} 0.332 \\ 0.265 \end{pmatrix}$$

Therefore the coordinates of P are

$$E_P = 7103 + 0.332 = 7103.332\mathrm{m}$$
$$N_P = 4096 + 0.265 = 4096.265\mathrm{m}.$$

These values would be used as improved approximate values for the coordinates of P, and the elements (coefficients and constant terms) of the two equations are re-established. The solution of equations is repeated to obtain new corrections.

Such a computational cycle is repeated until the corrections become negligible. In the case of the data supplied for this example, it will be clear that one cycle is sufficient.

The same principle may be applied to triangulation intersection, where the relationship between the observed bearing and the coordinates of the intersected point is non-linear. In Figure 2.23*b* the bearings α_{AP} and α_{BP} have been derived from observations at known points *A* and *B*, in order to fix the unknown point *P*. For either observation the following relationship exists:

$$\tan\alpha = \frac{\Delta E}{\Delta N} \tag{2.55}$$

which, on linearization, becomes

$$\frac{1}{\cos^2\alpha}\cdot d\alpha = \frac{1}{\Delta N}\cdot d(\Delta E) - \frac{\Delta E}{(\Delta N)^2}\cdot d(\Delta N)$$

or

$$d\alpha = \frac{\cos^2\alpha}{\Delta N}\cdot d(\Delta E) - \cos^2\alpha\cdot\frac{\Delta E}{(\Delta N)^2}\cdot d(\Delta N)$$

and since

$$\cos\alpha = \Delta N/D$$

therefore

$$d\alpha = \frac{\cos\alpha}{D}\cdot d(\Delta E) - \frac{\sin\alpha}{D}\cdot d(\Delta N)$$

or, by simplifying the notation,

$$d\alpha = p\cdot d(\Delta E) - q\cdot d(\Delta N) \tag{2.56}$$

where

$$p = \cos\alpha/D \quad \text{and} \quad q = \sin\alpha/D.$$

Again, for the line *AP*, $d(\Delta E)$ and $d(\Delta N)$ may be replaced by $(dE_P - dE_A)$ and $(dN_P - dN_A)$ respectively, therefore

$$d\alpha_{AP} = p_{AP}\cdot(dE_P - dE_A) - q_{AP}\cdot(dN_P - dN_A) \tag{2.57a}$$

and, for the line *BP*,

$$d\alpha_{BP} = p_{BP}\cdot(dE_P - dE_B) - q_{BP}\cdot(dN_P - dN_B) \tag{2.57b}$$

where dE_A, dN_A, dE_B, dN_B are 0 if the control points are considered to be fixed points.

Therefore the equations reduce to

$$\left.\begin{aligned} d\alpha_{AP} &= p_{AP}dE_P - q_{AP}dN_P \\ d\alpha_{BP} &= p_{BP}dE_P - q_{BP}dN_P \end{aligned}\right\} \tag{2.58}$$

where

dE_P, dN_P	are the unknowns—corrections to provisional coordinates of P;
$d\alpha_{AP}, d\alpha_{BP}$	are known discrepancies—the observed minus computed values—which must be derived to the same order of precision as the observations; and
$\left.\begin{matrix} p_{AP}, q_{AP} \\ p_{BP}, q_{BP} \end{matrix}\right\}$	are known coefficients, derived from the known coordinates of A and B, and the provisional coordinates of P; the number of digits used should be consistent with the size of the discrepancies.

In eqns (2.58) $d\alpha_{AP}$ and $d\alpha_{BP}$ are in radian measure and will be very small quantities. Similarly the coefficients p and q are also very small. The values may be converted into more manageable quantities by expressing the discrepancies in seconds of arc and multiplying the coefficients on the right-hand side by ρ, the factor for converting from radians to seconds of arc.

Thus, the equations become

$$\left.\begin{aligned} d\alpha_{AP} &= \rho p_{AP}\cdot dE_P - \rho q_{AP}\cdot dN_P \\ d\alpha_{BP} &= \rho p_{BP}\cdot dE_P - \rho q_{BP}\cdot dN_P \end{aligned}\right\} \tag{2.59}$$

where $\rho = 206265$ is normally sufficient.

Values for dE_P, dN_P are derived from a simultaneous solution of the equations; they are added to the initial provisional values of P to provide improved values, and these in turn are used as new provisional values. The process is repeated until discrepancies become negligible. With reasonably accurate initial values only one computational cycle is required.

Example 2.10

Point P is to be fixed from bearings, derived from observations made at two control points A and B. The same data as that used in Example 2.2 is used.

Coordinates:			*Observations:*	
Point	E	N	Line	Bearing
	(m)			° ′ ″
A	6091.13	3613.61	AP	126 08 26
B	5973.88	1561.00	BP	53 22 37

Solution

Let the approximate coordinates of P be

E	N
7437	2649

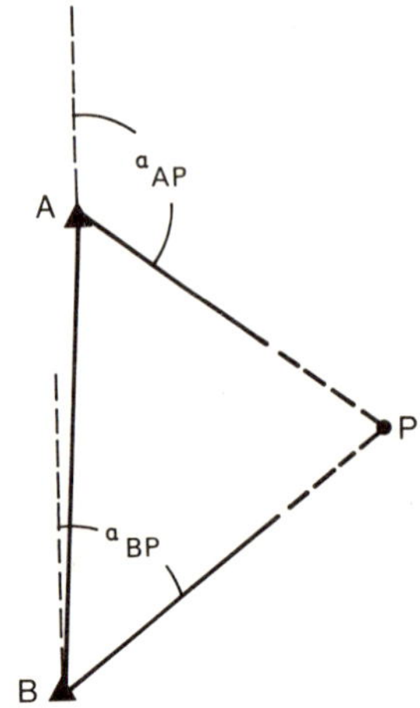

Figure 2.25

and when these are used to calculate 'joins' AP and BP the 'computed' values for bearing and distance are obtained. These values, together with the observed values, allow the determination of the required elements of eqns (2.59).

Line	Computed distance (m)	Computed bearing ° ′ ″	Observed bearing ° ′ ″	Observed— comp. bearing ($d\alpha''$)	$\frac{\rho \cdot \cos\alpha}{D}$ (p)	$\frac{\rho \cdot \sin\alpha}{D}$ (q)
AP	1666.400	126 07 58	126 08 26	28	− 72.99	99.97
BP	1823.311	53 21 54	53 22 37	43	67.50	90.78

Therefore, eqns (2.59) applied to this problem become

$$28 = -72.99\mathrm{d}E_P - 99.97\mathrm{d}N_P$$

$$43 = 67.50\mathrm{d}E_P - 90.78\mathrm{d}N_P$$

and the simultaneous solution provides values for $\mathrm{d}E_P$ and $\mathrm{d}N_P$. The equations may be expressed in matrix notation as

$$\boldsymbol{Ax} = \boldsymbol{b}$$

where

$$\boldsymbol{A} = \begin{pmatrix} -72.99 & -99.97 \\ 67.50 & -90.78 \end{pmatrix}, \quad \boldsymbol{x} = \begin{pmatrix} \mathrm{d}E_P \\ \mathrm{d}N_P \end{pmatrix}, \quad \boldsymbol{b} = \begin{pmatrix} 28'' \\ 43'' \end{pmatrix}$$

and the solution is found from

$$\boldsymbol{x} = \boldsymbol{A}^{-1}\boldsymbol{b} = \frac{1}{13374}\begin{pmatrix} -90.78 & 99.97 \\ -67.50 & -72.99 \end{pmatrix}\begin{pmatrix} 28 \\ 43 \end{pmatrix} = \begin{pmatrix} 0.131 \\ -0.376 \end{pmatrix}$$

Therefore the coordinates of P are

$$E_P = 7437 + 0.131 = 7437.131\mathrm{m}$$

$$N_P = 2649 + (-0.376) = 2648.624\mathrm{m}$$

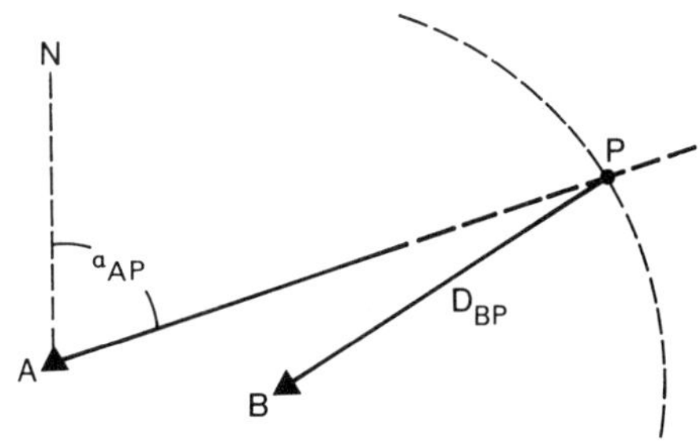

Figure 2.26 Mixed observation intersection.

These values may be used as improved approximate values for a further iteration; however, the new corrections are found to be negligible and so the coordinates above may be adopted as the final coordinates of P.

Equations (2.54) and (2.58) may also be used together to solve an intersection problem which involves both types of measurement. Thus in Figure 2.26 points A and B are control points with known coordinates; a bearing α_{AP} has been observed from A and a distance D_{BP} has been observed from B. Example 2.8 demonstrates how the problem may be solved directly. It may also be solved using linearized equations.

The two equations for these observations are

$$\left.\begin{aligned} d\alpha_{AP} &= p_{AP}dE_P - q_{AP}dN_P \\ dD_{BP} &= r_{BP}dE_P + s_{BP}dN_P \end{aligned}\right\} \quad (2.60)$$

and a simultaneous solution would provide values for dE_P, dN_P.

Example 2.11

Point P is to be fixed from a bearing derived from observations at control point A, and a distance observed from control point B. The same data as that given in Example 2.8 is used.

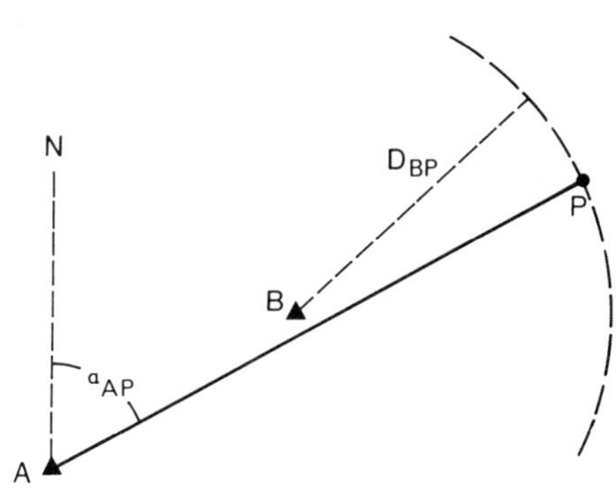

Figure 2.27

Coordinates:

Point	E (m)	N (m)
A	7589.69	3666.80
B	8246.18	4102.96

Observations:

$\alpha_{AP} = 60°31'20''$
$D_{BP} = 852.61\,\text{m}$

Solution

Let the approximate coordinates of P be

E	N
9015m	4472m

and when these are used to calculate the 'joins' AP and BP the 'computed' values for bearing and distance are obtained. These values, together with the observed values, allow the determination of the required elements of eqns (2.60)

Line	Computed bearing ° ′ ″	Computed distance (m)	Observed bearing ° ′ ″	Observed distance (m)	Observed— comp. bear. (dα)″	Observed— comp. dist. (dD)m	$\frac{\rho \cdot \cos\alpha}{D}$ (p)	$\frac{\rho \cdot \sin\alpha}{D}$ (q)	$\sin\alpha$ (r)	$\cos\alpha$ (s)
AP	60 32 12	1637.026	60 31 20	...	−52	...	61.98	109.7	...	...
BP	64 21 31	852.804	...	852.61	...	−0.194	...	...	0.9015	0.4327

Therefore, eqns (2.60) applied to this problem become

$$-52 = 61.98\text{d}E_P - 109.7\text{d}N_P$$

$$-0.194 = 0.9015\text{d}E_P + 0.4327\text{d}N_P$$

and the simultaneous solution provides values for $\text{d}E_P$ and $\text{d}N_P$. The equations may be expressed in matrix notation as

$$\boldsymbol{A}\boldsymbol{x} = \boldsymbol{b}$$

where

$$\boldsymbol{A} = \begin{pmatrix} 61.98 & -109.7 \\ 0.9015 & 0.4327 \end{pmatrix}, \quad \boldsymbol{x} = \begin{pmatrix} \text{d}E_P \\ \text{d}N_P \end{pmatrix}, \quad \boldsymbol{b} = \begin{pmatrix} -52'' \\ -0.194\text{m} \end{pmatrix}$$

and the solution is found from

$$\boldsymbol{x} = \boldsymbol{A}^{-1}\boldsymbol{b} = \frac{1}{125.713}\begin{pmatrix} 0.4327 & 109.7 \\ -0.9015 & 61.98 \end{pmatrix}\begin{pmatrix} -52 \\ -0.194 \end{pmatrix} = \begin{pmatrix} -0.348 \\ 0.277 \end{pmatrix}$$

Therefore the coordinates of P are

$$E_P = 9015 + (-0.348) = 9014.652\text{m}$$

$$N_P = 4472 + \quad 0.277 = 4472.277\text{m}$$

When these values are used as improved approximate coordinates, and the computational cycle repeated, the new corrections are found to be negligible. Therefore the values above may be adopted as the final coordinates for P.

Note that with this type of intersection problem, the control points A and B should subtend a fairly small angle at P to obtain a reliable fix for the point.

The solution of intersection problems using the principle of linearization may be extended to curves represented by higher-order functions. In practice however these functions seldom arise, and there is little point in considering the matter further.

2.10 Height determination

Traditionally the determination of height has been treated separately from the determination of planimetric position in surveying, and furthermore the commonly used datum surface (mean sea level) for heights is different from that used to represent planimetric information. Levelling instruments are designed specifically for the direct determination of height differences, and it is the accumulation of height differences between a point of known height and the unknown point that establishes the height of the latter. Therefore the mathematical model and the computational processes involved are extremely simple.

Figure 2.28 illustrates the field procedure for levelling operations. Point A is a benchmark of known height (H_A) above datum, and B is a point whose height is required. Point C, the final point sighted, is also of known height, and it is the misclosure at this point that gives some indication of the reliability of the observations. Let the observations (backsights and foresights) made to successive positions of the levelling staff be represented by $b_1 \ldots b_4$ and $f_1 \ldots f_4$. In order to avoid the problems caused by earth curvature and refraction, it is established practice to ensure that the foresight distance is roughly equal to the backsight distance for a particular instrument station.

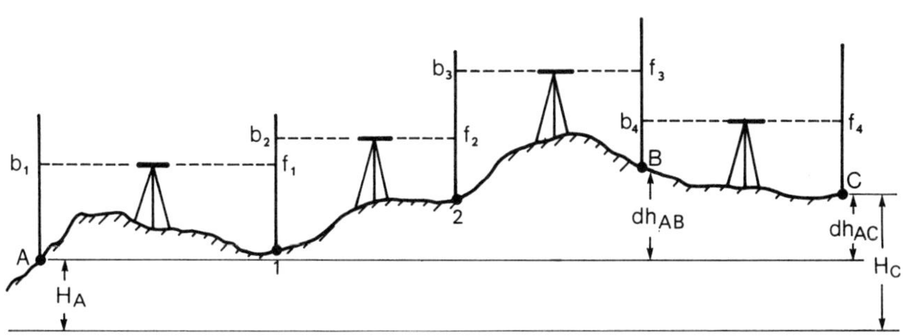

Figure 2.28 Levelling.

The height difference between A and B, expressed as a function of these observations, is

$$dh_{AB} = (b_1 - f_1) + (b_2 - f_2) + (b_3 - f_3) \tag{2.61}$$

Table 2.4 Rise and fall method

Back-sight	Inter-mediate	Fore-sight	Rise/fall	Reduced height	Remarks
b_1				H_A	benchmark A
b_2		f_1	$(b_1 - f_1)$	h_1 $(= H_A + (b_1 - f_1))$	1st change point
b_3		f_2	$(b_2 - f_2)$	h_2 $(= h_1 + (b_2 - f_2))$	2nd change point
	i_1		$(b_3 - i_1)$	h_{i_1} $(= h_1 + (b_3 - i_1))$	1st intermediate point
	i_2		$(i_1 - i_2)$	h_{i_2} $(= h_{i_1} + (i_1 - i_2))$	2nd intermediate point
b_4		f_3	$(i_2 - f_3)$	h_B $(= h_{i_2} + (i_2 - f_3))$	point B
		f_4	$(b_4 - f_4)$	h_C $(= h_B + (b_4 - f_4))$	benchmark C (observed height)
				H_C	benchmark C (known height)
$\sum b$		$\sum f$	$\sum \text{rise} + \sum \text{fall}$	$(H_C - h_C)$	Misclosure

and the height of B is found from

$$H_B = H_A + \mathrm{d}h_{AB} \tag{2.62}$$

If, in the process of levelling from A and B, other intermediate points have been observed, their heights may also be determined from an accumulation of height differences up to the points concerned. The arithmetic involved may be carried out systematically by arranging the observations in a tabular form, and two methods are shown in Tables 2.4 and 2.5.

In the Table 2.4 the sign of the difference $(b - f)$ will establish whether it is a rise or fall, and a check on the arithmetic may be carried out by summing the columns and verifying that

$$\sum b - \sum f = \sum \text{rise} + \sum \text{fall} = h_c - H_A \tag{2.63}$$

The misclosure is distributed between the observed height differences according to their relative weights, and the final reduced heights are then obtained. An intermediate point, which does not form part of the main levelling circuit, receives the same correction as its associated change point.

In Table 2.5 the recording of the observations (i.e. the first three columns) is identical to the first method. However, rises and falls are not deduced; instead, the height of collimation at each instrument station is calculated, and the staff reading for a particular point is subtracted from the height of collimation to obtain the height of the point.

Table 2.5 Height of collimation method

Back-sight	Inter-mediate sight	Fore-sight	Height of collimation	Reduced height	Remarks
b_1			$C_1 = H_A + b_1$	H_A	benchmark A
b_2		f_1	$C_2 = h_1 + b_2$	$h_1 = C_1 - f_1$	1st change point
b_3		f_2	$C_3 = h_2 + b_3$	$h_2 = C_2 - f_2$	2nd change point
	i_1			$h_{i_1} = C_3 - i_1$	1st intermed. pt.
	i_2			$h_{i_2} = C_3 - i_2$	2nd intermed. pt.
b_4		f_3	$C_4 = h_B + b_4$	$h_B = C_3 - f_3$	point B
		f_4		$h_C = C_4 - f_4$	benchmark C (observed height)
				H_c	benchmark C (known height)
$\sum b$		$\sum f$		$(H_C - h_C)$	

The arithmetic check is similar to the first method, namely

$$\sum b - \sum f = h_C - H_A \tag{2.64}$$

However, with this method there is no check that the reduced heights of the intermediate points have been derived correctly. A numerical example of levelling observations, and the corresponding results for the methods described above, are given in Tables 2.6 and 2.7.

Example 2.12

Both methods are used in practice, though each is suited to specific applications. The rise and fall method, which involves considerably more arithmetic, is more suitable when many intermediate sights are observed,

Table 2.6 Rise and fall method

Back sight	Inter-mediate sight	Fore-sight	Rise	Fall	Reduced height	Corr-ection	Adjusted height	Remarks
1.246					5.624		5.624	benchmark A
0.785		1.531		−0.285	5.339	+.002	5.341	1st change point
3.248		2.367		−1.582	3.757	+.004	3.761	2nd change point
	0.946		2.302		6.059	+.006	6.065	1st intermediate point
	3.512			−2.566	3.493	+.006	3.499	2nd intermediate point
0.816		2.804	0.708		4.201	+.006	4.207	point B
		3.542		−2.726	1.475	+.008	1.483	benchmark C (observed height)
					1.483			benchmark C (known height)
6.095		10.244	3.010	−7.159	+0.008			misclosure

$\Sigma b - \Sigma f = -4.149$
$\Sigma \text{rise} + \Sigma \text{fall} = -4.149$
$h_C - H_A = -4.149$

Table 2.7 Height of collimation method

Back-sight	Inter-mediate sight	Fore-sight	Height of collimation	Reduced height	Correc-tion	Adjusted height	Remarks
1.246			6.870	5.624		5.624	benchmark *A*
0.785		1.531	6.124	5.339	+.002	5.341	1st change point
3.248		2.367	7.005	3.757	+.004	3.761	2nd change point
	0.946			6.059	+.006	6.065	1st intermediate point
	3.512			3.493	+.006	3.499	2nd intermediate point
0.816		2.804	5.017	4.201	+.006	4.207	point *B*
		3.542		1.475	+.008	1.483	benchmark *C* (observed height)
				1.483			benchmark *C* (known height)
6.095		10.244		+0.008			misclosure

$$\Sigma b - \Sigma f = -4.149$$
$$h_C - H_A = -4.149$$

whereas the height of collimation method provides an adequate check if there are no intermediate sights.

Height differences may also be determined from a measurement of the vertical angle between the two endpoints of the line. This value, together with the horizontal or slope distance, provides a means of calculating the height difference. If the distance is determined by tacheometry the method is referred to as tacheometric levelling, whereas if the distance is observed directly (such as with EDM equipment), or deduced from known coordinates, it is referred to as trigonometric levelling.

Tacheometric levelling is based on the use of the stadia hairs found in most theodolites and, with reference to Figure 2.29, the formulae for horizontal distance (D) and height difference (Δh) are

$$\left.\begin{aligned} D &= 100s\cos^2 v \\ \Delta h &= 100s\sin v\cos v \end{aligned}\right\} \qquad (2.65)$$

where v is the vertical angle, and s is the stadia intercept. For details on the derivation of these equations see section 3.3.

Therefore the difference in height between the ground points A and B is

$$\mathrm{d}h = \Delta h + i_A - t_B \qquad (2.66)$$

The advantage of tacheometric levelling over direct levelling is the speed with which it may be conducted over rough ground. However, this is only achieved at the expense of accuracy, and with long sights or large values of v the measurement of s becomes imprecise. The results of tacheometric levelling are not normally considered reliable enough to warrant corrections for earth curvature and refraction. *Trigonometric levelling*, on the other hand, is conducted over longer distances and with greater precision, particularly if

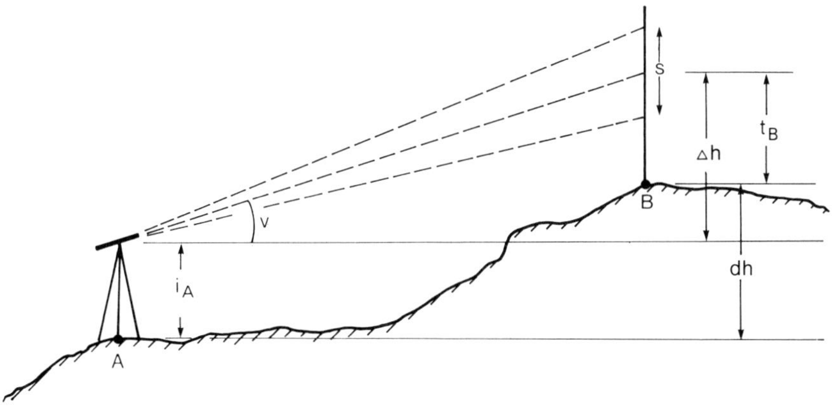

Figure 2.29 Tacheometric levellings.

EDM equipment is used to measure the distances. Therefore earth curvature and refraction must be accounted for.

The fundamental formulae for determining height difference from trigonometric levelling are

$$\left.\begin{aligned} &\Delta h = S \cdot \sin v, \quad \text{where } S \text{ is the slope distance} \\ \text{or} \quad &\Delta h = D \cdot \tan v, \quad \text{where } D \text{ is the horizontal distance} \end{aligned}\right\} \tag{2.67}$$

However, corrections are required for earth curvature and refraction. Figure 2.30*a* illustrates the curved line of sight *AB*, made from an instrument

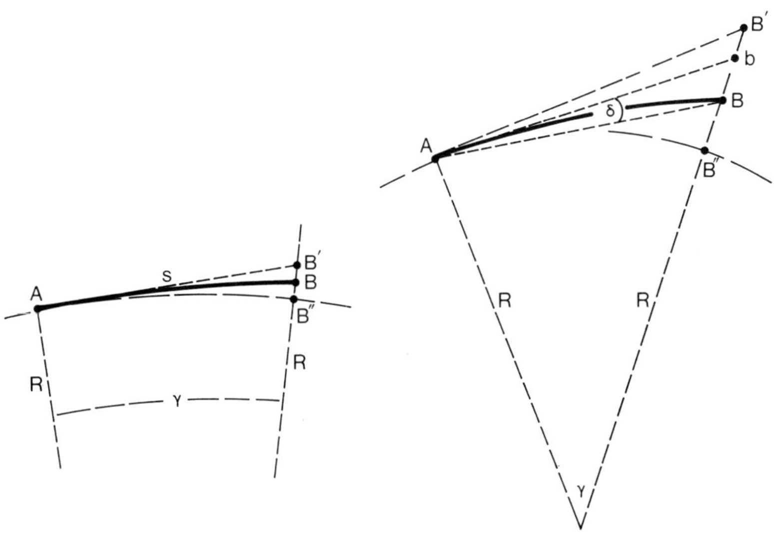

Figure 2.30 Earth curvature and refraction.

station A to a target B. AB' is the horizontal line at A, and γ is the angle subtended by the line at the centre of the earth. R is the mean radius of curvature of the earth for the line being considered, and s is the distance AB. Figure 2.30b is an exaggeration of the same diagram.

If c is the correction for earth curvature (i.e. the distance $B'B''$), then from the diagram,

$$R^2 + s^2 = (R + c)^2 = R^2 + 2Rc + c^2$$

therefore

$$c = \frac{s^2}{2R + c} \tag{2.68}$$

Since c is very small relative to R, the equation may be written

$$c = \frac{s^2}{2R} \tag{2.69}$$

without loss of accuracy.

Thus earth curvature correction varies according to the square of the distance sighted. With a value of R of 6370 km, and s given in km, the equation becomes

$$c = 0.0785s^2 \text{ m} \tag{2.70}$$

Figure 2.31 illustrates graphically how the earth curvature correction varies with distance.

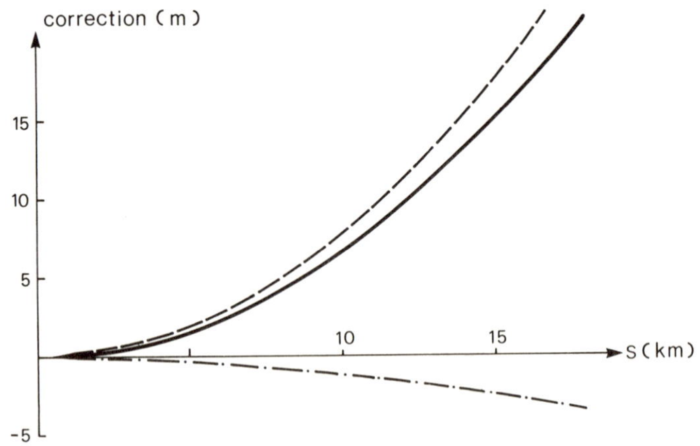

Figure 2.31 Earth curvature and refraction correction relative to distance. ——— earth curvature correction; —·—·— refraction correction; ——— combined correction.

Atmospheric refraction must also be accounted for when long lines are observed. Figure 2.30b shows how the line of sight is curved towards the earth, and the height difference, which should be based on the straight line AB will in fact be calculated using the tangent Ab. Thus the angle of refraction δ will give rise to a refraction correction r, being the distance bB in Figure 2.30a. The coefficient of refraction k may be defined as the ratio of the angles δ and γ as follows:

$$k = \frac{\delta}{\gamma} \tag{2.71}$$

and since the arc AB'' is approximately equal to the distance s, therefore

$$\gamma = \frac{s}{R} \qquad \text{and} \qquad k = \delta \cdot \frac{R}{s}$$

But $r = s \cdot \delta$, therefore, substituting for δ,

$$r = s^2 \cdot \frac{k}{R} \tag{2.72}$$

It should be mentioned that an alternative definition for the coefficient of refraction is R/R', where R' is the radius of curvature of the line of sight. In such a case the value of k will be twice that given above. This alternative definition is used in section 3.4, when the reduction of distances measured electronically is considered.

Equation (2.72) is similar in form to eqn (2.69), in that it is also a function of the distance squared, and a graph showing refraction (for $k = 0.07$) relative to distance is given in Figure 2.31. A significant difference however is the fact that the correction is of opposite sign—c is positive, whereas r is negative. Since r is usually smaller in absolute terms, the combined correction C—also shown graphically in Figure 2.31—may be obtained by combining the two equations (2.69) and (2.72). Thus

$$C = \frac{s^2}{2R} - \frac{s^2 k}{R} = \frac{s^2(1 - 2k)}{2R} \tag{2.73}$$

C will normally be positive, and it is the correction to be applied to the height difference obtained by trigonometric calculation.

The value of k will vary depending on the atmospheric conditions, but typically it has a mean value of approximately 0.07 over land. Atmospheric conditions, representative of the line as a whole, are difficult to monitor, and more reliable values of k are obtained from observations of reciprocal vertical angles, or from observations between points of known height. If simultaneous reciprocal vertical angles are observed the effect of earth curvature and refraction may be virtually eliminated.

Consider the diagram shown in Figure 2.32. Let H_A and H_B be the heights of A and B above sea level, and let α and β be the vertical angles observed at the two points. The curved line AB represents the line of sight, and the corresponding sea-level distance $A''B''$ is s.

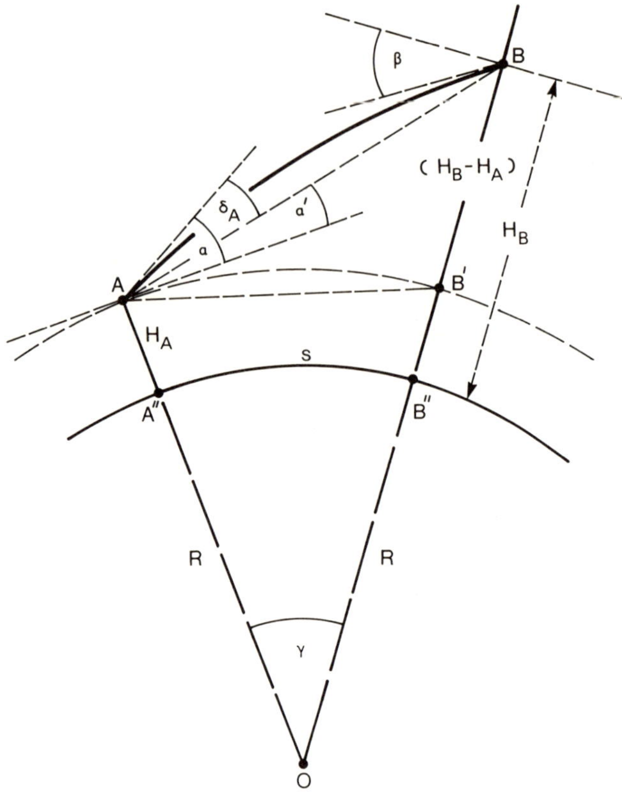

Figure 2.32 Trigonometric levelling.

In any triangle ABC the following trigonometrical relationship holds:

$$\frac{a-b}{a+b}=\frac{\tan[(A-B)/2]}{\tan[(A+B)/2]} \tag{2.74}$$

where A, B are two angles of the triangle, and a, b are the lengths of the opposite sides. When applied to the triangle ABO in Figure 2.32 the components are

$$a-b=(R+H_B)-(R+H_A)=H_B-H_A$$

$$a+b=(R+H_B)+(R+H_A)=2R+H_A+H_B$$

$$A-B=(90+(\alpha-\delta_A))-(90+(\beta-\delta_B))=\alpha-\beta-\delta_A+\delta_B$$

(note that β will be negative for an angle of depression)

$$A+B=180-\gamma$$

Let

$$\theta = \alpha - \beta - \delta_A + \delta_B \tag{2.75}$$

therefore

$$\frac{H_B - H_A}{2R + H_A + H_B} = \frac{\tan(\theta/2)}{\tan[90 - (\gamma/2)]} \tag{2.76}$$

and since $\tan(90 - (\gamma/2)) = \cot(\gamma/2))$ therefore

$$H_B - H_A = (2R + H_A + H_B)\tan(\theta/2)\tan(\gamma/2)$$

But $\gamma = s/R$, therefore $\tan(\gamma/2) = \tan(s/2R)$, and since γ is a small quantity

$$\tan\left(\frac{s}{2R}\right) = \frac{s}{2R} + \frac{1}{3}\left(\frac{s}{2R}\right)^3 + \cdots$$
$$= \frac{s}{2R}\cdot\left(1 + \frac{s^2}{12R^2} + \cdots\right)$$

Therefore

$$H_B - H_A = s\left(1 + \frac{H_A + H_B}{2R}\right)\left(1 + \frac{s^2}{12R^2} + \cdots\right)\tan\left(\frac{\theta}{2}\right)$$
$$= s\left(1 + \frac{H_m}{R} + \frac{s^2}{12R^2} + \frac{s^2 H_m}{12R^3} + \cdots\right)\tan\left(\frac{\theta}{2}\right) \tag{2.77}$$

where $H_m = (H_A + H_B)/2$ is the mean height.

The fourth term on the right-hand side is negligible for all practical purposes, therefore when θ is replaced by the expression in eqn (2.75) the equation may be written finally as

$$H_B - H_A = s\left(1 + \frac{H_m}{R} + \frac{s^2}{12R^2} + \cdots\right)\tan\left(\frac{\alpha - \beta - \delta_A + \delta_B}{2}\right) \tag{2.78}$$

and, in many applications the third term is also negligible.

The value of H_m is not required to any high degree of accuracy, and it may be found in the first instance by ignoring the correction for earth curvature and refraction.

It is normal practice to assume that the refraction angles are the same at each end of the line, particularly if simultaneous reciprocal angles are observed; therefore the quantities δ_A and δ_B cancel. When this assumption is made it becomes possible to determine a value for the coefficient of refraction as follows.

In Figure 2.32,

$$\text{angle } BAO = 90 + (\alpha - \delta) \quad \text{and} \quad \text{angle } ABO = 90 + (\beta - \delta)$$

therefore

$$\gamma = 180 - (90 + (\alpha - \delta)) - (90 + (\beta - \delta))$$

$$= 2\delta - \alpha - \beta$$

and so

$$\delta = \frac{\gamma + \alpha + \beta}{2}$$

Therefore

$$k = \frac{\delta}{\gamma} = \frac{\gamma + \alpha + \beta}{2\gamma} \tag{2.79}$$

The coefficient of refraction may also be determined from observations between points of known height. In Figure 2.32 the angle BAB' may be computed from the known heights of A and B, and after the correction for earth curvature has been applied in reverse a value α' will be obtained. The difference between the observed value α and the value α' represents the angle of refraction δ, and so a value of k may be found using eqn (2.71).

Note that height differences found using eqn (2.78) are based on observed values for α and β—i.e. vertical angles referred to the respective plumblines at A and B. If height differences with respect to the reference spheroid are required, the angles must be corrected for deviation of the vertical at these points. Note also that the derivations above are based on reciprocal observations where the theodolite and target at a point occupy the same position in the vertical sense as well as the planimetric sense. Any variation in instrument or target height during the interchange must be accounted for, as indicated in Chapter 3.

Height differences may also be determined using methods other than levelling or the measurement of vertical angles. Notable among these are methods which involve the measurement of pressure differences (barometric levelling), and parallax differences (photogrammetric heighting). Barometric levelling suffers from the uncertainty caused by changing atmospheric conditions, and is only considered when a lower order of accuracy may be tolerated. Photogrammetric heighting is dealt with in some detail later. In recent times, with the development of better instrumentation, the measurement of gravity and inertial differences has achieved greater significance; and hydrostatic levelling, involving the use of a long flexible pipe filled with some liquid, has proved useful for specific applications.

2.11 Two-dimensional coordinate transformations

One of the most frequently used mathematical models in surveying, photogrammetry and digital mapping, is the two-dimensional linear transformation from one cartesian coordinate system to another. In general a conformal or orthomorphic transformation (i.e. preserving shapc) is sufficient to satisfy the requirements. However, on certain occasions the more general affine transformation is used. The former—being more useful in practice—is dealt with in greater detail in this section.

The *linear conformal* (or *similarity*) transformation may be derived from a series of basic transformations. A simple rotation of a coordinate system (x, y) about the origin and through an angle θ, as shown in Figure 2.33, is represented by the equations

$$\left.\begin{aligned} E &= x\cos\theta - y\sin\theta \\ N &= x\sin\theta + y\cos\theta \end{aligned}\right\} \tag{2.80}$$

or in matrix notation

$$\begin{pmatrix} E \\ N \end{pmatrix} = \begin{pmatrix} \cos\theta & -\sin\theta \\ \sin\theta & \cos\theta \end{pmatrix}\begin{pmatrix} x \\ y \end{pmatrix}$$

or

$$\boldsymbol{E} = \boldsymbol{R}\cdot\boldsymbol{x} \tag{2.81}$$

where

$$\boldsymbol{R} = \begin{pmatrix} \cos\theta & -\sin\theta \\ \sin\theta & \cos\theta \end{pmatrix}$$

is an orthogonal matrix, and $\boldsymbol{E}$, $\boldsymbol{x}$ are coordinate vectors.

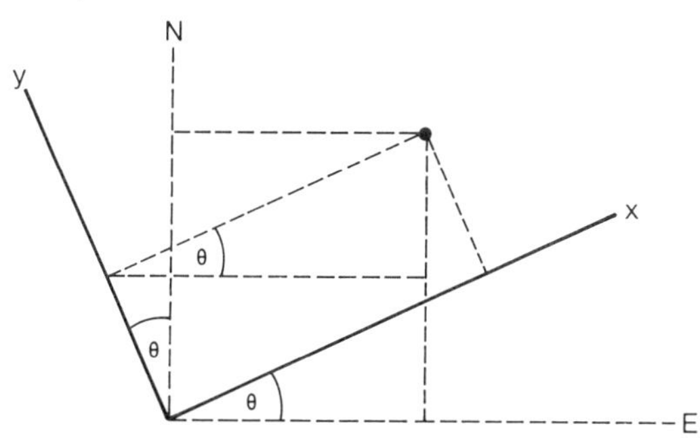

Figure 2.33 Rotation of coordinate system.

If there is a scale change (λ) between the two systems, the original coordinates must be multiplied by this factor. Thus the equations become

$$\begin{pmatrix} E \\ N \end{pmatrix} = \begin{pmatrix} \cos\theta & -\sin\theta \\ \sin\theta & \cos\theta \end{pmatrix} \lambda \begin{pmatrix} x \\ y \end{pmatrix} \tag{2.82}$$

and if both systems have the same origin, as indicated in Figure 2.34, then it is possible to determine values for the parameters θ and λ when only one point (excluding the origin) has known coordinates in both systems.

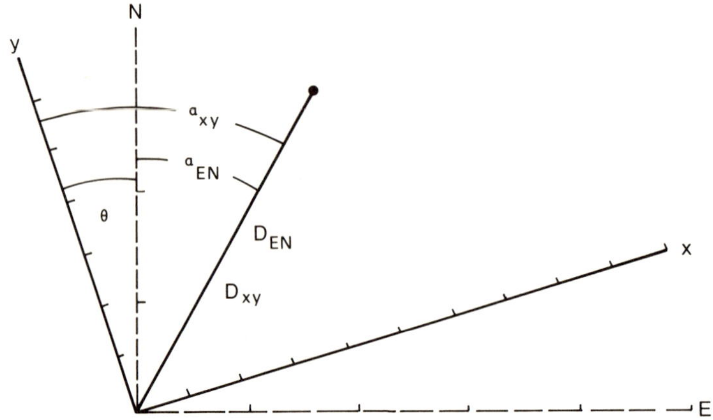

Figure 2.34 Determination of parameters θ and λ.

However, if the two systems do not have the same origin a shift of origin is required, and the equations become

$$\begin{pmatrix} E \\ N \end{pmatrix} = \begin{pmatrix} \cos\theta & -\sin\theta \\ \sin\theta & \cos\theta \end{pmatrix} \lambda \begin{pmatrix} x \\ y \end{pmatrix} + \begin{pmatrix} E_0 \\ N_0 \end{pmatrix} \tag{2.83}$$

where (E_0, N_0) are the coordinates of the origin of the old system, but expressed in terms of the new system, as shown in Figure 2.35.

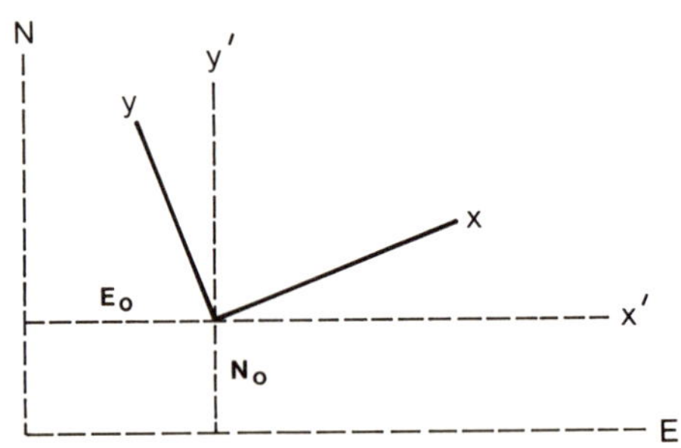

Figure 2.35 Shift of origin.

Equations (2.83) contain four independent transformation parameters—θ, λ, E_0, N_0—and these are the parameters usually required in practical applications.

The inverse transformation is also possible, and in terms of the same parameters it can be shown to be

$$\begin{pmatrix} x \\ y \end{pmatrix} = \begin{pmatrix} \cos\theta & \sin\theta \\ -\sin\theta & \cos\theta \end{pmatrix} \frac{1}{\lambda} \left(\begin{pmatrix} E \\ N \end{pmatrix} - \begin{pmatrix} E_0 \\ N_0 \end{pmatrix} \right) \tag{2.84}$$

Note that eqns (2.83) could be expressed in several different ways, to suit specific applications; such as

$$\begin{pmatrix} E \\ N \end{pmatrix} = \begin{pmatrix} \lambda\cos\theta & -\lambda\sin\theta & 1 & 0 \\ \lambda\sin\theta & \lambda\cos\theta & 0 & 1 \end{pmatrix} \cdot \begin{pmatrix} x \\ y \\ E_0 \\ N_0 \end{pmatrix} \tag{2.85}$$

or

$$\begin{pmatrix} E \\ N \end{pmatrix} = \begin{pmatrix} \lambda\cos\theta & -\lambda\sin\theta & E_0 \\ \lambda\sin\theta & \lambda\cos\theta & N_0 \end{pmatrix} \begin{pmatrix} x \\ y \\ 1 \end{pmatrix} \tag{2.86}$$

or

$$\begin{pmatrix} E \\ N \end{pmatrix} = \begin{pmatrix} a & -b \\ b & a \end{pmatrix} \begin{pmatrix} x \\ y \end{pmatrix} + \begin{pmatrix} c \\ d \end{pmatrix} \tag{2.87}$$

where $a = \lambda\cos\theta$, $b = \lambda\sin\theta$, $c = E_0$, $d = N_0$, or

$$\begin{pmatrix} E \\ N \end{pmatrix} = \begin{pmatrix} x & -y & 1 & 0 \\ y & x & 0 & 1 \end{pmatrix} \begin{pmatrix} a \\ b \\ c \\ d \end{pmatrix} \tag{2.88}$$

Equations (2.87) also contain four transformation parameters a, b, c, d, but a and b are not independent. However, if a and b are known, the independent parameters θ and λ may be derived as follows:

$$\left.\begin{aligned} \theta &= \arctan\left(\frac{b}{a}\right) \\ \lambda &= (a^2 + b^2)^{1/2} \end{aligned}\right\} \tag{2.89}$$

The determination of the four parameters requires at least two points with known coordinates in both systems and, in order to obtain reliable values, the points should be as far apart as possible. The parameters may be determined in stages, or by a simultaneous solution of four equations. A stage-by-stage determination involves:

(i) the calculation of bearing and distance for the line on each system, and the assessment of the *difference* in bearing and the *ratio* of distances, in order to establish θ and λ respectively; and

(ii) a comparison of coordinates for either point—those transformed by θ and λ, with those given in the new system. The difference establishes values E_0 and N_0.

The simultaneous solution involves establishing the four linear equations for the two points, in the form given by eqns (2.88), and then solving for a, b, c, d. The independent parameters may then be found using eqns (2.89). If more than two points have known coordinates in both systems the calculation of parameters which best suit these common points is desirable; thus a least-squares determination becomes the objective.

The example below is used to illustrate both methods when the minimum of two common points with known coordinates are provided.

Example 2.13

Points A, B, C, have been fixed by field survey, and the coordinates are given in a local x, y system. Points A and B also have known coordinates in the E, N-system. Both sets of coordinates, together with the xy-coordinates of point C are given below. Derive the parameters for a similarly transformation from the xy-system to the EN-system

(a) in stages, and

(b) simultaneously.

Hence transform the coordinates of point C:

Point	x (m)	y	E (m)	N
A	2000.000	2000.000	7319.13	6479.45
B	2033.719	1640.152	7355.90	6119.61
C	2047.622	1722.637		

Solution

(a) (i) The calculation of a 'join' for the line AB on each system provides a means of establishing the parameters θ and λ:

Line	xy-system α	D	EN-system α	D
AB	174°38′48″	361.424 m	174°09′56″	361.714 m

Therefore $\theta = 174°38'48'' - 174°09'56'' = 0°28'52''$
$\lambda = 361.714/361.424 = 1.000802$

(ii) Equation (2.83) is used with the coordinates of A to derive the translations

$$\begin{pmatrix} E_0 \\ N_0 \end{pmatrix} = \begin{pmatrix} E \\ N \end{pmatrix} - \begin{pmatrix} \cos\theta & -\sin\theta \\ \sin\theta & \cos\theta \end{pmatrix} \cdot \lambda \cdot \begin{pmatrix} x \\ y \end{pmatrix}$$

$$= \begin{pmatrix} 7319.13 \\ 6479.45 \end{pmatrix} - \begin{pmatrix} 1984.727 \\ 2018.341 \end{pmatrix} = \begin{pmatrix} 5334.403 \\ 4461.109 \end{pmatrix} \text{m}$$

(b) Equation (2.88) is the basis for the simultaneous solution of the four parameters. Thus, in matrix notation, let $\boldsymbol{E} = \boldsymbol{xA}$; where $\boldsymbol{E}$ and $\boldsymbol{x}$ are the matrices formed from the known coordinates, and $\boldsymbol{A}$ is the vector of unknown parameters; then

$$E = \begin{bmatrix} 7319.13 \\ 6479.45 \\ 7355.90 \\ 6119.61 \end{bmatrix}, \quad x = \begin{bmatrix} 2000 & -2000 & 1 & 0 \\ 2000 & 2000 & 0 & 1 \\ 2033.719 & -1640.152 & 1 & 0 \\ 1640.152 & 2033.719 & 0 & 1 \end{bmatrix}, \quad A = \begin{bmatrix} a \\ b \\ c \\ d \end{bmatrix}$$

The parameters A are found from

$$A = x^{-1}E = \begin{bmatrix} 1.000765 \\ 0.008407 \\ 5334.413 \\ 4461.106 \end{bmatrix}$$

and the parameters θ and λ are derived from eqns (2.89), thus

$$\theta = \arctan(b/a) = 0°28'53''$$
$$\lambda = (a^2 + b^2)^{1/2} = 1.000801$$

When the parameters derived in (b) are used to transform point C the following result is obtained:

$$\begin{pmatrix} E_c \\ N_c \end{pmatrix} = \begin{pmatrix} 2047.622 & -1722.637 & 1 & 0 \\ 1722.637 & 2047.622 & 0 & 1 \end{pmatrix} \begin{bmatrix} 1.000765 \\ 0.008407 \\ 5334.413 \\ 4461.106 \end{bmatrix} = \begin{pmatrix} 7369.119 \\ 6202.275 \end{pmatrix} \text{m}$$

Comment

Note that in (a) the sign of θ is given by subtracting the required bearing in the EN-system from the bearing in the xy-system.

The *linear affine* transformation does not preserve shape, and is designed to take account of some deformation between the two systems; however, parallel lines remain parallel after transformation. The change of shape arises because allowance is made for different scale factors in the x and y directions, and also for lack of orthogonality in the one system. Figure 2.36 illustrates the various stages in the transformation, and eqns (2.90) define the overall transformation from the xy-system to the EN-system.

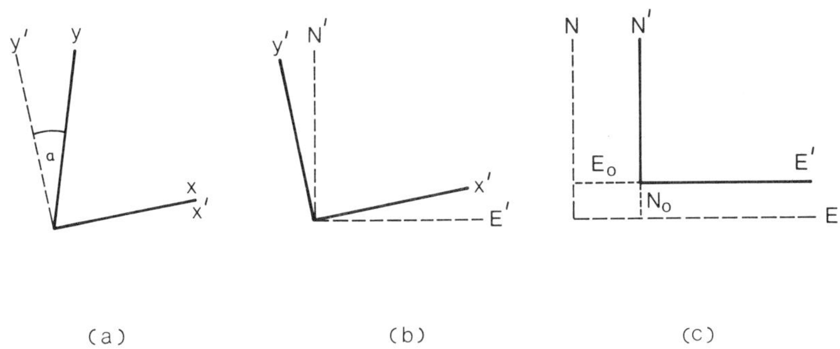

Figure 2.36 Affine transformation: *a*, lack of orthogonality; *b*, rotation; *c*, translation.

$$\begin{pmatrix} E \\ N \end{pmatrix} = \begin{pmatrix} \cos\theta & -\sin\theta \\ \sin\theta & \cos\theta \end{pmatrix} \begin{pmatrix} \lambda_x & \lambda_y \sin\alpha \\ 0 & \lambda_y \cos\alpha \end{pmatrix} \begin{pmatrix} x \\ y \end{pmatrix} + \begin{pmatrix} E_0 \\ N_0 \end{pmatrix} \tag{2.90}$$

where x, y is the original coordinate system, E, N is the final coordinate system, α is the lack of orthogonality, λ_x, λ_y are scale factors for x and y respectively, and E_0, N_0 are the translations.

α, θ, λ_x, λ_y, E_0, N_0 are the six independent parameters, which may be determined if at least three points have known coordinates on both systems. The equations may be expressed using a simpler notation as follows:

$$\left.\begin{aligned} E &= ax + by + c \\ N &= dx + ey + f \end{aligned}\right\} \tag{2.91}$$

or, in matrix notation,

$$\begin{pmatrix} E \\ N \end{pmatrix} = \begin{pmatrix} a & b \\ d & e \end{pmatrix} \begin{pmatrix} x \\ y \end{pmatrix} + \begin{pmatrix} c \\ f \end{pmatrix} \tag{2.92}$$

If more than three points have known coordinates in both systems then a least-squares determination of parameters is recommended, in the same manner as outlined for the four-parameter conformal transformation. The affine transformation has greater application in photogrammetry and digital mapping than in surveying, though possible uses could be in transforming control values from one projection system to another.

2.12 Calculation of area

The determination of the area of a closed figure is a common requirement in surveying. Examples are in applications such as cadastral surveys, engineering surveys, and topographic surveys for planning purposes, etc. Many different methods of determining area are possible, including analogue methods such as the planimeter, and mathematical methods which involve scaling certain dimensions of the figure and calculating the area from appropriate formulae.

Any closed polygon may be subdivided into a number of individual triangles, and an example of a mathematical method is to measure two sides and the enclosed angle of each triangle and calculate the area. In a triangle with angles A, B, C and side lengths a, b, c,

$$\text{area} = \tfrac{1}{2}ab\sin C \tag{2.93}$$

If only side lengths are given the area may be found from

$$\text{area} = (s(s-a)(s-b)(s-c))^{1/2} \tag{2.94}$$

where

$$s = \tfrac{1}{2}(a+b+c).$$

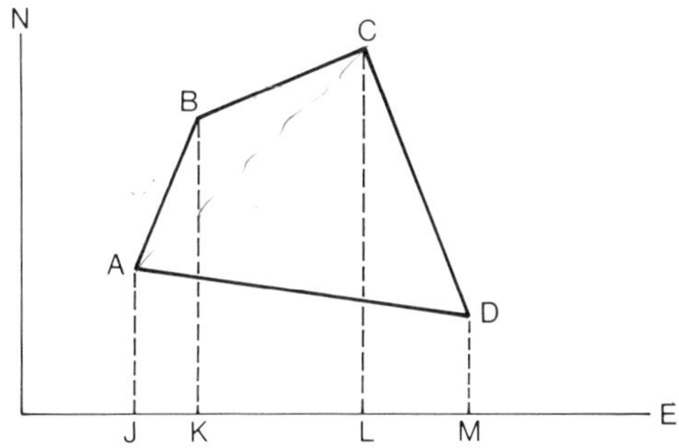

Figure 2.37 Area calculation.

If the positions of all corner points of the polygon are defined by coordinates in a rectangular coordinate system, then the area may be derived more conveniently by using these values. Consider the quadrilateral shown in Figure 2.37. The rectangular coordinates for points A, B, C, D are known, and when these points are projected orthogonally onto the eastings axis the trapeziums $ABKJ$, $BCLK$, $CDML$, and $ADMJ$ are defined.

The area of the figure $ABCD$ may be found from

$$\begin{aligned}\text{area}ABCD &= \text{area}ABCDMJ - \text{area}ADMJ \\ &= \text{area}ABKJ + \text{area}BCLK + \text{area}CDML - \text{area}ADMJ\end{aligned}$$

Now, since the area of any trapezium is the base length multiplied by the mean height, therefore

$$\begin{aligned}2 \times \text{area}ABKJ &= (N_A + N_B)(E_B - E_A) \\ 2 \times \text{area}BCLK &= (N_B + N_C)(E_C - E_B) \\ 2 \times \text{area}CDML &= (N_C + N_D)(E_D - E_C) \\ 2 \times \text{area}ADMJ &= (N_A + N_D)(E_D - E_A)\end{aligned}$$

Therefore

$$\begin{aligned}2 \times \text{area}ABCD = &(N_AE_B + N_BE_B - N_AE_A - N_BE_A) \\ &+ (N_BE_C + N_CE_C - N_BE_B - N_CE_B) \\ &+ (N_CN_D + N_DE_D - N_CE_C - N_DE_C) \\ &- (N_AE_D + N_DE_D - N_AE_A - N_DE_A)\end{aligned}$$

which, by cancelling certain terms and regrouping others, becomes

$$\begin{aligned}2 \times \text{area } ABCD = &N_A(E_B - E_D) + N_B(E_C - E_A) \\ &+ N_C(E_D - E_B) + N_D(E_A - E_C)\end{aligned} \tag{2.95}$$

Similarly, if the points A, B, C, D are projected onto the northings axis, the following formula for area may be derived.

$$2 \times \text{area} ABCD = E_A(N_B - N_D) + E_B(N_C - N_A) + E_C(N_D - N_B) + E_D(N_A - N_C) \quad (2.96)$$

Equations similar to (2.95) and (2.96) may be used for polygons defined by any number of points, and when expressed in this way the coordinates are arranged conveniently for electronic calculation. Therefore the general form of the equations would be

$$\left.\begin{aligned} 2 \times \text{area} &= \sum N_n(E_{n+1} - E_{n-1}) \\ \text{or} \quad 2 \times \text{area} &= \sum E_n(N_{n+1} - N_{n-1}) \end{aligned}\right\} \quad (2.97)$$

where $\sum$ indicates summation over all values of n, and the points are taken in a clockwise order.

The following example illustrates the method.

Example 2.14

Calculate the area enclosed by a parcel of land, defined by points A, B, C, D with the following coordinates:

Point	E (m)	N (m)
A	446.83	246.44
B	445.71	497.25
C	912.24	472.79
D	782.24	248.10

Solution

Equation (2.97) applied to this problem becomes

$$\begin{aligned} 2 \times \text{area} &= E_A(N_B - N_D) + E_B(N_C - N_A) + E_C(N_D - N_B) + E_D(N_A - N_C) \\ &= 446.83(249.15) + 445.71(226.35) + 912.24(-249.15) + 782.24(-226.35) \\ &= -192130.467 \text{ m}^2 \end{aligned}$$

Therefore, $\quad \text{area} = 96065.234 \text{ m}^2$

Comment

Note that the negative sign is simply ignored in the final answer.

It is often the case that a parcel of land is bounded by a curvilinear or irregular boundary in addition to straight-line boundaries. In these cases the objective should be to derive as much of the total area as possible using the coordinated values, and then apply suitable formulae or approximations to determine the area of the remaining segment. If the curve is a regular curve, such as the arc of a circle or parabola, the area of the segment will be defined by formulae. If on the other hand it is irregular, then either a suitable smoothing function may be used and the area calculated by formulae; or the irregular boundary may be specified by a series of coordinated points forming a polygon which approximates the true shape, and the area derived as outlined above.

A requirement often arises, particularly in cadastral applications, where a parcel of land must be subdivided according to a specific ratio of area. There might also be certain constraints imposed on the dividing line, such as

(i) it must pass through a given point; or
(ii) it must be oriented in a particular direction, or be parallel to a particular line; or
(iii) be a minimum or maximum distance from a particular point or structure;

and so on.

Figure 2.38 illustrates a situation involving case (i), where a parcel of land defined by the coordinated points *A*, *B*, *C*, *D*, *E* is to be subdivided into two equal areas. Point *F*, on the line *AE*, is chosen arbitrarily as one end of the dividing line, and its coordinates may be derived from a bearing and distance *AF*.

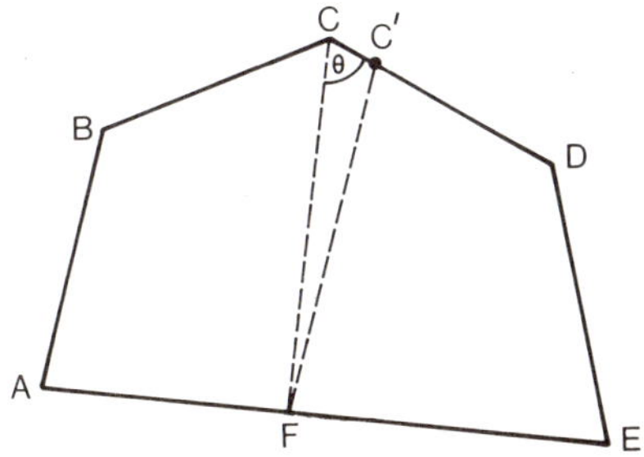

Figure 2.38

The area of the whole figure *ABCDE* may be found, and also that of the figure *ABCF*. A comparison of the two values will establish what discrepancy (Δ) in area exists. The discrepancy may be accounted for by calculating the position of a new subdivision line FC', where C' is established by calculating the distance CC' as follows.

Considering the triangle FCC'; the area (Δ) is given by

$$\Delta = \tfrac{1}{2}\cdot CF\cdot CC'\cdot \sin\theta$$

therefore

$$CC' = \frac{2\Delta}{CF\sin\theta} \qquad (2.98)$$

The coordinates of C' are then obtained using the known coordinates of C and the known bearing and distance CC'.

Figure 2.39 illustrates case (ii), where the line CF has been drawn in a specific direction. Therefore the coordinates of point F may be determined by intersection of the lines AF and CF, and the area of $ABCF$ established. A comparison with the total area will determine the discrepancy (Δ) in area. The discrepancy must be accounted for by a new line $C'F'$ which lies parallel to CF; hence the distance x must be found.

The area of the trapezium $CC'F'F$ is given by

$$\Delta = x \cdot CF - \tfrac{1}{2}x(x \cot\theta + x \cot\phi)$$

$$= x \cdot CF - \frac{x^2}{2}(\cos\theta + \cot\phi) \qquad (2.99)$$

where x is the unknown quantity.

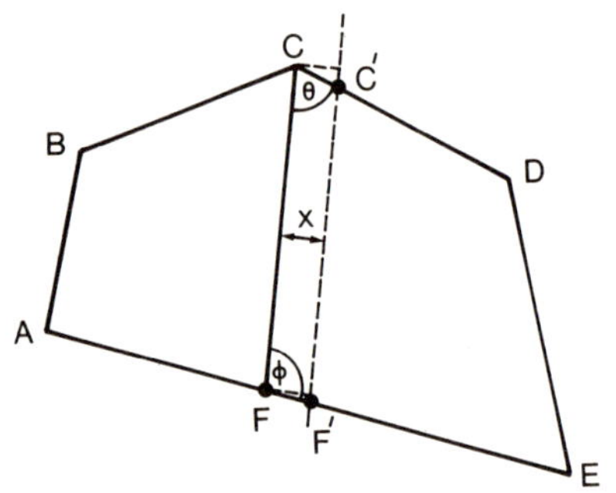

Figure 2.39

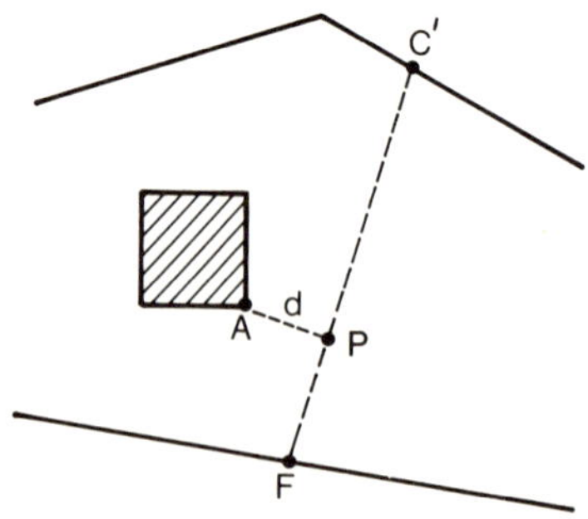

Figure 2.40

When the known values are substituted in eqn (2.99) it becomes a quadratic in x,

$$ax^2 + bx + c = 0$$

where a, b, c are known quantities.

The solution for x is therefore

$$x = \frac{-b \pm (b^2 - 4ac)^{1/2}}{2a}$$

where the positive root is used.

Figure 2.40 illustrates case (iii), where the subdivision line $C'F$ must be a minimum distance d from a point A.

The bearing $C'F$ must be specified in the first instance, and this establishes the bearing AP, which will be at right angles to it. Hence the coordinates of P may be calculated, followed by the calculation by intersection of points F and C'.

3 The reduction of observations in surveying

3.1 Introduction

The computations that take place in surveying are carried out with reference to some datum surface, such as the spheroid or the mean sea level surface. Observations on the other hand take place with respect to some local reference, such as the plumbline passing through the point of observation. Therefore, the actual observations are seldom used in their raw state for the computation of coordinates on or with respect to the datum surface. Invariably small corrections are applied in order to reduce them to equivalent values related to the reference surface. This chapter deals with the basic corrections which are applied to angle and distance observations, and emphasis is given to corrections generally required in plane surveying. The more refined corrections, which are the concern of geodetic surveying, are not considered. Neither are corrections for map projections considered in any great detail, as these are clearly related to the type of map projection—a subject which is beyond the scope of this text. The objective throughout is to regard the reference spheroid (or some approximation to it) as the reference surface for the reduction of observations.

3.2 The reduction of observed directions and angles

Section 2.3 has dealt with the matter of deriving horizontal angles or bearings from observed directions. The *orientation correction* required to convert directions to bearings may be large. Other corrections to horizontal directions are usually very small and are related either to imperfections in the construction or setting-up of the theodolite, or differences between the directions of the plumbline and the normal to the spheroid at the point. Modern well-constructed theodolites have reduced the optical and mechanical errors to a minimum, and those that remain are normally compensated by taking observations on both faces. Errors caused by faulty setting-up (e.g. incorrect levelling of the plate bubble) normally only affect the horizontal angles when steep lines of sight are involved. Differences between the plumbline and the normal to the spheroid have a negligible effect on horizontal angles, and the effect on vertical angles is significant only for geodetic work.

Corrections due to *refraction* of the observed ray may be significant in some cases, and the amount is associated with the temperature and pressure gradients. Vertical refraction has been dealt with already in section 2.10 on height determination. Horizontal or lateral refraction is very much less (by approximately one order of magnitude) than vertical refraction, but in extreme situations, such as when the ray passes close to some structure or the edge of a hill or cliff face, it may be significant. These are obviously situations to be avoided. Although it is possible to calculate the curvature of the ray based on measurements of temperature and pressure gradients, the difficulty of carrying out these measurements in practice makes it less than worthwhile.

Occasionally it may be necessary to calculate the amount of *spherical excess* (ε) in a triangle; that is, the amount by which the sum of the three angles in the spherical (or spheroidal) triangle exceeds 180°. The observed angles are those made on the spherical earth, and it is sometimes necessary to establish the misclosure when all three angles have been observed. Spherical excess is given by

$$\varepsilon = \frac{A}{R^2} \text{ radians} \tag{3.1}$$

where A is the area of triangle, and R is the radius of the sphere which best fits the triangle—i.e. R may be calculated from $R = (\rho\nu)^{1/2}$, where ρ and ν are the radii of curvature of the spheroid in the meridian and prime vertical planes respectively, and for a point which is approximately in the centre of the triangle. ε is normally a small quantity and therefore an approximate value of A will suffice. The graph shown in Figure 3.1*a* illustrates the variation in ε (in seconds of arc) with area, and Figure 3.1*b* shows the variation relative to side length when all three sides are equal.

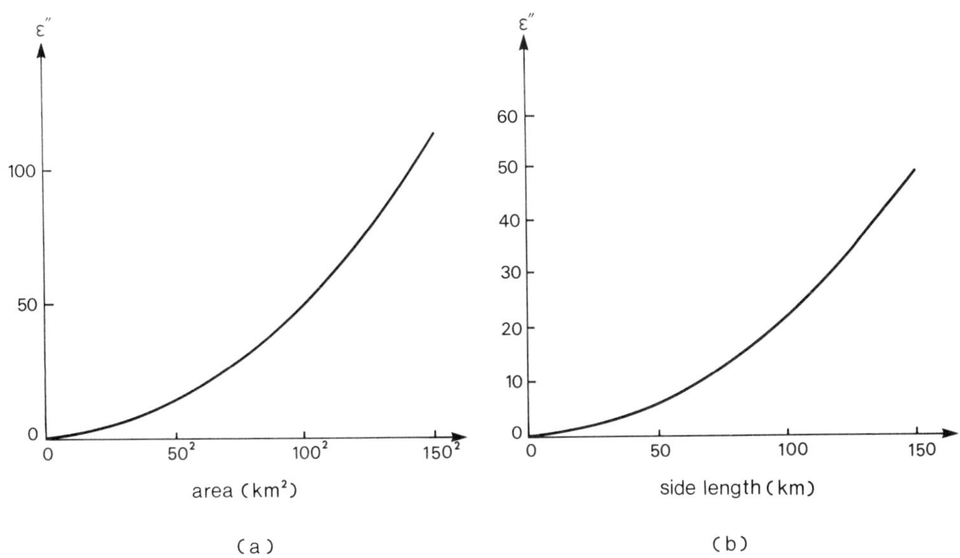

Figure 3.1 Variation of spherical excess.

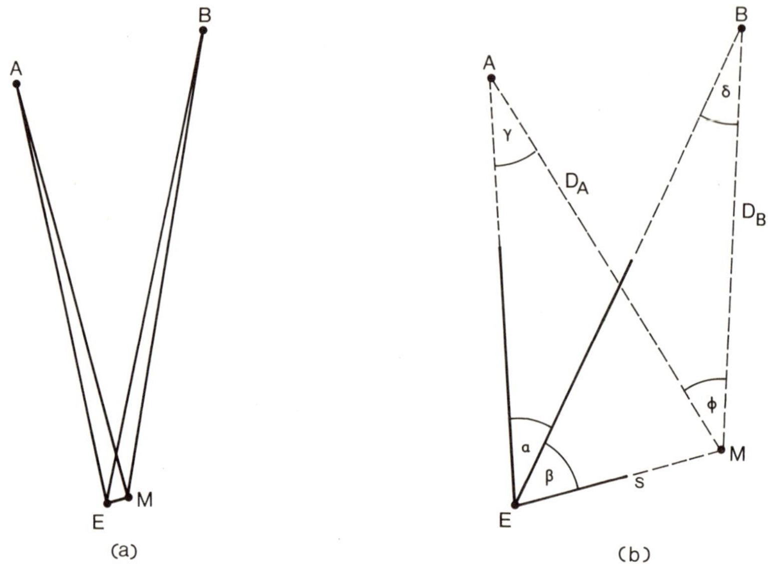

Figure 3.2 Eccentric station reduction.

A fairly common reduction to observed directions is the *eccentric station* (or *satellite station*) reduction. The problem arises when it is impossible or inconvenient to set a theodolite over the station, whereas it is possible to observe from a position close by. Examples are where it is difficult to remove the signal from a station prior to observing from it; or where the coordinated point is the corner of a building, which is clearly inaccessible to a theodolite. Figure 3.2 illustrates the situation where M is the station from which the observations are required (to give the angle ϕ), and E is the eccentric station close by, from which angles α and β are actually observed. Diagram (*a*) is a more realistic representation of the geometry of the situation, whereas diagram (*b*) is an exaggeration to clarify the theory. The eccentric distance s is very small relative to the lengths of the lines of sight to other points, and γ and δ are small angles at A and B subtended by s. They represent corrections to the directions AE and BE, and hence also to EA and EB. The distances AM and BM are represented by D_A and D_B, and they are approximately equal to AE and BE respectively.

From the sine rule,

$$\sin\gamma = \frac{s\sin(\alpha+\beta)}{D_A} \quad \text{and} \quad \sin\delta = \frac{s\sin\beta}{D_B}$$

but, since γ and δ are small angles, therefore

$$\left.\begin{aligned} \gamma'' &= \frac{s\sin(\alpha+\beta)}{D_A}\cdot\rho \\ \delta'' &= \frac{s\sin\beta}{D_B}\cdot\rho \end{aligned}\right\} \tag{3.2}$$

where $\rho = 206265$ in the sexagesimal system.

With γ and δ known, it is possible to derive a value for ϕ. If many points are observed from the same eccentric station the reduction may be set out systematically as follows.

Let θ be the angle (measured clockwise) from M to the particular point sighted, and if d_A, d_B, d_M are the observed directions to A, B, M respectively, then

$$\theta_A = d_A - d_M$$
$$\theta_B = d_B - d_M$$
$$\vdots$$
$$\text{etc.}$$

and

$$\gamma_A = \frac{\sin\theta_A}{D_A}\cdot k, \ldots \text{etc.} \tag{3.3}$$

where k is a constant ($= s \times 206265$).

If necessary the reduction may be tabulated as shown in the example given in Table 3.1.

Table 3.1 Eccentric station reduction

Point	Observed direction ° ′ ″	θ ° ′ ″	Approx. dist.(D)	$\frac{\sin\theta}{D}$	γ'' $\left(=\frac{\sin\theta}{D}\cdot k\right)$	Reduced direction ° ′ ″
A	342 29 00	285 56 00	8710	-110.4×10^{-6}	−83	342 27 37
B	83 03 30	26 30 30	5640	79.1×10^{-6}	60	83 04 30
C	124 50 59	68 17 59	3140	295.9×10^{-6}	224	124 54 43
D	203 35 03	147 02 03	13970	39.0×10^{-6}	29	203 35 32
E	235 02 26	178 29 26	8520	3.1×10^{-6}	2	235 02 28
M	56 33 00	00				

The sign of the correction will be determined by the sign of $\sin\theta$, i.e. it will be positive for $0 < \theta < 180$ and negative for $180 < \theta < 360$.

Eccentric station reduction should be avoided if possible, since mistakes in the reduction can go undetected. A useful field procedure which avoids the problem is to observe one round in the one position, and then observe a second round in a position diametrically opposite from M and the same distance from it. The corrections to the directions to a single point will be of the same absolute magnitude but have different signs, therefore when the mean of the two rounds is taken the observation is effectively reduced to that which would have been observed from M.

3.3 The reduction of observed distances

Distances may be measured in a variety of ways, including pacing, use of the odometer, time of travel, optical methods, taping, photogrammetry, inertial systems, and electronic distance measurement (EDM). All have an application in surveying, and are useful when the accuracy requirements, the cost, and the circumstances permit. Details of the circumstances are given in [15]. By far the most important to the practising surveyor however, are the three concerned with optical distance measurement, taping, and EDM.

In principle *optical distance measurement* involves the use of either a fixed base at the target end of the line being measured, or a fixed angle at the instrument end. The fixed base is usually a subtense bar, and the horizontal angle subtended by the targets at the extreme end of the bar is measured by a theodolite. Thus, with reference to Figure 3.3, if the theodolite at A measures the angle θ subtended by the bar with baselength b, then the horizontal distance D is given by

$$D = \frac{b}{2} \cdot \cot \frac{\theta}{2} \tag{3.4}$$

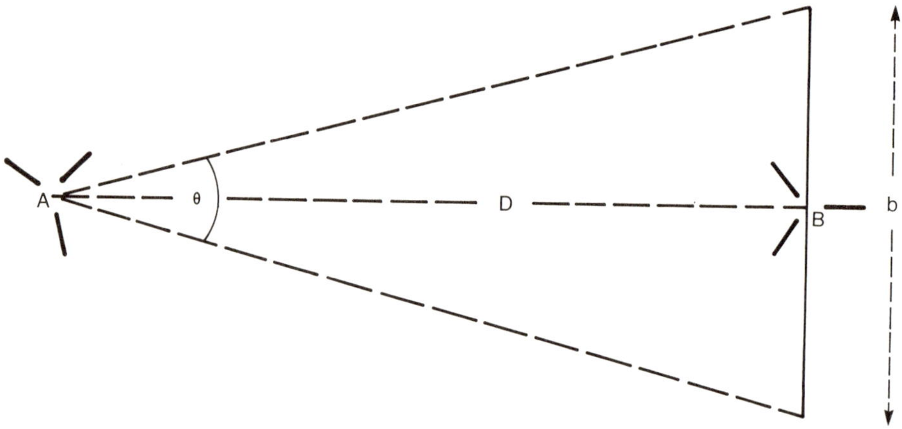

Figure 3.3 Observations to a subtense bar.

A subtense bar is normally 2m in length and, to maintain accuracy in the determination of D, the angle θ must not be too small and must be measured very precisely. Since the angle measured is a horizontal angle, eqn (3.4) produces the horizontal distance D directly. The height difference is found from

$$\Delta h = D \tan v \tag{3.5a}$$

where v is the vertical angle,

or

$$\Delta h = \frac{b}{2} \cdot \cot \frac{\theta}{2} \cdot \tan v \qquad (3.5b)$$

The alternative principle is tacheometry, which uses a fixed angle at the instrument end, and observations to a graduated staff. Two common methods of introducing this angle to the instrument are used; the first uses the principle of stadia hairs in addition to the normal cross-hairs of a surveying instrument, and the second uses a prism or deflection wedge to generate the fixed angle. An example of the way in which stadia hairs appear in the telescope is shown in Figure 3.4*a*, and an illustration of the deflection wedge is given in Figure 3.4*b*. By far the most common method used is that involving stadia

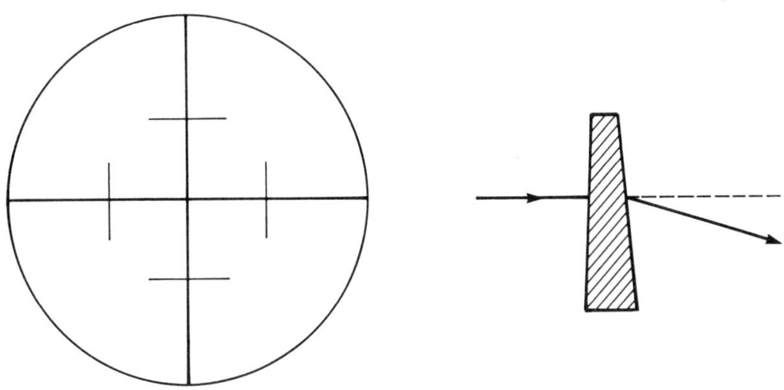

(a) (b)

Figure 3.4 Fixed angle at the instrument: *a*, stadia hairs; *b*, deflection wedge.

hairs, and if both vertical and horizontal stadia hairs are present then either a vertical or a horizontal graduated staff may be used. The principle of measurement involving a vertical staff is illustrated in Figure 3.5.

Images of A and B on the staff are brought to a focus on the outer stadia hairs a and b, which are a distance i apart. The observed stadia interval is the distance s, being the difference in the two readings at A and B. The focal length f is a value which may be regarded as constant, and c is a constant which relates to the point G on the telescopes axis, vertically above the ground point. Thus C is a constant equal to $(c + f)$. The stadia constant k (normally equal to 100) is defined as

$$k = \frac{f}{i} \qquad (3.6)$$

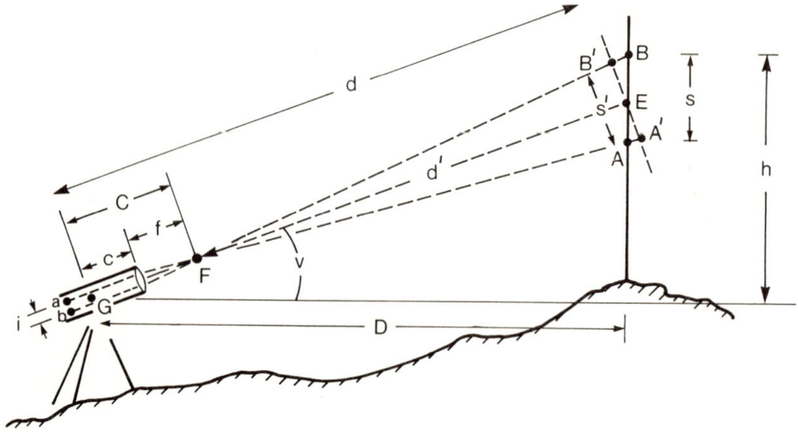

Figure 3.5 Vertical staff tacheometry.

If $A'B'$ is a line perpendicular to the line of sight, and if the distance $A'B'$ is s', then from similar triangles,

$$\frac{f}{i} = \frac{d'}{s'}$$

and so $d' = ks'$. Therefore the slope distance from E to the instrument axis is

$$d = ks' + (c + f) = ks' + C \tag{3.7}$$

The observed stadia interval is s, therefore s' must be expressed in terms of s. For all practical purposes the angles at A' and B' may be considered as right angles.

Therefore $s' = s \cos v$ and so eqn (3.7) becomes

$$d = ks \cos v + C \tag{3.8}$$

The horizontal distance and the height difference between G and E may therefore be found from

$$\left.\begin{aligned} D &= ks \cos^2 v + C \cos v \\ h &= ks \cos v \sin v + C \sin v \end{aligned}\right\} \tag{3.9}$$

Since

$$\cos v \sin v = \tfrac{1}{2} \sin 2v$$

the second equation in (3.9) may also be expressed conveniently as

$$h = \tfrac{1}{2} ks \sin 2v + C \sin v \tag{3.10}$$

Equations (3.9) and (3.10) are the general formulae for stadia tacheometry involving inclined sights. In many applications certain assumptions may be made: for example with small values of v the function $\cos v$ approximates to unity, and the first of eqns (3.9) could be written

$$D = ks + C \tag{3.11}$$

Another common assumption relates to the constant C, which is usually small or zero with modern theodolites; it is therefore usually neglected. Thus eqns (3.9) and (3.10) are normally used in the form

$$\left.\begin{aligned} D &= ks\cos^2 v \\ h &= ks\cos v\sin v \\ \text{or}\quad h &= \tfrac{1}{2}ks\sin 2v \end{aligned}\right\} \tag{3.12}$$

A variation of this type of tacheometry occurs when the staff is inclined so that it is perpendicular to the line of sight, as shown in Figure 3.6.

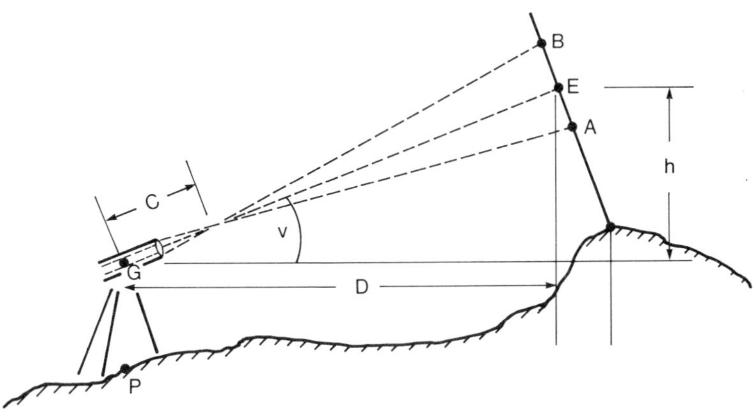

Figure 3.6 Inclined staff tacheometry.

In such a case s replaces s' in eqn (3.7), and the horizontal distance and height difference between G and E are given by

$$\left.\begin{aligned} D &= (ks + C)\cos v \\ h &= (ks + C)\sin v \end{aligned}\right\} \tag{3.13}$$

and again, the constant C is usually neglected.

Note that additional terms must be added to these equations if values with respect to the ground point P are required.

If the staff is used in a horizontal position and perpendicular to the line of sight the formulae are identical to eqns (3.13). The same formulae apply when a deflection wedge is used, and an increase in precision is achieved by using a parallel plate micrometer in addition to the deflection wedge.

Many manufacturers of instruments have incorporated the corrections represented in eqns (3.12) in an analogue fashion, by having curves etched on the vertical circle, and projecting these into the field of view. These are the self-reducing tacheometers, designed specifically for applications where a large amount of detail is to be surveyed, although, with the development of EDM, even these instruments are being phased out. More details on optical distance measurement and the related formulae may be found in [22]. The use of tapes and steel bands has also diminished with the introduction of EDM; however, their use is likely to remain for the measurement of shorter distances and, for the sake of completeness, a brief account of the main corrections is included. They may be listed as corrections for slope, sag, tension, temperature, and standardization.

The correction for *slope* may be applied, without actually being derived, by the formula,

$$D = s \cos v \tag{3.14}$$

where s is the slope distance, v is the vertical angle, and D is the horizontal distance.

Alternatively, the correction itself may be found from

$$\Delta_s = D - s$$

which, from eqn (3.14) becomes

$$\Delta_s = -s(1 - \cos v) \tag{3.15}$$

Or, if the heights of the terminal points are given,

$$D = (s^2 - h^2)^{1/2} = s\left(1 - \frac{h^2}{s^2}\right)^{1/2} \tag{3.16}$$

where h is the height difference between the points. Expansion of the right-hand side using the binomial series gives

$$D = s + \left(-\frac{h^2}{2s} - \frac{h^4}{8s^3} - \cdots\right) \tag{3.17}$$

where the term involving h^4 is usually insignificant.

Therefore the correction for slope, using a height difference, is

$$\Delta_s = -\frac{h^2}{2s} \tag{3.18}$$

Equations (3.14)–(3.18) are adequate even for long lines measured by tape, since the individual sections will be fairly short, and each section will be reduced according to the slope of that section. Thus the sum will follow the approximate shape of the spheroid. The reduction of much longer lines measured with EDM equipment is a different matter, however, and requires a more rigorous procedure, as outlined later in this section.

The errors caused by *misalignment* of the tape is of the same form as the slope error, except that it is in the horizontal sense. The correction therefore will be according to eqn (3.18), where h is the linear amount of misalignment at the one end.

The correction for *sag* (or *catenary*) is the correction for a length of tape suspended in catenary between two supports. The shape of the curve may be assumed to be a parabola, and is a function of the weight of the tape and the tension applied to hold it in position. The correction, sufficient for most purposes, is given as,

$$\Delta_c = -\frac{w^2 L^3}{24P^2} = -\frac{W^2 L}{24P^2} \tag{3.19}$$

where L is the length between the two supports, (m); w is the weight/unit length of the tape, (kg/m); W is the total weight of the tape between the supports, (kg); and P is the tension applied in newtons, (N).

It is clear from eqn (3.19) that the correction increases in absolute terms with the square of the total weight, and inversely with the square of the tension applied. There are advantages therefore in reducing the length of the bays between supports, and in using lighter tapes.

As all tapes are elastic to a certain extent, a change in *tension* will produce a change in length. The correction is associated with the change in tension relative to the tension used during the process of standardization, and the formula is

$$\Delta_P = \frac{(P - P_0)L}{AE} \tag{3.20}$$

where P is the tension applied in newtons, (N); P_0 is the tension used at standardization, (N); L is the measured length, (m); A is the cross-sectional area of the tape, (m^2); and E is the modulus of elasticity of the tape, (N/m^2).

It is important to note that an increase in tension relative to P_0 results in a positive correction.

The correction for change in *temperature* is also associated with the temperature used during standardization, and is given by the formula

$$\Delta_T = \alpha L(T - T_0) \tag{3.21}$$

where α is the coefficient of thermal linear expansion for the material of the tape, (m/° C); L is the measured length, (m); T is the temperature of the tape during measurement, (° C); and T_0 is the temperature of the tape during standardization, (° C).

Note that, as in the case of tension, an increase in temperature relative to T_0 results in a positive correction.

The process of *standardization* of the tape is linked with tension and temperature, and values for the two parameters must be specified if the standardization correction is to have any meaning. The correction is then given as

$$\Delta_{st} = \frac{l}{L} \cdot \Delta_L \tag{3.22}$$

where l is the measured length; L is the total length of the tape; and Δ_L is the standardization correction for the total length at a specified temperature, tension, and support arrangement.

A tape that is too long will give a positive value for Δ_L; and hence Δ_{st} will also be positive.

Instead of quoting a standardization correction Δ_L, it is often convenient to quote a temperature (or tension) at which the nominal length of the tape agrees exactly with the true length, and any correction to a field measurement will be based on a variation with respect to this value.

The very precise measurement of distances with steel or invar tapes is still carried out, although to a much more limited extent; the main application is to very short ranges where the uncertainty in the constant error of EDM equipment becomes unacceptable. Further details on the application and accuracy of taping are given in [15]. Corrections to distances measured with EDM equipment are far more important to the practising surveyor, and these are considered next.

3.4 The reduction of distances measured electronically

Distances measured electronically often extend to much greater lengths than the methods discussed so far; and the principles of measurement are so different that many additional errors arise or become significant. Details concerning these principles and the equipment used are contained in [10] and [25], and will not be pursued in this text. What will be considered are the corrections that are necessary in order to reduce the measured value to the equivalent value on the spheroid.

Most EDM instruments display a distance (D') which is obtained from the fundamental equation

$$D' = V \cdot t \tag{3.23}$$

where V is the velocity of the signal along the path and t is the time taken for the signal to travel the required distance.

The corrections may be classified broadly as velocity corrections and geometrical corrections. The velocity (V_0) of electromagnetic waves in a vacuum is a physical constant, which has been determined many times in the past [18]. Historically, its determination has been associated with waves in the visible part of the spectrum; however, in a vacuum all electromagnetic waves travel at the same speed. The most recent value, recommended at the XVIth General Assembly of the International Union of Geodesy and Geophysics in 1975 [23], is

$$V_0 = 299\,792\,458 \text{ m s}^{-1} \qquad \pm 1.2 \text{ m s}^{-1}$$

The precision of its determination represents 0.004 parts per million, which is well within the requirements of most operations in surveying and geodesy, and therefore it may be regarded as a constant which is free of error.

A *velocity correction* is required because the signal is not travelling in a vacuum, but usually in some medium which has the effect of reducing the speed of travel. The amount of reduction is determined by the *refractive index* (n), which is defined as

$$n = \frac{V_0}{V} \tag{3.24}$$

where V_0 is the velocity of electromagnetic waves in a vacuum, and V is the velocity of the particular wave in the medium of the measurement—usually the atmosphere.

Therefore if D' is the distance displayed by the EDM instrument—referred to as the 'instrument' distance—and D is the corrected distance, then

$$D = \frac{D'}{n} \tag{3.25}$$

The velocity V, in the atmosphere, depends on the frequency (f) of the signal, and on the physical properties of the atmosphere—i.e. the temperature, pressure, and relative humidity. Therefore measurements of these atmospheric conditions must be made simultaneously with the distance measurement, in order to determine a value of n. The relationship between refractive index and wavelength, referred to as *dispersion*—is given by the Cauchy equation, viz.

$$n = A + \frac{B}{\lambda^2} + \frac{C}{\lambda^4} \tag{3.26}$$

where A, B, C are constants which relate to specific atmospheric conditions.

As far as the application is concerned, it is more convenient to establish a correction (Δ_r) to the distance due to refractive index; this may be considered as having two components, referred to as first and second velocity corrections.

The *first velocity correction* (Δ_{r1}) is

$$\begin{aligned}\Delta_{r1} = D - D' &= \frac{D'}{n} - D' \\ &= D'\left(\frac{1}{n} - 1\right) \\ &= D'\left(\frac{1-n}{n}\right)\end{aligned}$$

and since $n \approx 1$, the equation may be written

$$\Delta_{r1} = D'(1 - n) \tag{3.27}$$

with sufficient accuracy.

A value of n for standard air (0° C temperature, 1013.25 mbar pressure, and dry air with 0.03% carbon dioxide) is given by Barrel and Sears [7] to be

$$(n_s - 1) = \left(287.604 + \frac{1.6288}{\lambda^2} + \frac{0.0136}{\lambda^4}\right)10^{-6} \tag{3.28}$$

where λ is expressed in μm.

In practice, modulated waves are propagated by EDM instruments. Therefore a velocity known as group velocity (V_g) determines the distance, and so group refractive index (n_g) becomes the required quantity. Group velocity is always smaller than the individual (carrier) wave velocities, therefore

$$V_g = V - \frac{dV}{d\lambda}\cdot\lambda \tag{3.29}$$

From eqn (3.24),

$$V = \frac{V_0}{n} = \frac{n_g V_g}{n}$$

therefore, by substituting in eqn (3.29),

$$V_g = \frac{n_g V_g}{n} - \frac{dV}{d\lambda}\cdot\lambda$$

and so

$$n_g = n + \frac{n}{V_g}\cdot\frac{dV}{d\lambda}\cdot\lambda \tag{3.30}$$

But

$$n = \frac{V_0}{V},$$

and

$$\frac{dn}{d\lambda} = -\frac{V_0}{V^2}\cdot\frac{dV}{d\lambda}$$

$$\approx -\frac{n}{V_g}\cdot\frac{dV}{d\lambda}$$

therefore, substituting in eqn (3.30),

$$n_g = n - \frac{dn}{d\lambda}\cdot\lambda \tag{3.31}$$

By differentiation of eqn (3.26),

$$\frac{dn}{d\lambda} = -\frac{2B}{\lambda^3} - \frac{4C}{\lambda^5} \tag{3.32}$$

Therefore, substituting eqns (3.26) and (3.32) in (3.31),

$$n_{\dot{g}} = \left(A + \frac{B}{\lambda^2} + \frac{C}{\lambda^4} \right) + \frac{2B}{\lambda^2} + \frac{4C}{\lambda^4}$$

$$= A + \frac{3B}{\lambda^2} + \frac{5C}{\lambda^4} \tag{3.33}$$

where λ is the individual (carrier) wavelength.

Therefore, for group velocity in standard air, and λ expressed in μm, eqn (3.28) becomes

$$(n_{gs} - 1) = \left(287.604 + \frac{4.8864}{\lambda^2} + \frac{0.0680}{\lambda^4} \right) 10^{-6} \tag{3.34}$$

For example, in the case of instruments using infrared and light waves,

$n_{gs} = 1.0002936$, for an instrument with $\lambda = 0.91\ \mu$m
$n_{gs} = 1.0003039$, for an instrument with $\lambda = 0.56\ \mu$m

In the case of microwave instruments ($\lambda = 0.03$ m to $\lambda = 1.00$ m), the second and third terms in eqns (3.34) and (3.28) are negligible; therefore at these frequencies the atmosphere may be considered to be non-dispersive, and n_{gs} would be equal to 1.0002876.

For atmospheric conditions other than standard, the following equation for group refractive index for *light waves* is given by Barrell and Sears [7]:

$$(n_{gl} - 1) = \frac{(n_{gs} - 1)}{1 + \alpha t} \cdot \frac{p}{1013.25} - \frac{4.1 \times 10^{-8}}{1 + \alpha t} \cdot e \tag{3.35}$$

where n_{gl} is the group refractive index for light waves, which corresponds to p, t and e; n_{gs} is the group refractive index for standard air; p is the atmospheric pressure, (mbar); t is the 'dry bulb' temperature, (°C); e is the partial water vapour pressure, (mbar); and α is the coefficient of thermal expansion of air ($= 1/273.16$ per °C).

Equation (3.35) may be rearranged as

$$(n_{gl} - 1) = \frac{(d_{gs} - 1)}{\alpha(1/\alpha + t)} \cdot \frac{p}{1013.25} - \frac{4.1 \times 10^{-8}}{\alpha(1/\alpha + t)} \cdot e$$

and when the value for α is introduced the equation becomes

$$(n_{gl} - 1) = \frac{273.16(n_{gs} - 1)p}{(273.16 + t)1013.25} - \frac{11.2 \times 10^{-6}}{(273.16 + t)} \cdot e \tag{3.36}$$

Equation (3.36) is suitable for visible light and infrared waves with carrier wavelengths in the range 0.5 μm, to 0.93 μm and is reliable to ± 0.2 ppm. It is valid for temperatures within the range $-40°$ C to $+50°$ C, and for pressures within the range 533 mb to 1067 mb. An error analysis may be carried out by differentiation of eqn (3.36), and for conditions of $n_{gs} = 1.000300$, $t = 12°$ C, $p = 1000$ mbar and $e = 15$ mbar. When the coefficients are evaluated the differential equation becomes

$$dn_{gl} \times 10^6 = 0.28\,dp - 0.99\,dt - 0.04\,de \qquad (3.37)$$

Equation (3.37) reveals that n_{gl}, and hence the distance, is altered in the following way by errors in the determination of the three parameters:

(i) a 1 mbar error in p increases n_{gl} by 0.28 ppm
(ii) a 1° C error in t decreases n_{gl} by 0.99 ppm
(iii) a 1 mbar error in e decreases n_{gl} by 0.04 ppm.

Therefore, it is clear that the determination of temperature is critical, whereas the determination of partial water vapour pressure is of little practical value, and the second term in eqn (3.36) is usually neglected. Temperature may be measured very precisely, and with little effort, at both ends of the line. However, the mean derived from these measurements does not represent the integral value for the line as a whole; and the accuracy in the determination of n_{gl} is therefore limited to ± 1 ppm. With the longer lines this will degenerate even further.

Microwaves are more sensitive to humidity changes, and the following formula is given by Essen and Froome [17]:

$$(n_{gm} - 1)10^6 = \frac{77.64(p-e)}{(273.16+t)} + \frac{64.68}{(273.16+t)}\left(1 + \frac{5748}{(273.16+t)}\right)e \qquad (3.38)$$

where n_{gm} is the group refractive index for microwaves, which correspond to p, t and e; p is the atmospheric pressure (mbar); t is the 'dry bulb' temperature (°C); and e is the partial water vapour pressure (mbar).

Equation (3.38) is suitable for carrier wavelengths in the range 0.03 m to 1.00 m. It is reliable to ± 0.1 ppm under normal conditions, and ± 1 ppm under abnormal conditions. An error analysis may be carried out, based on eqn (3.38) and for conditions of $t = 12°$ C, $p = 1000$ mbar, $e = 15$ mbar; the differential equation becomes

$$dn_{gm} \times 10^6 = 0.27dp - 1.46dt + 4.59de \qquad (3.39)$$

By inspection of eqn (3.39),

(i) a 1 mbar error in p increases n_{gm} by 0.27 ppm
(ii) a 1° C error in t decreases n_{gm} by 1.46 ppm
(iii) a 1 mbar error in e increases n_{gm} by 4.59 ppm.

Therefore, it is clear that the most critical of the parameters for microwave distance measurement is e—the partial water vapour pressure. Values of wet and dry bulb temperatures may be measured easily enough in order to determine e at the end points, but the mean derived from these values is a weak approximation for the line as a whole—particularly for the longer lines. Thus an accuracy better than ± 3 ppm for n_{gm} is difficult to achieve.

The value of e may be determined from wet and dry bulb readings taken with an instrument such as an aspiration psychrometer, and the equation [34] for this determination is

$$e = E - 0.000662(t - t')p \tag{3.40}$$

where E is the saturation water vapour pressure (mbar), for temperature t' over water; t is the dry bulb temperature (°C); t' is the wet bulb temperature (°C); and p is the atmospheric pressure (mbar).

Equation (3.40) is reliable to $\pm 1\%$ and valid for values of t and t' above 0° C. In the case of a frozen wet bulb wick, eqn (3.40) becomes

$$e = E' - 0.000583(t - t')p$$

where E' is the saturation water vapour pressure (mbar) for temperature t' over ice.

The values of E and E' may be obtained from tables given in [26], and [34], or calculated from the following formulae given in [34]:

$$\left.\begin{aligned} E &= 10\exp\left(\frac{7.5t'}{237.3 + t'} + 0.7858\right) \\ E' &= 10\exp\left(\frac{9.5t'}{265.5 + t'} + 0.7858\right) \end{aligned}\right\} \tag{3.42}$$

which are valid for a range of t' from -70° C to $+50$° C.

Partial water vapour pressure may also be determined using a known value for relative humidity (h), obtained for example from hygrometer readings, or weather bulletins. The results will be considerably less precise; however, in some applications (such as for short ranges) the value may satisfy the requirements. The equation is

$$e = \frac{Eh}{100} \tag{3.43}$$

where h is relative humidity, expressed as a percentage. For further discussion on the determination of e and n, the reader is referred to [34] and [10].

Once E and e have been established a value for n_g may be found, and hence Δ_{rl} may be derived from eqn (3.27). The value of Δ_{rl} may be reduced substantially if the manufacturer adjusts the main oscillator to give a modulation frequency which corresponds with a *reference refractive index* (n_{ref}), where n_{ref} is a value approximately equal to the average value of n

found under field conditions. Thus, instead of eqn (3.27), which refers to a reference index of unity, the correction becomes

$$\Delta_{r1} = D'(n_{ref} - n) \tag{3.44}$$

The value of n_{ref} depends on the type of equipment (i.e. carrier wavelength), and also on what are considered 'average' conditions, but in general it will be approximately 1.0003. It should be made clear however, that although the magnitude of Δ_{r1} is much smaller, the precision of its determination remains the same.

Example 3.1

A microwave instrument ($n_{ref} = 1.0003200$) was used to measure a distance, and the following readings were obtained: $D' = 5674.32\,\text{m}$; $t = 18°\,\text{C}$; $t' = 14°\,\text{C}$; $p = 1010\,\text{mbar}$. Calculate a value for the first velocity correction.

Solution

From eqn (3.42),

$$\begin{aligned} E &= 10\exp\left(\frac{7.5 \times 14}{237.3 + 14} + 0.7858\right) \\ &= 16.0\,\text{mb} \end{aligned}$$

From eqn (3.40),

$$\begin{aligned} e &= 16.0 - (0.000662 \times (18 - 14) \times 1010) \\ &= 13.3\,\text{mb} \end{aligned}$$

From eqn (3.38),

$$\begin{aligned} (n_{gm} - 1)10^6 &= \frac{77.64(1010 - 13.3)}{(273.16 + 18)} + \frac{64.68}{(273.16 + 18)}\left(1 + \frac{5748}{273.16 + 18}\right)13.3 \\ &= 265.78 + 61.28 \\ &= 327.06 \end{aligned}$$

From eqn (3.44),

$$\begin{aligned} \Delta_{r1} &= 5674.32(1.0003200 - 1.0003271) \\ &= 5674.32(-0.0000071) \\ &= -0.040\,\text{m} \end{aligned}$$

The *second velocity correction* (Δ_{r2}) is applied in an attempt to compensate for the fact that the radius of curvature of the signal path is not the same as that of the spheroid (or the atmospheric layers). It can be shown that this additional correction is given by

$$\Delta_{r2} = -(K - K^2)\cdot\frac{D'^3}{12R^2} \tag{3.45}$$

where K, the coefficient of refraction, is defined as

$$K = \frac{\text{curvature of signal path}}{\text{curvature of spheroid}} = \frac{1/r}{1/R} = \frac{R}{r} \tag{3.46}$$

and R is the mean radius of curvature of the spheroid for the measured line; r is the radius of curvature of the signal path; and D' is the instrument distance.

Since r is approximately $8R$ and $4R$ for light waves and microwaves respectively, values of K for the respective waves are therefore

$$K_1 = 0.13 \qquad \text{and} \qquad K_{\text{m}} = 0.25$$

In practice the second velocity correction is likely to be more important for microwaves than light waves, because values of D' are likely to be greater for these waves. For example, for a distance of 50 km and $R = 6370$ km, the value of Δ_{r2} for microwaves is -48 mm.

Most manufacturers supply environmental correction nomograms for the determination of the first velocity correction, and this is usually with respect to a fixed value n_{ref}. Some instruments also allow a value of n to be introduced before measurement takes place, in order to compensate for environmental conditions.

The objective in applying *geometrical corrections* is to reduce the measured distance (corrected for refractive index) to the equivalent distance on the spheroidal surface. The corrections may be split into two groups—those that are always present in a theoretical sense, and those that arise because of problems related to the design or location of the equipment. Some corrections are more significant than others, depending on the type of equipment and the length of line being considered.

A measured distance will invariably be between points which are not on the spheroid. Therefore a reduction to the spheroid becomes necessary. Figure 3.7 shows a section containing the measured distance AB, and the required spheroid distance A' B': s_1 represents the measured distance which has been corrected for refractive index, s_2 is the corresponding chord distance, s_3 is the spheroidal chord, and s_4 is the surface distance on the spheroid. The final value s_4 is the length of the normal section which, for most practical purposes, is the same as the length of the geodesic. h_A and h_B are the heights above the spheroid of points A and B. γ is the angle subtended by the points at the centre of the equivalent sphere (of radius R) which passes through the points A' and B' ; and δ is the angle between the tangents to the curved ray path at A and B, and also between the normals to the ray path at these points.

A value for R is found from

$$R = \frac{\rho\nu}{\nu\cos^2\alpha + \rho\sin^2\alpha} \tag{3.47}$$

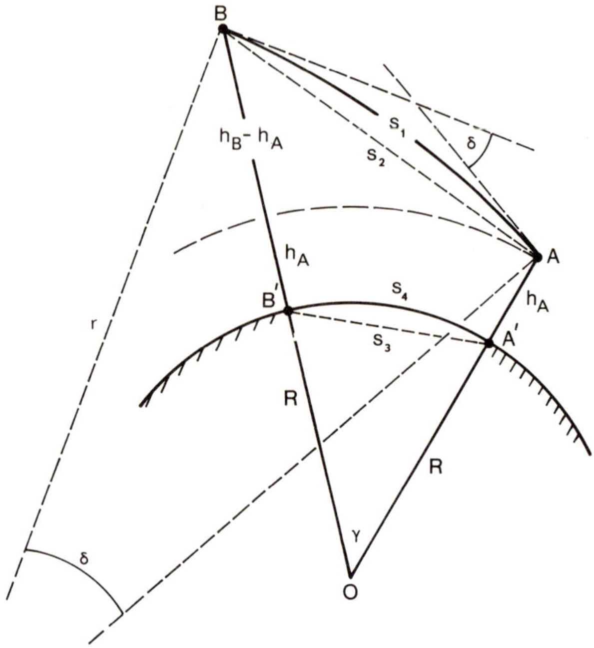

Figure 3.7 Reduction to the spheroid, using heights.

where ρ is the radius of curvature of the spheroid in the meridian plane; ν is the radius of curvature of the spheroid in the prime vertical plane; and α is the azimuth of the line.

Values of ρ and ν are based on latitude; the latitude of the centre of the line, or even the centre of the survey area, will be sufficient for these purposes. The reduction may be carried out in stages as follows:

(i) The reduction from s_1 to s_2 is found from:

$$s_2 = 2r \sin \frac{\delta}{2} \tag{3.48}$$

which, by series expansion, becomes

$$s_2 = 2r\left(\frac{\delta}{2} - \frac{1}{6}\left(\frac{\delta}{2}\right)^3 + \frac{1}{120}\left(\frac{\delta}{2}\right)^5 - \cdots\right)$$

and since $\delta = \dfrac{s_1}{r}$

therefore

$$s_2 = 2r\left(\frac{s_1}{2r} - \frac{1}{48}\left(\frac{s_1}{r}\right)^3 + \frac{1}{3840}\left(\frac{s_1}{r}\right)^5 - \cdots\right)$$

$$= s_1 - \frac{s_1^3}{24r^2} + \frac{s_1^5}{1920r^4} - \cdots$$

The third term is negligible, and so the equation reduces to

$$s_2 = s_1 - \frac{s_1^3}{24r^2} \tag{3.49}$$

or, in terms of a correction Δ_1,

$$\Delta_1 = s_2 - s_1 = -\frac{s_1^3}{24r^2} \tag{3.50a}$$

which, by using eqn (3.46), may also be expressed as

$$\Delta_1 = -\frac{s_1^3 K^2}{24R^2} \tag{3.50b}$$

(ii) The reduction from S_2 to S_3 is found by considering the triangle ABO. From the cosine rule

$$s_2^2 = (R + h_A)^2 + (R + h_B)^2 - 2(R + h_A)(R + h_B)\cos\gamma \tag{3.51}$$

and since $\cos\gamma = 1 - 2\sin^2\frac{\gamma}{2}$

therefore eqn (3.51) becomes

$$\begin{aligned} s_2^2 &= R^2 + 2Rh_A + h_A^2 + R^2 + 2Rh_B + h_B^2 - 2(R^2 + h_Ah_B + Rh_A + Rh_B) \\ &\quad + 4(R + h_A)(R + h_B)\sin^2\frac{\gamma}{2} \\ &= h_A^2 - 2h_Ah_B + h_B^2 + 4(R + h_A)(R + h_B)\sin^2\frac{\gamma}{2} \\ &= (h_B - h_A)^2 + 4R^2\left(1 + \frac{h_A}{R}\right)\left(1 + \frac{h_B}{R}\right)\sin^2\frac{\gamma}{2} \end{aligned}$$

But $s_3 = 2R\sin\frac{\gamma}{2}$

therefore

$$s_2^2 \; (h_B - h_A)^2 + s_3^2\left(1 + \frac{h_A}{R}\right)\left(1 + \frac{h_B}{R}\right)$$

Therefore

$$s_3 = +\left\{\frac{s_2^2 - (h_B - h_A)^2}{[1 + (h_A/R)][1 + (h_B/R)]}\right\}^{1/2} \tag{3.52a}$$

or

$$s_3 = +\left(\frac{R^2(s_2^2 - \Delta h^2)}{(R + h_A)(R + h_B)}\right)^{1/2} \tag{3.52b}$$

where $\Delta h = h_B - h_A$.

Equations (3.52) are rigorous inasmuch as they apply to a sphere which best fits the spheroidal line $A'B'$.

(iii) The reduction from s_3 to s_4 is of the same type as in (i), but with opposite sign, and therefore may be written as

$$\Delta_3 = \frac{s_3^2}{24R^2} \tag{3.53}$$

Equations (3.46), (3.48) and (3.52) may be combined to provide a single equation for the geometrical reduction of a measured distance to the spheroid. In Figure 3.7

$$s_4 = 2R \arcsin \frac{s_3}{2R}$$

but, substituting the second equation in (3.52) for s_3 gives

$$s_4 = 2R \arcsin\left(\frac{s_2^2 - \Delta h^2}{4(R + h_A)(R + h_B)}\right)^{1/2}$$

and when eqn (3.48) is substituted for s_2 it becomes

$$s_4 = 2R \arcsin\left[\frac{r^2 \sin^2(\delta/2) - (\Delta h^2/4)}{(R + h_A)(R + h_B)}\right]^{1/2}$$

Now $$\frac{\delta}{2} = \frac{s_1}{2r}$$

therefore

$$s_4 = 2R \arcsin\left[\frac{r^2 \sin^2(s_1/2r) - (\Delta h^2/4)}{(R + h_A)(R + h_B)}\right]^{1/2}$$

and from eqn (3.46), $r = \dfrac{R}{K}$

therefore

$$s_4 = 2R \arcsin\left[\frac{R^2 \sin^2(s_1 K/2R) - (K^2 \Delta h^2/4)}{K^2(R + h_A)(R + h_B)}\right]^{1/2} \tag{3.54}$$

where s_4 is the spheroidal distance; R is the mean radius of curvature of the spheroid for the measured line; s_1 is the measured distance, after the velocity correction has been applied; K is the coefficient of refraction $(=R/r)$, where r is the radius of curvature of the signal path; Δh is the difference in spheroidal height between the stations; and h_A, h_B are the spheroidal heights of the two stations.

Equation (3.54) is rigorous, accepting that the spheroid is replaced by a local sphere. Due care should be taken, however, to ensure that enough significant digits are used, in particularly with regard to the trigonometric functions where small angles are involved.

Very often the vertical angle (α) at one end of the line is known, and the reduction from s_2 to s_3 is carried out in two stages as shown by Figure 3.8. If α' is the vertical angle at A which has been corrected for refraction (i.e. $\alpha' = \alpha - \delta$; see Figure 2.32), and AB'', the intermediate distance, is represented by s_5, the stages are

(i) a reduction for slope from s_2 to s_5, and
(ii) a reduction for height above the spheroid, from s_5 to s_3.

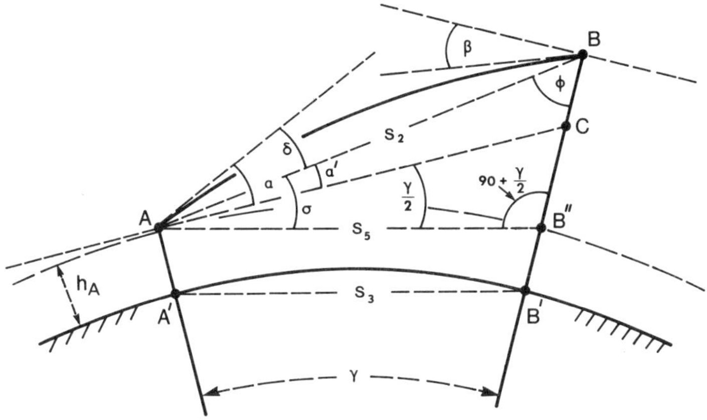

Figure 3.8 Reduction for slope, and height above the spheroid.

In the triangle ABB'',

$$\sigma = \alpha' + \frac{\gamma}{2}$$

and

$$\phi = 180 - \left(\alpha' + \frac{\gamma}{2}\right) - \left(90 + \frac{\gamma}{2}\right) = 90 - (\alpha' + \gamma)$$

also

$$\frac{s_5}{\sin\phi} = \frac{s_2}{\sin\left(90 + \dfrac{\gamma}{2}\right)}$$

Therefore

$$s_5 = s_2 \sin\phi \sec\frac{\gamma}{2} = s_2 \cos(\alpha' + \gamma) \sec\frac{\gamma}{2} \tag{3.55}$$

and the slope correction (Δ_2') is

$$\begin{aligned} \Delta_2' &= s_5 - s_2 \\ &= -s_2\left(1 - \cos(\alpha' + \gamma)\sec\frac{\gamma}{2}\right) \end{aligned} \tag{3.56}$$

For short lines γ is often assumed to be 0, in which case eqn (3.56) reduces to

$$\Delta_2' = -s_2(1 - \cos\alpha') \tag{3.57}$$

as in eqn (3.15).

However, for the longer EDM lines γ will be significant, and eqn (3.56) should be used. A value for γ may be found from

$$\gamma = \frac{s_5}{R + h_A} \tag{3.58}$$

where $s_5 \approx s_2$.

Figure 3.9 illustrates the difference in the slope correction between eqns (3.56) and (3.57), for a range of values of s_2 and α; and for $h_A = 0$.

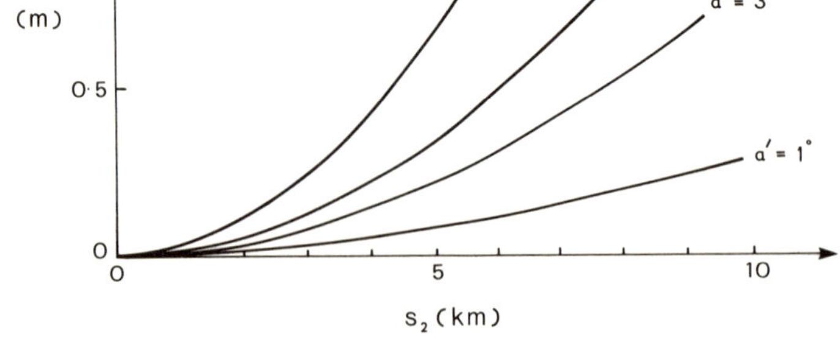

Figure 3.9

If the angle of refraction δ is not known, but reciprocal vertical angles α and β have been observed, then in the triangle ABB'',

$$\sigma = \alpha' + \frac{\gamma}{2} = (\alpha - \delta) + \frac{\gamma}{2}$$

but, from section 2.10,

$$\delta = \frac{(\gamma + \alpha + \beta)}{2}$$

therefore,

$$\left.\begin{aligned} \sigma &= \alpha - \frac{\gamma + \alpha + \beta}{2} + \frac{\gamma}{2} \\ &= \frac{\alpha - \beta}{2} \end{aligned}\right\} \tag{3.59}$$

where β is negative for an angle of depression. Therefore

$$\phi = 180 - \left(\sigma + 90 + \frac{\gamma}{2}\right) = 90 - \left(\sigma + \frac{\gamma}{2}\right)$$

Now

$$\frac{s_5}{\sin\phi} = \frac{s_2}{\sin\left(90 + \frac{\gamma}{2}\right)}$$

therefore

$$s_5 = s_2 \cos\left(\sigma + \frac{\gamma}{2}\right) \sec\frac{\gamma}{2} \tag{3.60}$$

and the slope correction Δ_2' is given as

$$\begin{aligned} \Delta_2' &= s_5 - s_2 \\ &= -s_2\left(1 - \cos\left(\sigma + \frac{\gamma}{2}\right)\sec\frac{\gamma}{2}\right) \end{aligned} \tag{3.61}$$

where $\sigma = (\alpha - \beta)/2$.

The second reduction is for the height (h_A) of the chord above the spheroid, and is found from

$$\frac{s_3}{s_5} = \frac{R}{(R + h_A)} \tag{3.62}$$

Therefore

$$s_3 = s_5\left(1 + \frac{h_A}{R}\right)^{-1}$$

which may be expanded as a series to give

$$s_3 = s_5\left(1 - \frac{h_A}{R} + \left(\frac{h_A}{R}\right)^2 - \cdots\right)$$

Therefore the spheroid correction (Δ_2'') is

$$\Delta_2'' = s_3 - s_5$$

$$= -s_5\left(\frac{h_A}{R} - \left(\frac{h_A}{R}\right)^2 + \cdots\right)$$

and since $\left(\frac{h_A}{R}\right)^2$ is negligible

$$\Delta_{2''} = -s_5\frac{h_A}{R} \qquad (3.63)$$

The more important geometrical corrections which are related to instrument design or location in the field may be classified as follows:

(i) calibration and standardization corrections;
(ii) eccentric station corrections; and
(iii) corrections for unequal instrument/target/reflector heights.

Calibration and standardization corrections arise because of changes in the components of EDM equipment. The corrections include those necessary to compensate for scale error, zero error, and cyclic error. Scale errors are caused mainly by errors in the modulation frequencies, but other sources have a similar effect —e.g. errors in assessing the environmental parameters, which have a greater impact on microwave distance measurement than electro-optical distance measurement. The zero error (or index error) on the other hand is often more important when electro-optical distance measurement is carried out; it is the difference between the mechanically defined centres of the instruments (or reflectors) at each end, and their electrically (or optically) defined centres, and is considered constant over a period of time. Cyclic error becomes important when very precise measurement is required over short ranges, and the variation is usually a function of the phase difference between the reference and measurement signals. Cyclic error, if it exists at all, tends to have a pattern resembling that of a sine curve.

A scale-factor correction may be determined by measuring the frequency of the oscillator with a calibrated frequency counter, or by measuring a distance whose value is known. Standardization with a frequency counter has the advantage that the problems of reduction of a measured distance are avoided. If values for the other corrections are also required then more than one distance must be measured. The standard procedure is to measure all possible distances between a number of stations on one straight line, and determine the correction parameters by least squares. If the stations are arranged in such a way that a range of values over a modulation wavelength is recorded then a determination of cyclic error is possible. After scale and zero-error corrections have been applied, the residual errors that remain will indicate whether or not cyclic error is present. Details of these procedures are given in [37], [34] and [10].

Calibration and standardization correction parameters will change as the equipment becomes older, and so they should be determined periodically. They will also change if either the instrument or reflector/remote components of a pair is changed.

Thus the corrected distance (s) may be expressed as

$$s = c_1 s' + c_2 + c_3 \tag{3.64}$$

where s' is the recorded distance; c_1 is the scale-factor correction; c_2 is the zero-error correction; and c_3 is the cyclic error correction—a function of the remaining fraction of the unit wavelength in the measured distance.

Although the scale-factor c_1 might have the largest effect of all three corrections, it is not necessarily the most important. In many applications involving surveys between existing control points the method of computation of coordinates is likely to involve procedures which compensate for scale error. Examples are the Bowditch adjustment of a traverse, and a variation of coordinates adjustment of a control network. Furthermore, the instrument is not the only source of scale error; most national networks have scale errors in excess of those found in modern EDM instruments. The cyclic error correction c_3 is normally very small and usually neglected, except for the most precise work. If the error becomes large the defect in the instrument may be rectified by the manufacturer.

Electro-optical instruments normally use glass corner-cube retro-reflectors to return the signal from the far end of the line. When the reflector is aligned correctly the zero correction will take account of any constant eccentricity between the effective centre of the reflector and the point vertically above the ground station. If the reflector face is at an angle, however, a small error in the distance will occur, due to the fact that a greater portion of the ray path is through glass. Details of the amounts involved are given in [34], but they become significant at angles of about 10° and greater.

Some microwave instruments suffer from a problem where the beam is wide enough to be reflected from objects adjacent to the line, thus producing incorrect distances. The phenomenon is often referred to as 'ground swing', and may be reduced by adjusting the position of the instrument so that the peripheral part of the beam is screened off. Alternatively, the carrier frequency may be varied by small amounts and a series of fine readings taken, from which the cyclic variation in the measured distance may be revealed, and a suitable mean adopted. Thus a field procedure rather than a reduction is used to compensate for the error. Modern instruments minimize the problem by using higher frequencies and narrower beamwidths.

Eccentric station reductions are more complicated for distance measurement than angle measurement since the eccentricity is likely to be in a three-dimensional sense. Therefore the derivation of the appropriate equations is best handled in three dimensions, and spherical trigonometry is used. The eccentricity may also arise at either the instrument end or the reflector/remote end of the line, or at both ends.

In the more common case of eccentricity at the instrument end, the angles may be observed from either the main station or the eccentric station. Consider first of all the case where both angles and distances are observed from the eccentric station. In Figure 3.10 E, the eccentric station, is at the origin of a three-dimensional system, and M is the main or permanent station. The xy-plane is the horizontal plane, and the line EM lies in the xz-plane. Also, T is the target (reflector/remote) point; T' is the point where the extension of the line ET meets the unit sphere; M' is the point where the extension of the line EM meets the unit sphere; e is the eccentric distance; α is the horizontal angle (measured clockwise) between the main station and the target; β is the plane angle MET; v_M, v_T are the vertical angles for M and T respectively; and s', s are the observed and reduced distances respectively.

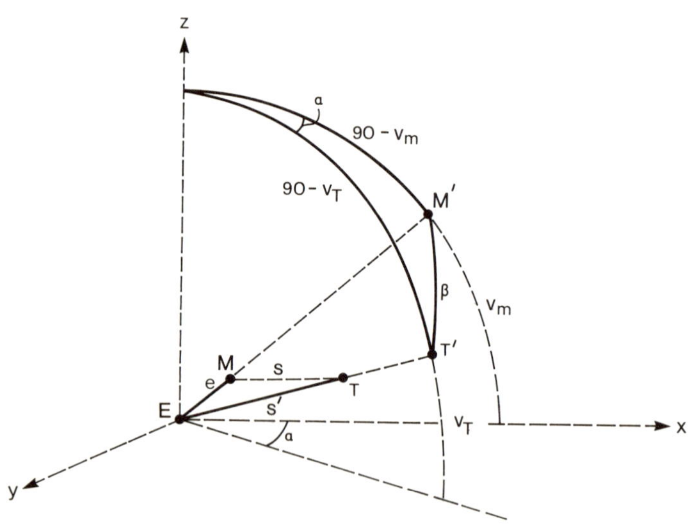

Figure 3.10 Reduction for eccentric angle and distance measurement.

From the cosine formula in spherical trigonometry,

$$\cos\beta = \cos(90 - v_M)\cos(90 - v_T) + \sin(90 - v_M)\sin(90 - v_T)\cos\alpha \tag{3.65}$$

and, from the cosine formula in plane trigonometry,

$$s^2 = s'^2 + e^2 - 2s'e\cos\beta \tag{3.66}$$

Therefore, substituting eqn (3.65) in (3.66), and taking the positive square root,

$$s = s'\left(1 + \frac{e^2}{s'^2} - \frac{2e}{s'}(\sin v_M \sin v_T + \cos v_M \cos v_T \cos\alpha)\right)^{1/2} \tag{3.67}$$

The vertical angle at M may be derived by considering the height differences:

$$\Delta h_{MT} = \Delta h_{ET} - \Delta h_{EM}$$

i.e.

$$s\sin v_{MT} = s'\sin v_T - e\sin v_M$$

therefore

$$\sin v_{MT} = \frac{s'}{s}\sin v_T - \frac{e}{s}\sin v_M \tag{3.68}$$

where e, s', v_T, v_M are measured, and s is derived from eqn (3.67).

When angles are measured at the main station and distances from the eccentric station, the situation is as indicated in Figure 3.11.

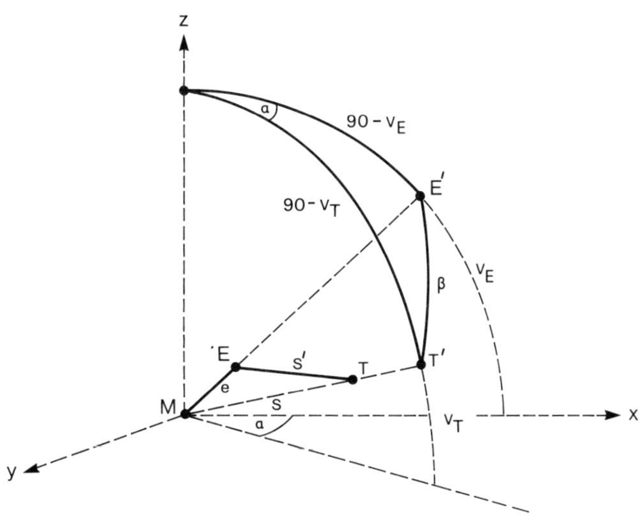

Figure 3.11 Reduction for eccentric distance measurement.

A similar procedure is used to derive an equation for s, with the following result:

$$s = s'\left(1 - \frac{e^2}{s'^2} + \frac{2es}{s'^2}(\sin v_E \sin v_T + \cos v_E \cos v_T \cos\alpha)\right)^{1/2} \tag{3.69}$$

In this case s also appears on the right-hand side. However, an approximate value ($= s'$) may be used, and if necessary it may be improved by further iterations.

Details concerning the other cases of eccentricity are not given here, but may be found in [34].

Corrections for *unequal instrument/target/reflector heights* are all too common in electronic distance measurement, particularly with the earlier EDM equipment. Modern instruments are designed to make such corrections unnecessary. However, if the theodolite observations are made separately, the likelihood is that the theodolite axis will be at a different height from the EDM instrument. The same applies to the target end of the line. Therefore the equation for such a correction becomes useful.

The correction may be applied before the slope correction, and Figure 3.12 illustrates a typical situation. E is the EDM instrument and E' is the reflector

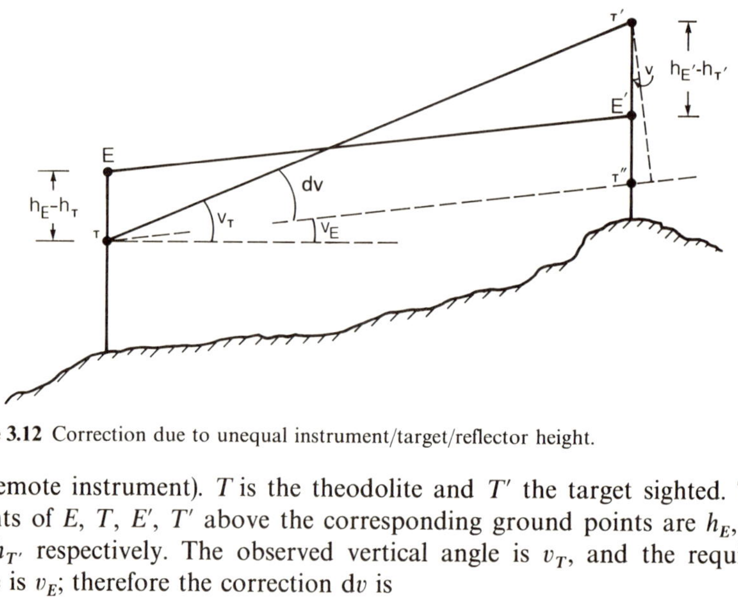

Figure 3.12 Correction due to unequal instrument/target/reflector height.

(or remote instrument). T is the theodolite and T' the target sighted. The heights of E, T, E', T' above the corresponding ground points are h_E, h_T, $h_{E'}$, $h_{T'}$ respectively. The observed vertical angle is v_T, and the required angle is v_E; therefore the correction dv is

$$\begin{aligned} \mathrm{d}v &= v_E - v_T \\ \mathrm{d}v &= \frac{T'T''}{TT'} \cos v \end{aligned} \tag{3.70}$$

where v is an approximate value for the vertical angle; for convenience it may be the observed value v_T.

Let

$$\Delta h = T'T'' = (h_E - h_T) - (h_{E'} - h_{T'}) \tag{3.71}$$

and since $TT' \approx EE' = s$, the measured distance, therefore

$$\left.\begin{aligned} dv &= \frac{\Delta h}{s} \cdot \cos v \text{ radians} \\ v_E &= v_T + dv \end{aligned}\right\} \tag{3.72}$$

The sign of dv will be determined by the sign of Δh; and for maximum precision in the determination of the reduced distance, Δh should be as small as possible in absolute terms. With steep lines of sight the determination of Δh (i.e. the measurement of heights h_E, h_T, $h_{E'}$, $h_{T'}$) becomes critical.

Equations (3.71) and (3.72) apply also to the situation where the EDM instrument is mounted on the trunion axis supports of a theodolite, in which case $(h_E - h_T)$ becomes a constant for a particular EDM/ theodolite combination. An alternative situation is where the EDM instrument is mounted on the telescope of the theodolite, so that the axes are offset and parallel to each other. Therefore if suitable corrections are not applied the distances will be incorrect. If on the other hand a special tiltable target/reflector with the same offset is provided then a correction becomes unnecessary. The same comment applies to EDM equipment which has the transmitter and receiver located on opposite sides of the theodolite telescope. A problem does arise, however, with instruments that require two pointings to the reflector—one for the angular measurements and another for the distance measurement. Figure 3.13 is an exaggerated diagram that illustrates the problem. Point T

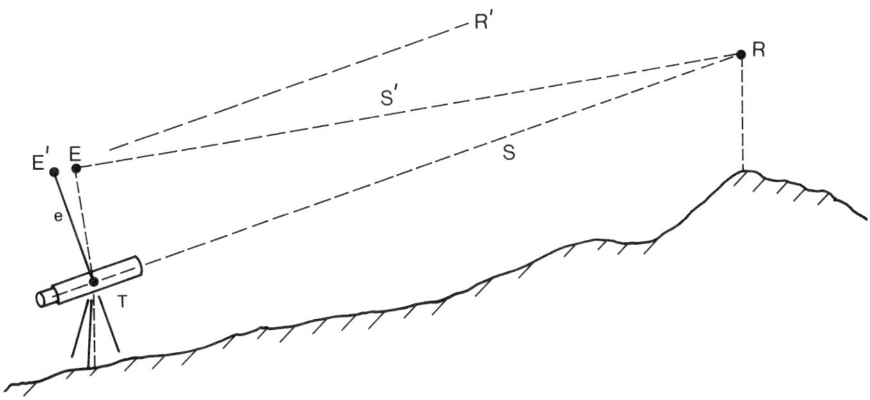

Figure 3.13 Telescope-mounted EDM.

represents the theodolite, R the reflector, and E and E' the successive positions of the EDM instrument. Angular measurements are made with the theodolite telescope pointing to R and the EDM in the position E', with axis $E'R'$. This is followed by the second pointing with the EDM in position E and axis ER. Thus if e is the eccentric distance, s' the measured distance ER, and s the required distance TR,

$$s^2 = s'^2 + e^2$$

therefore

$$s = s'\left(1 + \frac{e^2}{s'^2}\right)^{1/2} \tag{3.73}$$

which, on expansion of the term in brackets, becomes

$$s = s'\left(1 + \frac{e^2}{2s'^2} - \cdots\right)$$

$$= s' + \frac{e^2}{2s'} - \cdots \tag{3.74}$$

or, the correction (Δs) to the measured distance, to obtain the distance from the horizontal axis of the theodolite, is

$$\Delta s = \frac{e^2}{2s'} \tag{3.75}$$

In practice Δs is usually small enough to be neglected.

3.5 Projection corrections

Observations which have been reduced to the surface of the spheroid must have further corrections applied if the computations are to be carried out on the plane projection. Many different projections are used worldwide, and it is beyond the scope of this text to consider them in detail; on the other hand it is appropriate to give some indication of the magnitudes of the corrections involved. Orthomorphic (or conformal) projections are usually used, and of these the transverse Mercator projection is the most common. In this section some simplified formulae are considered which are appropriate to this projection, and which may be used to correct observations made for general (lower-order) surveying purposes—where the accuracy requirements are of the order of one second of arc and five parts per million for angle and distance measurements respectively. Further details concerning the more rigorous formulae may be found in [8], and concerning the projection system for the UK in [31]. The most commonly applied corrections are the 't-T' (or 'arc-chord') and scale-factor corrections for directions and distances respectively.

The *t-T correction* (δ) arises because of the fact that the geodesic (or normal section)—which represents the observation on the spheroid—appears as a curved line on the projection. Therefore, for computation using plane trigonometry, a small correction is needed to derive the direction of the chord between the end points. Figure 3.14 illustrates how geodesics, represented by full lines, appear on the projection. Two important characteristics are evident:

(i) curvature of the line as always concave towards the central meridian, thus the sign of the correction is immediately evident; and

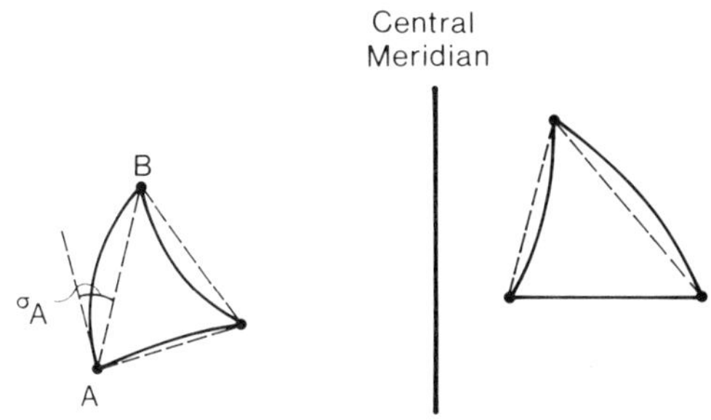

Figure 3.14 *t-T* or arc-chord correction.

(ii) the amount of curvature (and hence correction) is a function of the distance (i.e. eastings coordinate) from the central meridian, and the difference in the northings coordinates for the endpoints.

The formula for *t-T* is not a closed formula, but rather in the form of a series, where the number of terms adopted depends on the application and the required precision. In most survey operations, with a precision requirement of 1″, the use of only the first term is sufficient. Thus, for a linc *AB*, the direction would be corrected by an amount obtained using the formula

$$\delta_A = -\frac{(N_B - N_A)(E_B + 2E_A)}{6\rho v} \tag{3.76}$$

where δ_A is the *t-T* correction (in radians) at *A*; ρ is the radius of curvature of the spheroid in the meridian plane; v is the radius of curvature of the spheroid in the prime vertical plane (perpendicular to the meridian plane); and E_A, N_A, E_B, N_B are eastings and northings coordinates for the two endpoints.

Note that the eastings values are distances with respect to the central meridian. If the eastings coordinate includes a constant (E_0)—as in the UK grid coordinate system—it is subtracted before the value is introduced in eqn (3.76). Thus E' would be used, where

$$E' = E - E_0 \tag{3.77}$$

It is clear from eqn (3.76) that δ_A and δ_B will differ by a very small amount, and in many applications the difference is negligible for practical purposes. Therefore for surveys of limited extent a simpler formula may be used, such as

$$\delta_A = -\frac{\Delta N_{AB} \cdot E}{2R^2} \tag{3.78}$$

where *E* is the eastings value for the midpoint of the line; *R* is an approximate value of radius of curvature of the spheroid for the area—derived, for example, for a point in the centre of the survey area; and ΔN_{AB} is the difference in northings for the line *AB*.

Examples which illustrate the varying magnitude of *t-T*, for lines of different length and location on the transverse Mercator projection for the UK, are given in Table 3.2.

Table 3.2

Distance (km) E or W of the CM	ΔN (km)	δ ($''$)
50	10	1
250	10	6
100	30	8

Scale-factor (F) is the multiplying factor which is used to convert a 'true' distance (S) on the reference spheroid to the equivalent grid (chord) distance (s) on the projection.

Thus

$$\left.\begin{aligned} s &= F \times S \\ \text{or} \qquad S &= s/F \end{aligned}\right\} \tag{3.79}$$

Scale-factor in the vicinity of a point is also mainly dependent on the distance from the central meridian. It changes slowly from point to point, particularly for areas close to the central meridian; and very often a single value is sufficient for all measurements within a 5km radius of the point. The formula for scale-factor is also expressed in the form of a series, but for surveys of limited extent the following equation will generally suffice:

$$F = 1 + \frac{E^2}{2\rho\nu} \tag{3.80}$$

Thus the variation of scale-factor with distance from the central meridian is according to a parabolic curve, as indicated by the dotted line in Figure 3.15*a*.

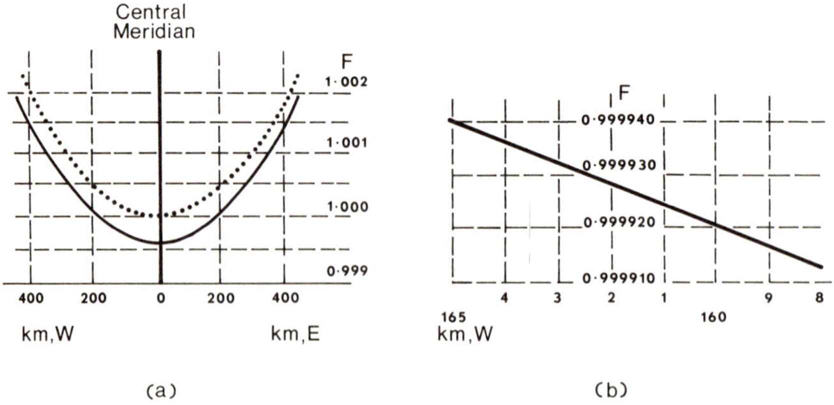

Figure 3.15 Projection scale factor.

Equation (3.80) gives a value for scale-factor where the value on the central meridian is unity; and the value increases going east and west of the meridian. In order to have a more acceptable range of values across what is considered the useful zone of the projection, it is common practice to assign a value (F_0), not equal to unity, for the central meridian. Therefore eqn (3.80) becomes

$$F = F_0\left(1 + \frac{E^2}{2\rho v}\right) \tag{3.81}$$

and the parabola which relates to this equation (where $F_0 = 0.9996$) is shown as a full line in Figure 3.15*a*.

Thus for example in the UK projection, where F_0 is approximately 0.9996, the scale factor is unity on lines nearly parallel with the central meridian, and approximately 180 km on either side of it. Figure 3.15*b* shows the variation in scale factor in the vicinity of Loch Lomond, Scotland, and it is clear that over a few kilometres the change is almost linear.

Again, ρv may usually be replaced by R^2; and if the eastings coordinate contains a constant (E_0) it must be subtracted before the coordinate value is introduced into eqns (3.80) and (3.81).

Table 3.3 lists values for scale factor for a 10 km line at various locations on the UK projection; the corresponding corrections to the observed distances are also given.

Table 3.3

Distance (km) E or W of the CM	Scale-factor	Correction (m) to a 10 km line
10	0.9996025	− 3.98
100	0·9997238	− 2.76
180	0·9999984	−0·02
200	1.0000916	+ 0.92
250	1.0003674	+ 3.67

As indicated earlier, very often a single value may be used for a particular survey. For larger surveys a value for each specific line may be calculated, based on the eastings coordinate for the midpoint of the line. Lines up to 30 km may be corrected in this way, with errors not exceeding 2 ppm. The scale factor for longer lines may be established from values calculated for both ends and the midpoint, using Simpson's Rule, viz.

$$\frac{1}{F} = \frac{1}{6}\left(\frac{1}{F_A} + \frac{4}{F_M} + \frac{1}{F_B}\right) \tag{3.82}$$

where A and B are the endpoints, and M is the midpoint.

The calculated value for scale-factor should have a precision consistent with—or one order better than—the observed distance; in many cases this implies a value calculated to six or seven decimal places.

4 Computation of rectangular coordinates using redundant data

4.1 Introduction

Chapter 2 was concerned with the computation of rectangular coordinates of single points when the minimum number of observations was made. In practice the coordinates of a *network* of control points is normally required, and many more observations than the minimum are made—i.e. redundant observations are present. Not only does the inclusion of additional observ-ations provide a means of identifying gross errors in observations or control data, but it also improves the accuracy of the result; furthermore, it provides a means of establishing the accuracy of the result. Therefore, as far as control points are concerned, it is rare that the absolute minimum number of observations is made. This chapter considers some computational methods for the calculation of plane rectangular coordinates of points in a control network when redundant observations are present. The assumption from the outset is that all measured values have been reduced to the computational surface. This includes the corrections outlined in Chapter 3, and the additional projection corrections necessary to convert a spheroidal value to the equivalent value on the projection.

The basic problem associated with redundant observations is the fact that each observation is simply an estimate of a quantity, and there will be some variation between successive observations of the same quantity. A more reliable estimate may be derived by considering all the observations which relate to that quantity; and the best estimate is obtained when the relative weights and correlation between observations is taken into account. This latter and more rigorous treatment of observations is considered in Chapters 9 and 10. The purpose of the present chapter, however, is to consider less rigorous methods, which do not require the same amount of computation but which nevertheless provide satisfactory solutions to many position-fixing problems.

Two different approaches to the problem are considered: the first establishes small corrections to the observations in order to obtain better estimates, which are then used to derive final coordinates, whereas the second uses the actual observations to establish final coordinates, and better estimates for the observed quantities are obtained as a secondary phase. The methods described involve a series of individual computational steps, and the stochastic (or statistical) properties of the observations are not given the same close attention as in the more rigorous least-squares methods described later. They are therefore considered more appropriate to networks of third order or lower.

4.2 The adjustment of figures using condition equations

The objective of the first of the two approaches is to apply small corrections to the observations so that they are consistent with one another; in other words so that any subset containing the required minimum number of observations will produce the same coordinates for the points as another subset. The objective is achieved by ensuring that the corrected observations satisfy certain geometrical conditions, and with each redundant observation there is one more condition to be satisfied.

Thus in the case of a simple plane triangle, if all three angles A, B, C have been observed, the condition to be satisfied is

$$A + B + C = 180^\circ \tag{4.1}$$

A similar expression may be derived for any closed polygon, where the sum of the interior angles must equal $(2n - 4)$ right angles, n being the number of sides. Other conditions arise also, and the *angle conditions* in a network may be classified as:

(i) polygon (normally triangle) conditions—as specified above;
(ii) horizon (or centrepoint) conditions—if all the angles about a point have been observed, their sum must equal 360°; and
(iii) station conditions—if one or more observed angles at a station may be expressed as a function of other observed angles at that station.

It is usually possible to establish morc conditions than the number (r) of redundant observations in a network; however, the requirement is that *independent* condition equations are used—that is, equations which cannot be derived from a combination of any of the other equations adopted.

All the angle conditions of a network may be satisfied without the network attaining the required consistency. For example, consider the network shown in Figure 4.1*a*. Angles 1 to 8 have been observed, and the following three independent triangle conditions have been satisfied:

$$\left.\begin{aligned} 1 + 2 + 3 + 4 &= 180 \\ 5 + 6 + 7 + 8 &= 180 \\ 1 + 2 + 7 + 8 &= 180 \end{aligned}\right\} \tag{4.2}$$

AB is assigned a distance, and the coordinates of points A and B are established; the coordinates of points C and D are then calculated via triangles ABC and ABD.

However, the coordinates of C may also be calculated via the triangle ADC, and a second position C' would be obtained. Values $5'$ and $6'$ would also satisfy

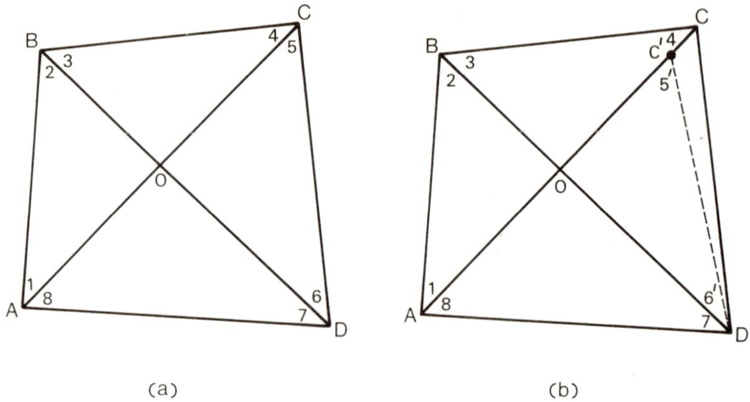

Figure 4.1 Braced quadrilateral.

the three conditions of eqns (4.2), but would nevertheless produce a different position for C. Therefore the fourth condition is one which makes the figure as a whole geometrically consistent. The condition is known as the *side condition*, and is required when the length of any side of the network, such as CD, can be computed via more than one route. The condition is enforced by ensuring that the side has the same length irrespective of the route taken to compute it.

Thus in Figure 4.1*a* the side condition is established as follows:

$$\left.\begin{aligned}\frac{AB}{\sin 7}&=\frac{AD}{\sin 2}; &\quad \frac{BC}{\sin 1}&=\frac{AB}{\sin 4}\\ \frac{CD}{\sin 3}&=\frac{BC}{\sin 6}; &\quad \frac{AD}{\sin 5}&=\frac{CD}{\sin 8}\end{aligned}\right\} \tag{4.3}$$

therefore

$$\frac{AB}{AD}\cdot\frac{BC}{AB}\cdot\frac{CD}{BC}\cdot\frac{AD}{CD}=\frac{\sin 7}{\sin 2}\cdot\frac{\sin 1}{\sin 4}\cdot\frac{\sin 3}{\sin 6}\cdot\frac{\sin 5}{\sin 8}=1 \tag{4.4}$$

An equation such as (4.4) is formed from the observations opposite the centrepoint O, and it may be written down immediately by inspection of the figure. It is non-linear, and the problem of satisfying the condition is eased considerably if it is converted into a linear function.

It may be written as

$$F=\frac{\sin 1 \sin 3 \sin 5 \sin 7}{\sin 2 \sin 4 \sin 6 \sin 8}-1=S-1=0 \tag{4.5}$$

which on linearization becomes

$$F_{o}+\frac{\partial S}{\partial 1}\cdot \mathrm{d}1+\frac{\partial S}{\partial 2}\cdot \mathrm{d}2+\cdots \text{etc.}=0 \tag{4.6}$$

where $F_o = S_o - 1$ and the subscript o implies that observed values of the angles are used.

Now

$$\frac{\partial S}{\partial 1} = \left(\frac{\sin 3 \sin 5 \sin 7}{\sin 2 \sin 4 \sin 6 \sin 8}\right)\cdot \cos 1$$

$$= \left(\frac{\sin 1 \sin 3 \sin 5 \sin 7}{\sin 2 \sin 4 \sin 6 \sin 8}\right)\cdot \frac{\cos 1}{\sin 1}$$

$$= S_o \cot 1$$

But $S_o \approx 1$, therefore $\dfrac{\partial S}{\partial 1} = \cot 1$

and

$$\frac{\partial S}{\partial 2} = -\cot 2, \quad \frac{\partial S}{\partial 3} = \cot 3, \ldots \text{etc.}$$

Therefore eqn. (4.6) becomes

$$F_o + \cot 1 \cdot d1 - \cot 2 \cdot d2 + \cdots \text{etc.} = 0 \tag{4.7}$$

where d1, d2,....etc. are the unknown corrections (v's), and F_o is a known constant—the amount by which the condition equation is not satisfied when the observations are introduced. Therefore the condition equation may be written

$$\cot 1 \cdot v_1 - \cot 2 \cdot v_2 + \cot 3 \cdot v_3 - \cdots \text{etc.} = C_4 \tag{4.8}$$

where $C_4 = -F_o$.

The angle conditions (eqns (4.2)), which are linear functions, may also be written so that they include the observed values and the unknown corrections to these values; thus for the first equation,

$$\left.\begin{aligned} (1+v_1)+(2+v_2)+(3+v_3)+(4+v_4) &= 180 \\ \text{or} \qquad v_1+v_2+v_3+v_4 &= C_1 \end{aligned}\right\} \tag{4.9}$$

where C_1 represents the misclosure for the triangle.

The other two conditions may be arranged in a similar fashion. Therefore, the final set of linearized condition equations becomes

$$\left.\begin{aligned} v_1 + v_2 + v_3 + v_4 &= C_1 \\ v_5 + v_6 + v_7 + v_8 &= C_2 \\ v_3 + v_4 + v_5 + v_6 &= C_3 \\ \cot 1 \cdot v_1 - \cot 2 \cdot v_2 + \cdots + \text{etc.} &= C_4 \end{aligned}\right\} \tag{4.10}$$

where $C_1 \ldots C_4$ are known constants—the misclosures.

Once the condition equations for the figure have been established, as set out in eqns (4.10), they are satisfied one at a time, in a specified sequence. The values used for each new condition equation will be those derived after completion of the previous stages, and appropriate steps must be taken after satisfying a particular condition to ensure that the earlier conditions remain satisfied. This might involve applying additional corrections to the angles in these conditions. The example of the braced quadrilateral given below illustrates the procedure. In the process of satisfying the condition equations the assumption is made that angles obtain equal corrections. Therefore in the first of eqns (4.10), the correction to each angle will be

$$v = \frac{C_1}{4}, \qquad \text{where} \quad C_1 = 180 - (1 + 2 + 3 + 4) \tag{4.11}$$

Similarly for the second and third equations where, if necessary, the C's are obtained using angles derived after previous conditions have been satisfied.

In the side condition alternate angles must necessarily have corrections of opposite sign. Therefore, if all v's are to have the same absolute magnitude but alternate signs, the absolute value must be derived from

$$v(\cot 1 + \cot 2 + \cot 3 + \cdots \text{etc.}) = C_4 \tag{4.12}$$

and the appropriate sign added in order to satisfy the equation in (4.10). Therefore

$$v = \frac{\rho \cdot C_4}{\sum \cot \theta} \tag{4.13}$$

where v is in seconds; $\rho = 206265$; $C_4 = 1 - S_0$; and $\sum \cot \theta = \cot 1 + \cot 2 + \cot 3 + \cdots + \text{etc.}$

The sines in S_0 should be evaluated to six decimal places, but the cotangents are only required to two or three places. Note that for angles between 90° and 180° the cotangents will be negative. The v's are applied to the angles in such a way that when S is recalculated it is found to be unity. The side equation should not affect the other conditions, since the corrections to the peripheral angles are applied with alternate signs. After completion of the adjustment all the condition equations should be satisfied. However, different results will be obtained depending on the order of application of the condition equations, and this demonstrates the approximate nature of the solution. Nevertheless, consistency will have been achieved, and the coordinates of the points in the network may be calculated using any convenient route. The method is suitable for very simple networks involving angular measurements for the most part, and with few overlapping triangles. It becomes complicated with larger networks, or networks involving redundant distance measurement. Although the latter may be incorporated as additional side conditions, they are prone to error and are not considered worthwhile in a method which is after all approximate.

The number of independent condition equations for any network is equal to the number of redundant observations (r), and may be established by building up a network from known data. Observations are added one at a time until finally all points in the network are established, and the minimum number of observations (m) which are required is known. The number of redundant observations is then

$$r = n - m \tag{4.14}$$

where n is the total number of observations.

A variety of formulae have been devised to establish the number of condition equations directly, and examples may be found in [15] and [32]. The method described above, however, is considered sufficient for the networks of limited size which are considered here.

A numerical example of the adjustment of a braced quadrilateral is given below, using the diagram and equations established above. The process is carried out systematically with the aid of a table of corrections (Table 4.1), and the steps in the process are as follows:

(i) List the observations in column 2 and derive the misclosures C_1 and C_2 for the first two conditions. Derive the corrections, enter them in column 3, and add them to the observations to give the first adjusted values in column 4.

(ii) Derive the misclosure C_3 for the third condition and enter the corresponding corrections in column 5. Re-establish the closure for the first two conditions by compensating the other two angles in each triangle. Apply the corrections and derive the second adjusted values in column 6.

(iii) Derive the corrections for the side condition and enter these in column 7. Add the corrections to the values in column 6 to obtain the final adjusted values in column 8. Note that the corrections for the even and odd angles should have opposite signs.

(iv) Check that the values in column 8 satisfy all the conditions—in particular the side condition.

Example 4.1

Table 4.1 Adjustment of a braced quadrilateral

Angle	Observed value			$C_1/4$ $C_2/4$	1st adjusted value			$C_3/4$	2nd adjusted value			$\frac{\rho \cdot C_4}{\Sigma \cot \theta}$	Final adjusted angle		
	°	′	″	″	°	′	″	″	°	′	″	″	°	′	″
1	84	07	59	+3	84	08	02	−4	84	07	58	+2.5	84	08	00.5
2	40	10	36	+3	40	10	39	−5	40	10	34	−2.5	40	10	31.5
3	16	11	20	+3	16	11	23	+4	16	11	27	+2.5	16	11	29.5
4	39	29	53	+3	39	29	56	+5	39	30	01	−2.5	39	29	58.5
	179	59	48												
$C_1 =$			+12												
5	75	03	02	−4	75	02	58	+5	75	03	03	+2.5	75	03	05.5
6	49	15	28	−4	49	15	24	+5	49	15	29	−2.5	49	15	26.5
7	30	05	21	−4	30	05	17	−5	30	05	12	+2.5	30	05	14.5
8	25	36	25	−4	25	36	21	−5	25	36	16	−2.5	25	36	13.5
	180	00	16												
$C_2 =$			−16												
1					84	08	02	−4							
2					40	10	39	−5							
7					30	05	17	−5							
8					25	36	21	−5							
					180	00	19								
				$C_3 =$			−19								

Side condition

$$S_o = 0.99986996$$

$$C_4 = 1 - S_o = 0.00013004$$

$$\Sigma \cot \theta = 10.885$$

$$v = \frac{\rho C_4}{\Sigma \cot \theta} = 2.46''$$

Check after satisfying the side condition:

$$C_4 = -0.0000019$$

which is negligible.

4.3 Traverse adjustment

In section 2.5 an open traverse was defined, but dismissed as being of little practical use because it is unchecked. The minimum requirements for checking

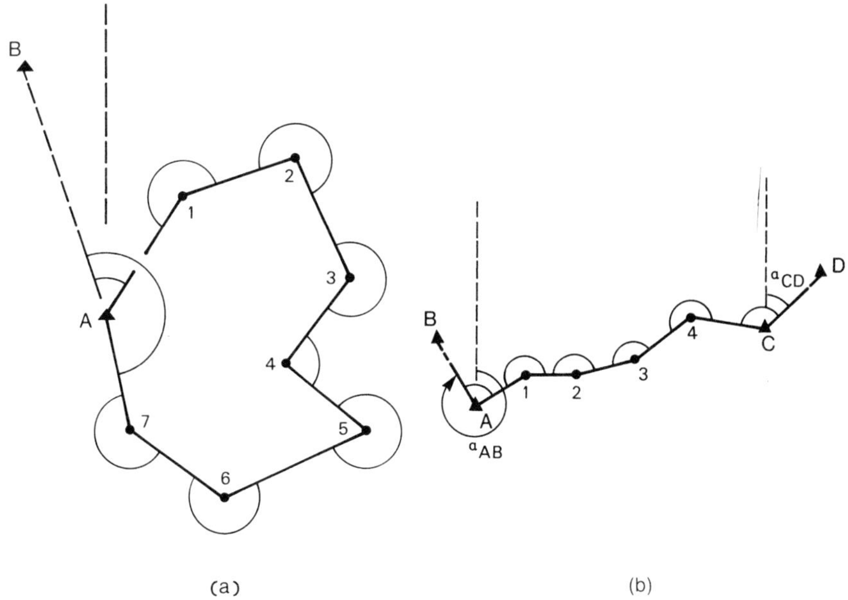

Figure 4.2 closed traverses.

purposes are that it closes either on the starting point, or on another point which has known coordinates. The two situations are shown in Figure 4.2*a* and 4.2*b* respectively.

In the case of Figure 4.2*a* the sum of the interior angles of the polygon should equal the required amount, and in (*b*) the derived bearing α_{CD} should equal the value computed from known coordinates. Any angular misclosure may be distributed among the observed angles in order to achieve provisionally adjusted values. These are used to derive provisional bearings which, together with the observed distances, allow the computation of coordinate differences ΔE and ΔN between traverse points. The sum of the respective differences should be 0 in case (*a*), and equal to a known difference in case (*b*). Because of observational error, or faulty coordinates for the known points, it is very rare that this happens in practice. Therefore the misclosures must be distributed amongst the traverse legs in order to obtain final coordinates for the points. This is referred to as *a traverse* adjustment, and it may be carried out in a number of ways.

It should be noted here that although the loop traverse shown in (*a*) may be closed by an appropriate adjustment, certain systematic errors in the observations may go undetected. An error in determining the orientation of the traverse, based on the single observation *AB*, will not be detected. Also, if there is a scale difference between the unit of measurement and the coordinate system it will not show in the results. For these reasons the loop traverse is considered to be inferior.

The objective of a traverse adjustment is to obtain the best estimates of the coordinates of the traverse points; thus the chosen method should be consistent with the relative precision of the observed quantities. When the quantities are dissimilar, as in the case of angle and distance measurement, relative precision may be assessed by comparing the error in distance measurement with the distance subtended by an angular error. Alternatively, the ratio of the distance error to the distance measured may be compared with the angular error expressed in radians.

Three alternative situations are possible in traversing:

(i) when the precision of angular measurement exceeds that of distance measurement;
(ii) when the precision of angle and distance measurement is approximately equal; and
(iii) when the precision of distance measurement exceeds that of angle measurement.

Although the objective is normally (ii), the circumstances in practice do not always support such conditions. Therefore approximate methods have been devised to suit each case. However, if the number of additional observations is limited to only one or two at the end of the traverse it is questionable which method would give the best results. Also, the quality of the results of a particular adjustment depends to a large extent on the geometry of the traverse—i.e. whether the direction of the traverse approaches a straight line, or whether it deviates considerably from it. Therefore in practice it is normal to adopt one method and stick to it. Further discussion on the relative merits of approximate methods is contained in [15] and [33]. In this text only the Bowditch adjustment—also referred to as an adjustment by the compass rule—will be considered, as it is simple to apply and the results are usually adequate for lower-order surveys. The more rigorous approach to traverse adjustment is to use a process of least-squares estimation, and this is considered in Chapter 10, where the subject is dealt with in terms of a control network with mixed observations.

The stages in the computation, based on a Bowditch adjustment, are as follows:

(i) The angles (or bearings) are adjusted so that the accumulated value agrees with a specified value—which could be the sum of the angles for a closed polygon, or a calculated bearing between points with known coordinates.
(ii) Coordinate differences ΔE and ΔN for all traverse legs are calculated, based on the bearings derived in (i) and the observed distances (corrected for slope, height above spheroid, projection scale-factor, etc.).
(iii) The coordinate differences are accumulated to establish the total misclosure corrections $\mathrm{d}E$ and $\mathrm{d}N$, and these are then distributed

proportionally amongst the individual coordinate differences according to the rule

$$\left.\begin{aligned}\delta E_{ij} &= \frac{l_{ij}}{L}\cdot \mathrm{d}E \\ \delta N_{ij} &= \frac{l_{ij}}{L}\cdot \mathrm{d}N\end{aligned}\right\} \qquad (4.15)$$

where $\delta E_{ij}, \delta N_{ij}$ are corrections to the coordinate differences ΔE, ΔN for the traverse leg ij; l_{ij} is the length of the traverse leg ij; and L is the total length of the traverse. The adjusted differences are accumulated in order to verify that the result satisfies the requirements; and any error in rounding must be accounted for at this stage.

(iv) The adjusted coordinate differences are accumulated to obtain final values for the coordinates of the traverse points, and from these values final bearings and distances may be derived.

An example is given below to demonstrate the application of the Bowditch adjustment.

Example 4.2

Figure 4.2*b* shows a traverse which starts at point *A* and closes on point *C*. Use the observed angles and distances listed on p. 112, and the given coordinates of *A*, *B*, *C* and *D*, to calculate the coordinates of points 1, 2, 3 and 4.

Solution

The known coordinates of *A*, *B*, *C* and *D* are used to calculate the bearings α_{AB} and α_{CD}, which are 330°27′06″ and 49°15′07″ respectively. The value for α_{AB} is used to establish the bearing α_{A1}, and the other observed angles are added progressively in order to establish a value $\alpha'_{CD'}$. The angular misclosure correction is found from $(\alpha_{CD} - \alpha'_{CD'})$, and in this case it is + 39″. The correction is distributed equally between the observed angles; or, in other words linearly between the derived bearings, as indicated in the fourth column. The coordinate differences (ΔE and ΔN) are calculated and adjusted for any misclosure, as indicated by step (iii) above. Finally, the adjusted differences are added to the coordinates of *A* to obtain the coordinates of points 1 to 4.

4.4 Semigraphic methods of fixing position

The second approach to computation involving redundant observations uses the observations more directly, and derives final coordinates of points before establishing final values for the observed quantities. Instead of satisfying condition equations in a particular sequence the method derives the coordinates of points sequentially. Therefore it is also an approximate solution, in the

Trav. stn.	Observed angle and distance	Bearing (α) °	′	″	Corr. ″	Adjusted bearing °	′	″	ΔE ΔN	Corr. δE δN	Adjusted ΔE ΔN	Final Coordinates E	N
B												672.450	690.150
	$\alpha_{AB}=$	330	27	06									
A	86 12 50 361.00	56	39	56	+6	56	40	02	301.613 198.370	+0.012 −0.019	301.625 198.351	898.360	291.640
1	216 10 35 301.33	92	50	31	+12	92	50	43	300.958 −14.958	+0.010 −0.016	300.968 −14.974	119.985	489.991
2	162 40 37 352.17	75	31	08	+18	75	31	26	340.989 88.034	+0.012 −0.018	341.001 88.016	1500.953	475.017
3	160 17 51 425.49	55	48	59	+25	55	49	24	352.012 239.018	+0.014 −0.022	352.026 238.996	1841.954	563.033
4	219 05 45 432.67	94	54	44	+32	94	55	16	431.075 −37.116	+0.014 −0.023	431.089 −37.139	2193.980	802.029
C	134 19 44	49	14	28	+39	49	15	07				2625.070	764.890
	$\alpha_{CD}=$	49	15	07									
	$L=\Sigma l=1872.66$								$\Sigma(\Delta E)=1726.648$ $E_C-E_A=1726.710$	$\Sigma(\Delta N)=473.348$ $N_C-N_A=473.250$			
D												2986.620	1076.400
Misclosure correction					+39				+0.062	−0.098			

sense that a different sequence of computation in a network will produce a different answer.

The basic principle of the method is to obtain, at a large scale, a graphical representation of the interaction between observations at the point to be fixed. Once such an image is available an assessment of the most probable position of the point may be made, and hence final coordinates for the point may be adopted. These in turn lead to final estimates for the observed quantities.

The first stage in the process of fixing the position of a single point is to derive provisional (approximate) coordinates using a minimum number of observations. Thus the first step is a polar calculation, triangulation intersection calculation, or trilateration intersection calculation, based on eqns (2.8), (2.19), or (2.21) respectively. It is unlikely that any additional observation would be consistent with these provisional coordinates, and the second stage is to determine the extent of this inconsistency. Once the inconsistencies for all observations have been established the observations may be represented graphically, and at a suitable scale, producing what is referred to as an *error figure*. The calculation required to establish the inconsistency is often referred to as a 'cut' calculation, and the procedure will depend on whether the observation is a bearing or a distance. A *cut point* is a reference point established by the cut calculation, which allows the graphical construction of the line representing the observation. It may be on an eastings or northings axis for an angle observation, or on the line representing the direction of observation for a distance measurement.

Consider first of all the case where angular observations have been made from points A, B, C in order to fix the position of point P, as illustrated in Figure 4.3a. The coordinates of A, B and C are known and therefore the bearings and distances between them may be calculated. The bearings α_{AP}, α_{BP},

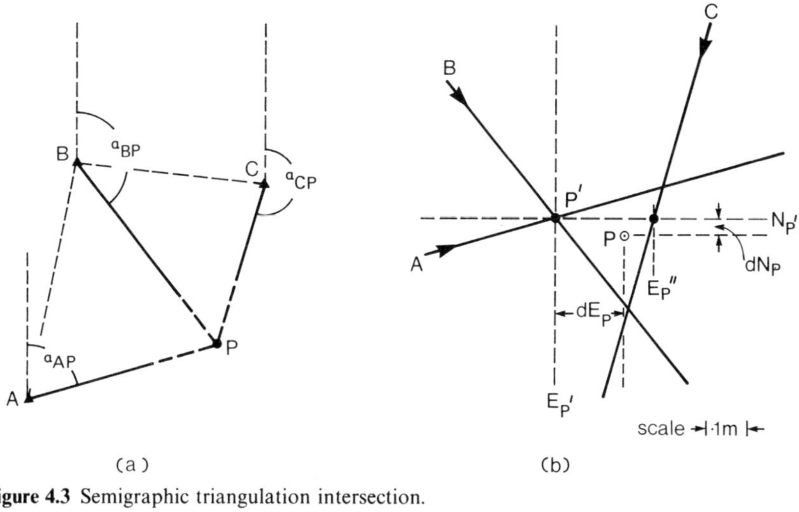

Figure 4.3 Semigraphic triangulation intersection.

α_{CP} may be derived from the observations at these points and from the computed bearings α_{AB} and α_{BC}.

Figure 4.3*b* is an enlarged view of the situation at point *P*, and represents the error figure for the fixation of *P*. Its construction is carried out in a number of stages, as follows:

(i) Calculate provisional coordinates of *P*, by triangulation intersection (eqn (2.19)), using the known coordinates of *A* and *B*, and the bearings α_{AP} and α_{BP}. Let these values be $E_{P'}, N_{P'}$.

(ii) Using $E_{P'}$, $N_{P'}$ as a local origin, construct lines to represent the observations α_{AP} and α_{BP}. These may be drawn by means of a protractor centred on the local origin.

(iii) Derive the differences $\Delta E_{CP'}$ and $\Delta N_{CP'}$ for the line CP', from

$$\Delta E_{CP'} = E_{P'} - E_C; \qquad \Delta N_{CP'} = N_{P'} - N_C \tag{4.16}$$

(iv) Calculate the cut point on the appropriate axis passing through P', using either of the following formulae:

$$E_{P''} = E_C + \Delta N_{CP'} \tan \alpha_{CP} \tag{4.17a}$$

when $|\Delta N_{CP'}|$ is the larger, or

$$N_{P''} = N_C + \Delta E_{CP'} \cot \alpha_{CP} \tag{4.17b}$$

when $|\Delta E_{CP'}|$ is the larger

where $|\quad|$ indicates that the absolute value is to be considered.

Figure 4.3*b* shows the position of $E_{P''}$ when the first of eqns (4.17) is used. Centre the protractor on this point and construct the line representing the observation α_{CP}.

(v) Repeat steps (iii) and (iv) for all remaining angular observations to or from known points.

(vi) Complete the annotation of the error figure by entering the scale, point names, and approximate length of each line of sight. The annotation should indicate the end of the line with known coordinates, and also the direction of observation of each line.

(vii) Select the final position of *P* by giving due consideration to all the factors involved—i.e. observational circumstances, reliability of control, etc. Scale off the corrections dE_P and dN_P, and derive the final coordinates of *P* from

$$\left.\begin{aligned} E_P &= E_{P'} + dE_P \\ N_P &= N_{P'} + dN_P \end{aligned}\right\} \tag{4.18}$$

(viii) The coordinates E_P, N_P are used to obtain final bearings and distances between *P* and the other known points.

Example 4.3

Use the data given below, and a semigraphic method, to derive final coordinates for the point *P*.

Coordinates:			*Observations*:	
Point	*E* (m)	*N*	Line	Bearing ° ′ ″
A	6091.13	3631.61	*AP*	126 08 26
B	5973.88	1561.00	*BP*	53 22 37
C	6947.21	178.43	*CP*	11 13 10
D	7597.31	2063.49	*DP*	344 41 38

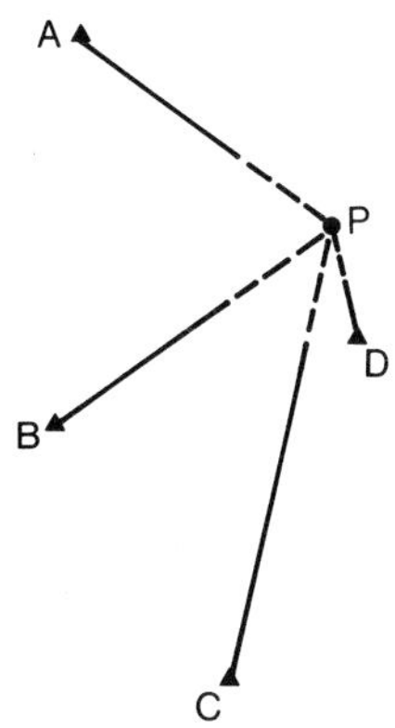

Figure 4.4

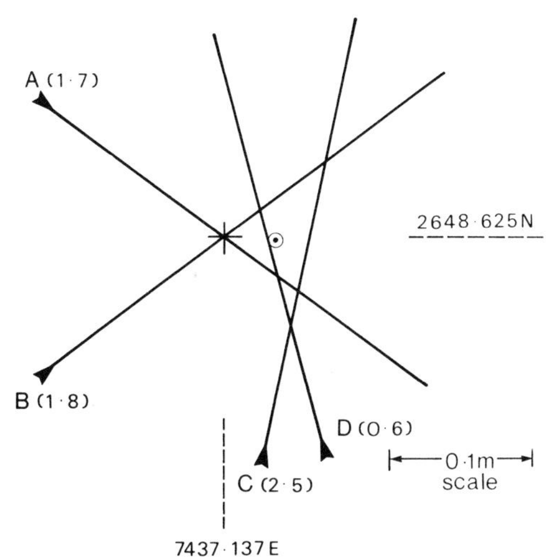

Figure 4.5

Solution

(i) The provisional coordinates of *P*, calculated by triangulation intersection and using the observations *AP* and *BP*, are

	E	N
P'	7437.137	2648.625

(ii) The 'cuts' for the observations *CP* and *DP* are calculated as follows:

For *CP*: $\Delta E_{CP'} = 489.927$; $\Delta N_{CP'} = 2470.195$

$$E_{P''} = E_C + \Delta N_{CP'} \tan \alpha_{CP} = 7437.193$$

For *DP*: $\Delta E_{DP'} = -160.173$; $\Delta N_{DP'} = 585.135$

$$E_{P''} = E_D + \Delta N_{DP'} \tan \alpha_{DP} = 7437.168$$

(iii) The error figure is constructed as shown in Figure 4.5, and the final position of *P* is selected by inspection.

(iv) The final coordinates of *P* are scaled off the error figure and found to be

	E	N
P	7437.17	2648.62

The procedure for trilateration is similar in principle. Figure 4.6*a* illustrates a trilateration intersection based on distances *AP*, *BP* and *CP*, measured from known points *A*, *B*, and *C*, and Figure 4.6*b* is the corresponding error figure.

The stages in the construction of the error figure are as follows:

(i) Calculate provisional coordinates of *P* by trilateration intersection

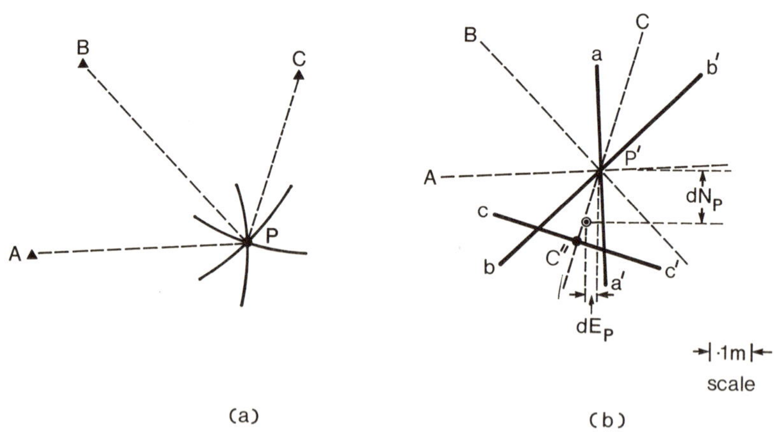

Figure 4.6 Semigraphic trilateration intersection. ——— locus of measured distance; – – – – direction of measured line

(eqn (2.21)), using the known coordinates of A and B and the observed distances AP and BP. Let these values be $E_{P'}$, $N_{P'}$.

(ii) Using $E_{P'}$, $N_{P'}$ as a local origin, construct lines passing through the point which represent the approximate bearings for all observed distances to P. The loci for the observed distances AP and BP are then constructed as straight lines (aa' and bb') at right angles to the corresponding bearings, and also passing through the point P'. These lines represent the observations and are in fact the tangents to the circular arcs shown in Figure 4.6*a*.

(iii) Derive the differences $\Delta E_{CP'}$ and $\Delta N_{CP'}$ for the line CP' from

$$\left.\begin{aligned} \Delta E_{CP'} &= E_{P'} - E_C \\ \Delta N_{CP'} &= N_{P'} - N_C \end{aligned}\right\} \qquad (4.19)$$

and the distance CP' from

$$CP' = (\Delta E_{CP'}^2 + \Delta N_{CP'}^2)^{1/2} \qquad (4.20)$$

(iv) Derive the discrepancy (Δ) in the distance from

$$\Delta = CP - CP' \qquad (4.21)$$

(i.e. the observed value minus the computed value)

and scale off this distance from P', and along the line CP'. If Δ is positive it will be located away from C along CP' extended (as shown in Figure 4.6*b*); if it is negative it will be located towards C. This represents the cut point (indicated by C'' in the diagram), and the locus (cc') representing the observation is drawn at right angles through it.

(v) Steps (iii) and (iv) are repeated for all remaining distance observations to or from known points.

(vi) Complete the annotation of the error figure to aid the selection of the final position. This includes scale, point names, approximate distances, etc.

(vii) Select the final position of P and scale off the corrections $\mathrm{d}E_P$ and $\mathrm{d}N_P$. The final coordinates of P are then obtained from

$$\left.\begin{aligned} E_P &= E_{P'} + \mathrm{d}E_P \\ N_P &= N_{P'} + \mathrm{d}N_P \end{aligned}\right\} \qquad (4.22)$$

(viii) The coordinates E_P, N_P are used to obtain final bearings and distances between P and other known points.

Example 4.4

Use the data given below, and a semigraphic method, to derive final coordinates for the point *P*.

Coordinates: Point	*E* (m)	*N* (m)	*Observations*: Line	Distance (m)
A	5997.73	3912.56	*PA*	1120.76
B	7279.56	4643.10	*PB*	574.53
C	7781.63	2593.70	*PC*	1648.73
D	9002.47	4011.21	*PD*	1901.22

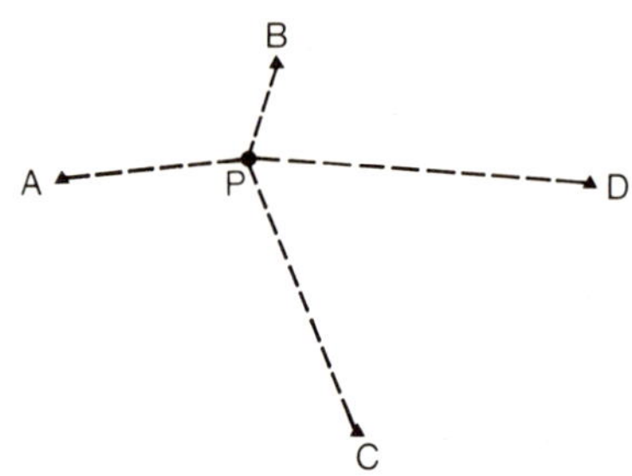

Figure 4.7

Solution:

(i) The provisional coordinates of *P*, calculated by trilateration intersection and

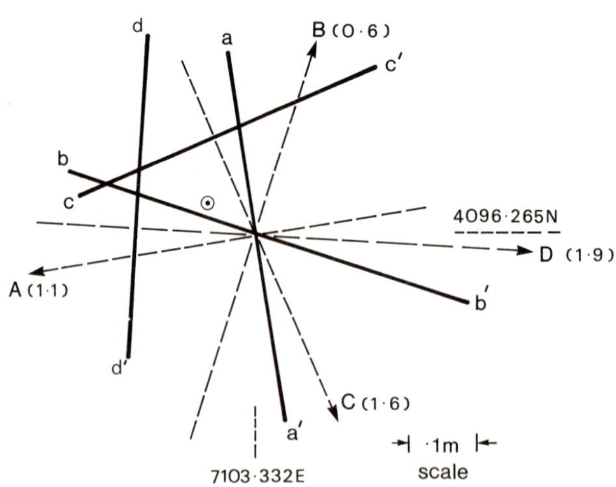

Figure 4.8

using the observations PA and PB, are

	E	N
P'	7103.332	4096.265

(ii) The calculation of 'joins' from P' to all four control points gives the following results:

Line	Computed Bearing °	′	″	Computed Distance (m)
AP'	80	33	58	1120.760
BP'	197	51	45	574.530
CP'	335	42	16	1648.572
DP'	272	33	52	1901.042

(iii) The 'cuts' for the observations PC and PD are calculated by subtracting the computed distances from the observed distances. In the case of points A and B these will be 0.

For PC: $\Delta_{CP} = 1648.73 - 1648.572 = 0.158$
For PD: $\Delta_{DP} = 1901.22 - 1901.042 = 0.178$

(iv) The error figure is constructed as shown in Figure 4.8 and the final position of P is selected by inspection.

(v) The final coordinates of P are scaled off the error figure and found to be

	E	N
P	7103.26	4096.31

The semigraphic procedures described so far are based on the assumption that inconsistencies between observations are of a random nature. In practice this is not always the case, and situations occur where systematic error is present to a greater or lesser extent. The obvious examples are orientation error in a round of observed directions, and scale error in distance measurement. If sufficient care is taken in the planning and execution of the fieldwork the effects may be reduced to a minimum, until they become negligible. However circumstances in the field may prevent this, and it is therefore necessary to include procedures in the computation which allow recognition and compensation for systematic error. Both orientation error and scale error are considered here.

The classic example of *orientation error* arises in resection, where the initial error is usually large, and is accounted for by the standard resection calculation. However, a situation such as that shown in Figure 4.9*a* could give rise to a much smaller but nevertheless significant orientation error.

A, B, C, are points with known coordinates, and directions have been observed from A and P in order to fix the position of P. The point P is a good deal closer to A than B and C, and if the orientation of the observations

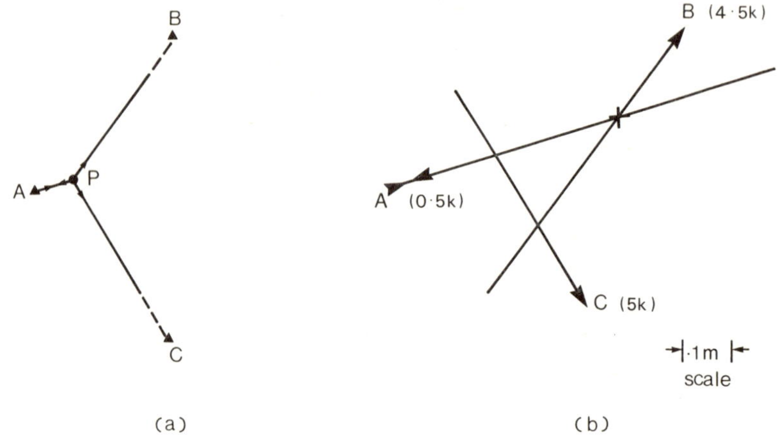

Figure 4.9 Effect of orientation error.

is based on those made at A there could be orientation error present. This is reflected in the error figure shown in Figure 4.9*b* where, in order to reduce the size of the figure, all the rays must be rotated in a clockwise sense about the fixed points. If the angular rotation (correction) is to be the same the linear shift on the error figure will vary, and will depend on the length of the line being considered. Diagram (*b*) is drawn at a large scale, and after a small rotation of a line about the fixed control point it will appear in a position nearly parallel to the original position, and it may be drawn as such. The problem is that the amount of rotation (orientation correction) is unknown at this stage; it is solved in the following way:

(i) Adopt a convenient orientation correction (say 10″), and calculate the amount of the parallel displacement x on the error figure that corresponds with this value, using the formula

$$x = D \cdot \frac{\theta}{\rho} \tag{4.23}$$

where x is the displacement; D is the approximate length of the line; θ is the arbitrary orientation correction (10″) in seconds; and $\rho = 206265$.

Apply eqn (4.23) to all the observations from P to give x_A, x_B, and x_C.

(ii) The displacements are introduced into the error figure (as shown in Fig. 4.10*a*) by drawing lines parallel to the original lines, and at distances x_A, x_B, and x_C away. The new positions must be such that they indicate a rotation in the same sense (clockwise in this case) about each fixed point.

(iii) The final stage is to join the original intersection point to the new intersection point, for each pair of observations, as shown in Figure 4.10*b*. These construction lines are in fact a representation of the tangents to the three circles passing through P, which were shown earlier in Figure 2.13 in the context of a full resection. The intersection of these tangents at P represents a position where orientation error has been eliminated. The amount of the orientation correction may be determined by scaling a distance x_P and calculating a value of θ from eqn (4.23). When only three known points have been observed, as in the case above, the tangents will meet in a single point which may or may not be selected as the final position. Due consideration should also be given to the observation from A to P, which would lead to a choice of position close to but not coincident with the intersection of the tangents. If more than three known points are observed the tangents will not all intersect in one point, and an error figure formed by the tangents will arise. This should be considerably smaller than the original error figure–assuming orientation error was present; and the final position is selected by considering such a figure and the forward observation AP. A numerical example which involves observations to four control points is given below.

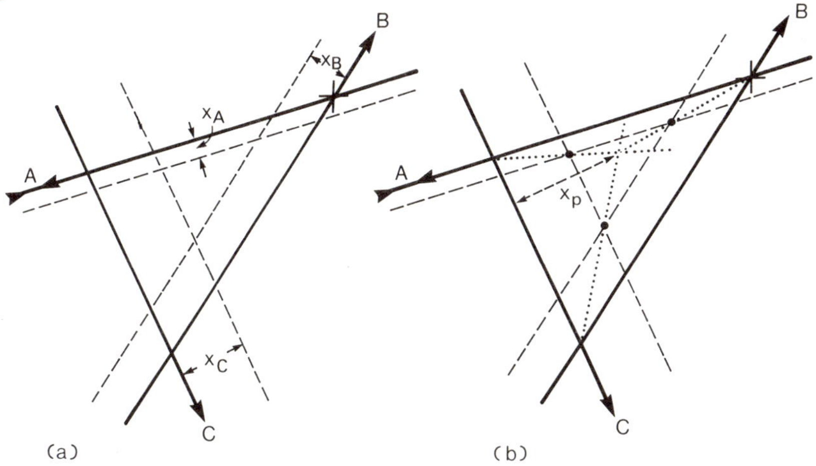

Figure 4.10 Compensation for orientation error. ———— original observation; – – – – displaced observation; ········ locus of intersection.

Example 4.5

Directions have been observed from P to the control points A, B, C and D. The known coordinates and the set of theodolite observations made at P are:

Coordinates:

Point	E (m)	N (m)
A	8246.18	4102.96
B	7781.63	2593.70
C	5997.73	3912.56
D	7279.56	4643.10

Observations at P:

To	Direction ° ′ ″
A	57 45 00
B	171 13 11
C	280 08 12
D	343 44 11

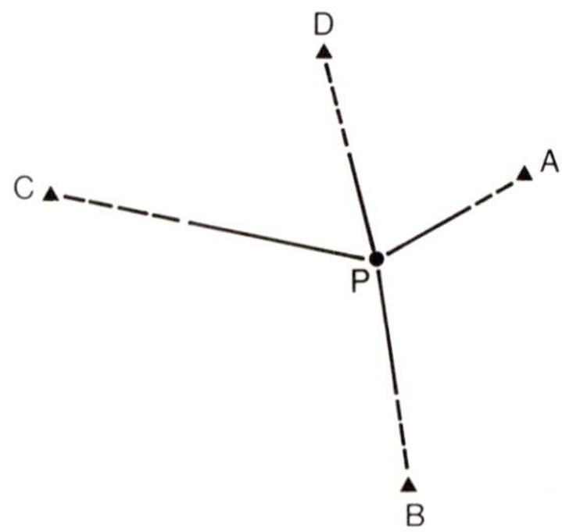

Figure 4.11

Derive final coordinates for the point P, and a mean orientation correction for the observations at P.

Solution

(i) An approximate orientation correction of $-1°21'$ is applied to the observations to give the following values:

	Approx. Bearing °	′	″
A	56	24	00
B	169	52	11
C	278	47	12
D	342	23	11

The observations to A and B are used to calculate the following provisional coordinates of P by triangulation intersection:

E	N
7589.876	3666.913

(ii) The calculation of 'cuts' CP and DP, based on the bearings derived in (i), produce discrepancies of

$$\delta N_{CP} = -0.451, \quad \text{and} \quad \delta E_{DP} = -0.396$$

(iii) The error figure for the observations is constructed as shown in Figure 4.12, and the loci of the intersections established after introducing a convenient orientation correction—in this case $-15''$. The displacements derived from such a correction are

$$x_A = 0.057\,\text{m}, \quad x_B = 0.079\,\text{m}, \quad x_C = 0.117\,\text{m}, \quad x_D = 0.074\,\text{m}$$

(iv) The intersection of the loci indicate an area which is free of orientation error, and a position of P is selected within this area. Hence the final coordinates of P are

	E	N
P	7589.74	3666.69

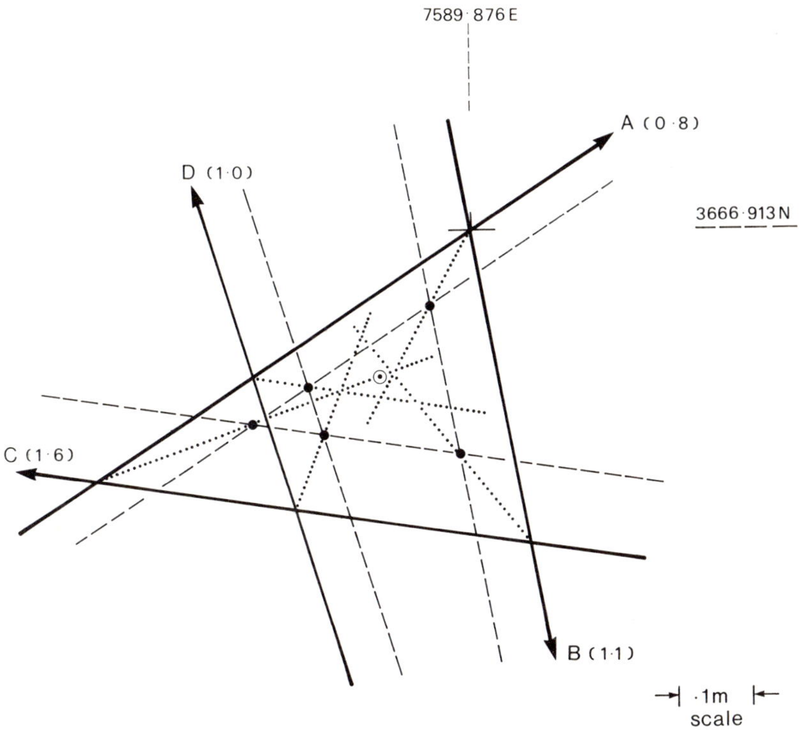

Figure 4.12

(v) The final bearings to control points are calculated from 'joins' *PA*, *PB*, *PC* and *PD*, and from these results a mean final orientation correction of $-1°21'28.5''$ is derived, as indicated below.

	Observed direction ° ′ ″	Computed bearing ° ′ ″	Orientation correction ° ′ ″
A	57 45 00	56 23 31	−1 21 29
B	171 13 11	169 51 38	−1 21 33
C	280 08 12	278 46 46	−1 21 26
D	343 44 11	342 22 35	−1 21 36
		Mean =	$-1° 21' 31''$

As far as distance measurement is concerned, *scale error* could be due to any of a number of causes, including frequency drift in the EDM instrument, neglect of projection scale factor, or scale error in the control system. Identification of, and compensation for, scale error follows a similar pattern as orientation error. Figure 4.13 illustrates the situation when distance have been measured from *P* to three control points *A*, *B*, and *C*.

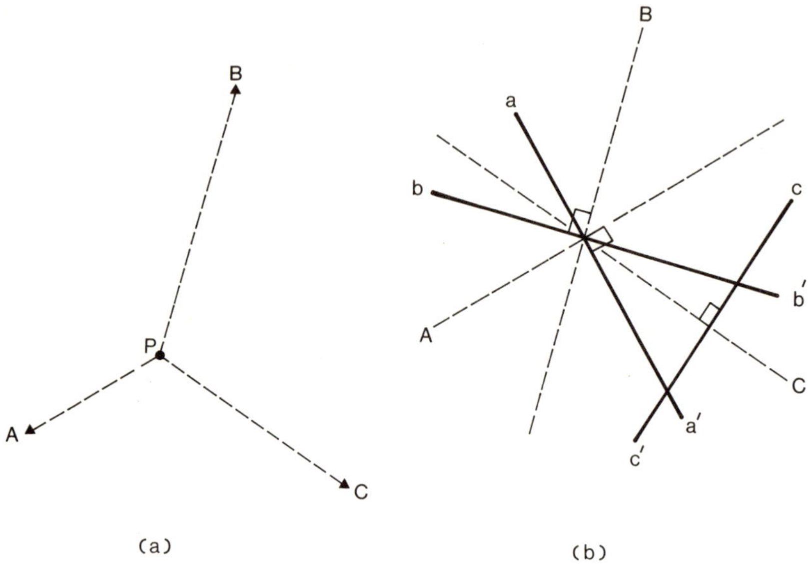

Figure 4.13 Effect of scale error. ———— locus of distance measurement.

The stages involved in compensating for scale error are as follows:

(i) The error figure shown in Figure 4.13*b* may be reduced by increasing each of the measured distances. Therefore if each observed value is multiplied by an arbitrary scale factor (λ), greater than unity, the corrected distance D' would be found from,

$$D' = \lambda \cdot D \tag{4.24}$$

where D is the measured distance.

The scale error is likely to be small, with λ close to unity, therefore the difference ($\Delta\lambda$) from unity is a more convenient value to use in the application. The correction x to the distance is therefore found from

$$\begin{aligned} x &= D' - D = \lambda D - D \\ &= (\lambda - 1)D = \Delta\lambda.D \end{aligned} \tag{4.25}$$

A convenient value for $\Delta\lambda$ is adopted and used to calculate values of x for each observation.

(ii) The displacements are introduced into the error figure (as shown in Figure 4.14*a*) by constructing parallel lines at distances x_A, x_B, x_C from the original positions, and such that they represent (in this case) an increase in each distance.

(iii) Finally, the original intersection points are joined with the new and the loci intersect at the point P, as shown in Figure 4.14*b*. This may be adopted as the final position of the unknown point; it is the position consistent with full compensation for scale error. The amount of this scale error may be determined by measuring the distance x_P and relating it to the corresponding observed distance to the control point.

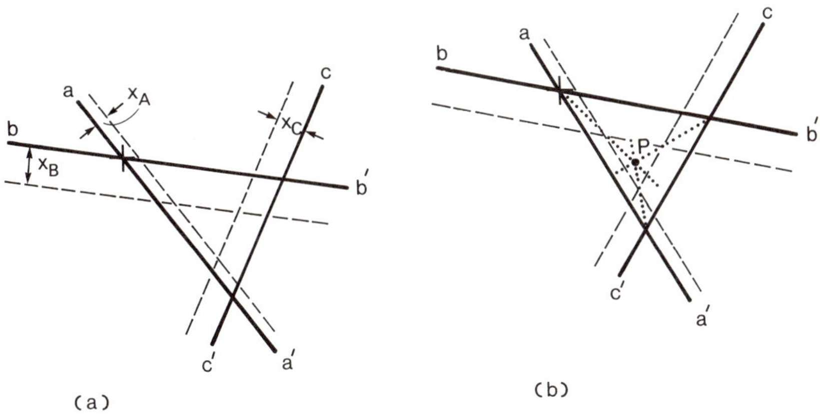

Figure 4.14 Compensation for scale error. ———— locus of observed distance; – – – – locus of corrected distance; ········ locus of intersection point.

If distances to more than three control points have been observed the loci of the intersection points will not all meet in a single point, and a secondary error figure will be formed instead. However, it should be smaller than the original, since scale error will have been compensated, and selection of the final position should present little difficulty. A numerical example involving observations to four control points is given below.

Example 4.6

The data supplied with Example 4.4 is used again for this problem.

Coordinates: Point	*E* (m)	*N*	*Observations*: Line	Distance (m)
A	5997.73	3912.56	*PA*	1120.76
B	7279.56	4643.10	*PB*	574.53
C	7781.63	2593.70	*PC*	1648.73
D	9002.47	4011.21	*PD*	1901.22

Derive final coordinates for the point *P*, and a mean value for the scale-factor λ.

Solution

(i) The observations *PA* and *PB* are used as before to calculate provisional coordinates of *P*,

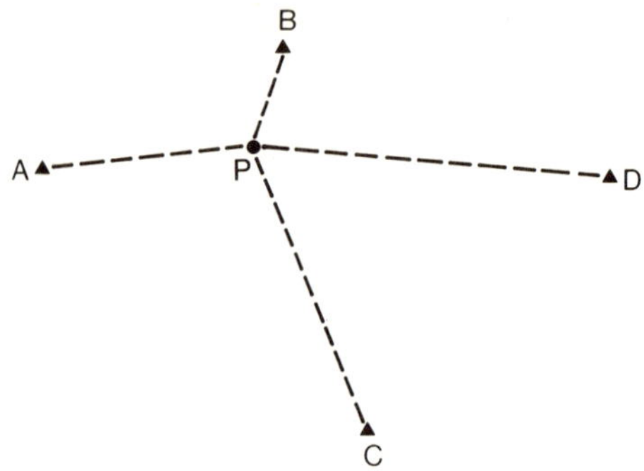

Figure 4.15

	E	N
	7103.332	4096.265

and the same discrepancies relating to observations PC and PD are obtained, viz.

$$\Delta_{CP} = 0.158, \quad \text{and} \quad \Delta_{DP} = 0.178$$

(ii) The error figure is constructed as shown in Figure 4.16, and it is clear from the relative positions of the observations that a negative scale-factor correction ($\Delta\lambda$) is required. A convenient value is applied to the observations—in this case -3×10^{-5}—and the displacements are found to be

$$x_A = -0.034, \quad x_B = -0.017, \quad x_C = -0.049, \quad x_D = -0.057$$

(iii) The loci of the intersection points are found to converge to form a smaller error figure, from which the final position may be chosen by inspection to be

	E	N
P	7103.26	4096.31

which is identical to the result in Example 4.4.

(iv) The final distance to the control points are derived from 'joins' PA, PB, PC, and PD, and from these results a mean value for the scale-factor λ is derived, as indicated below.

Line	Observed distance (m)	Computed distance (m)	Scale-factor (λ) $\left(=\dfrac{\text{computed distance}}{\text{observed distance}}\right)$
PA	1120.76	1120.696	0.9999429
PB	574.53	574.509	0.9999634
PC	1648.73	1648.643	0.9999472
PD	1901.22	1901.116	0.9999453
		Mean =	0.9999497

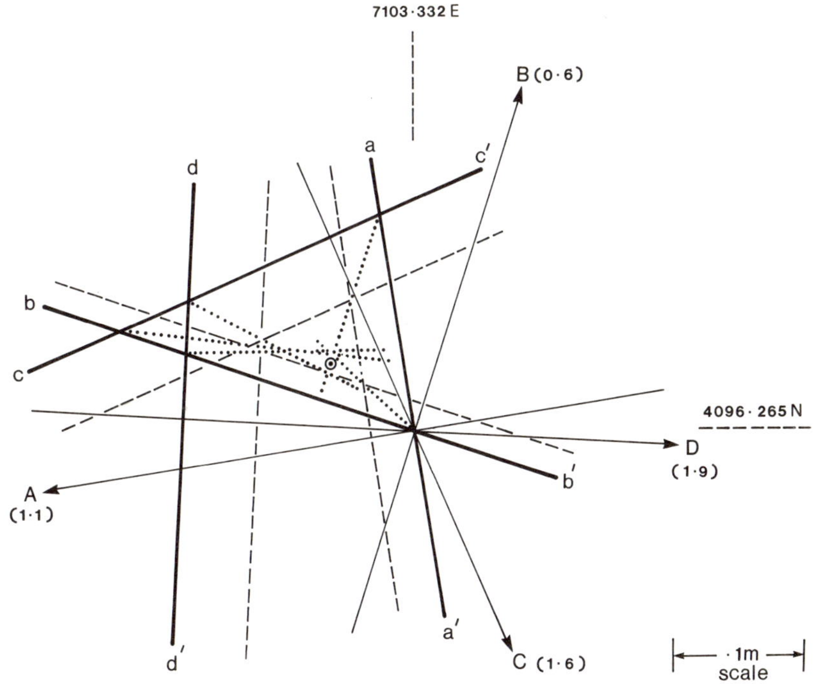

Figure 4.16

Most survey networks established in modern times will have a mixture of angle and distance observations, and the semigraphic method is well able to deal with these networks. Angle and distance observations may be represented on the same error figure; the systematic errors may be compensated if necessary, and the final position assessed after consideration of all matters affecting the observations. The points are still coordinated successively rather than simultaneously, and the order of computation must be specified beforehand. The sequence will be based on what elements of the network have been observed, and on the proximity of the control points to the new points. Thus the order of computation will be from the point which is potentially well fixed to that which would appear less so. As each new point is fixed it serves as additional control for other new points. In a network involving several new points the problem essentially is one of using the many observations in a tidy and systematic way. This is achieved by making use of a *direction sheet*, on which the observed directions are recorded. These directions are the original observations at each station, but rotated by means of an orientation correction so that the listed values are approximate grid bearings. Thus if the observer had sufficient foresight in the field to orient the observations the values for the direction sheet could be taken directly from the field book. Orientation of observations at each station is carried out either by using the observations to other known stations if the occupied station is known, or on the basis of reverse bearings if the occupied station is not known. Thus an orientation correction may be derived from observations to more than one station; in fact this is preferable, in order to isolate gross errors and improve the accuracy of the result. The actual computation of coordinates and cuts for the construction of the error figure is carried out using the mean of forward and back bearings. The procedure is best illustrated by means of an example.

Example 4.7

Figure 4.17 shows a small network involving known control points *A*, *B*, *C*, *D*, *E*, *F*, and new points *X* and *Y*. Lines whose directions have been observed in both the forward and reverse sense are shown as full lines, and those observed in one direction only are shown in full at the end occupied. Measured distances *DX*, *AY*, and *XY* are shown as lines encased in two parallel lines. The fixed control points that have been occupied for the purpose of measuring angles are *A*, *B*, and *C*. The coordinates of all control points are given below.

Point	*E* (m)	*N* (m)
A	8557.13	5934.72
B	13462.82	7930.45
C	13014.54	11142.59
D	7074.36	16410.66
E	1877.30	516.41
F	14404.00	451.79

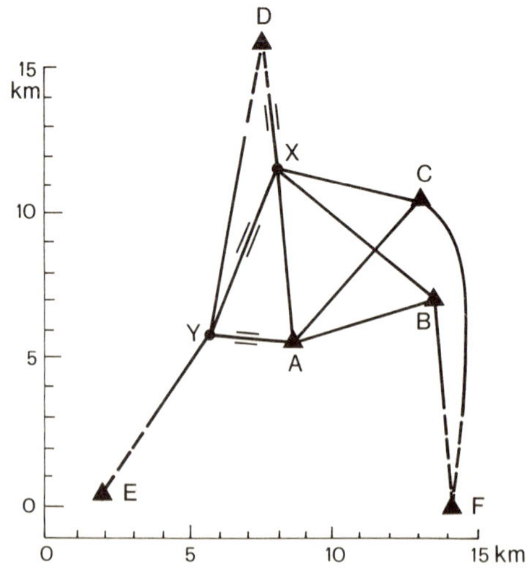

Figure 4.17

Table 4.2 shows the direction sheet containing all the observations, which have been oriented approximately (within 20″ say) to grid north, and listed in column 3; thus it becomes unnecessary to record degrees and minutes in every column. The measured distances are listed in column 2. By inspection of the diagram in Figure 4.17 a decision is made that X will be the first point computed. Values in italics in Table 4.2 are only derived after the final coordinates for point X have been established.

The stages in the computation are as follows:

(i) Calculate grid bearings between the known control points and enter the values in column 8. Establish final corrections to observations made between known point and known point and enter the value in column 7. The mean orientation correction for a particular set of observations is derived and applied to the observation to X. Thus for observations at A the final correction to the observations to C and B is $+0$ and $+1$ respectively, and the adopted mean ($+0$) is applied to the observation to X, to give the forward direction in column 5. Similarly, for the observations at B, the mean ($+6$) of $+9$ and $+3$ is applied to the observations to X, to give a value 52″ in column 5; and at C a value of 33″ is obtained in column 5.

(ii) The three oriented observations to X in column 5 are transferred to the set observed from X, and entered in column 6 as back directions. Values in column 6 minus those in column 3 establish individual orientation corrections $+9$, $+0$, and $+4$, and the mean ($+4$) is applied to the observations to known points C, B, A, D, to give the values 28″, 56″, 45″, and 13″ respectively. The means of forward and back directions are then used to fix the position of X semigraphically, and an observation made in one direction only (such as to D – 13″) is used as its stands. The distance measurement DX is also included in the set of observations and constructed on the error figure, which is shown in Figure 4.18*a*. Final coordinates for X are derived from the error figure, and these are used as a basis for calculating final directions to other known points. These values are entered appropriately in column 8 for the sets of observations made from A, B, C, and X.

(iii) The calculation of Y follows a similar pattern except that X is now an additional control point, and is used as such. Thus the forward direction from X to Y of 47″ is derived by applying a mean orientation correction of $+3''$ instead of $+4''$. The forward directions from A and X to Y are transferred to the set observed from Y, and entered in column 6; these enable a mean orientation correction of $+7''$ to be applied to the observations from Y to give the forward directions in column 5. The means of the forward and back directions to X and A, and the single observations to E and D, together with the distance measurements AY and XY, are used to construct the error figure shown in Figure 4.18*b*. Once the final coordinates for Y have been determined, the final bearings may be calculated and entered in columns 8.

A slight difference in the result would have been obtained had Y been selected as the first point. Also, the choice of final position in an error figure has an element of subjectivity about it. Therefore, on these grounds, the method is regarded as approximate. However, provided sufficient observations are made

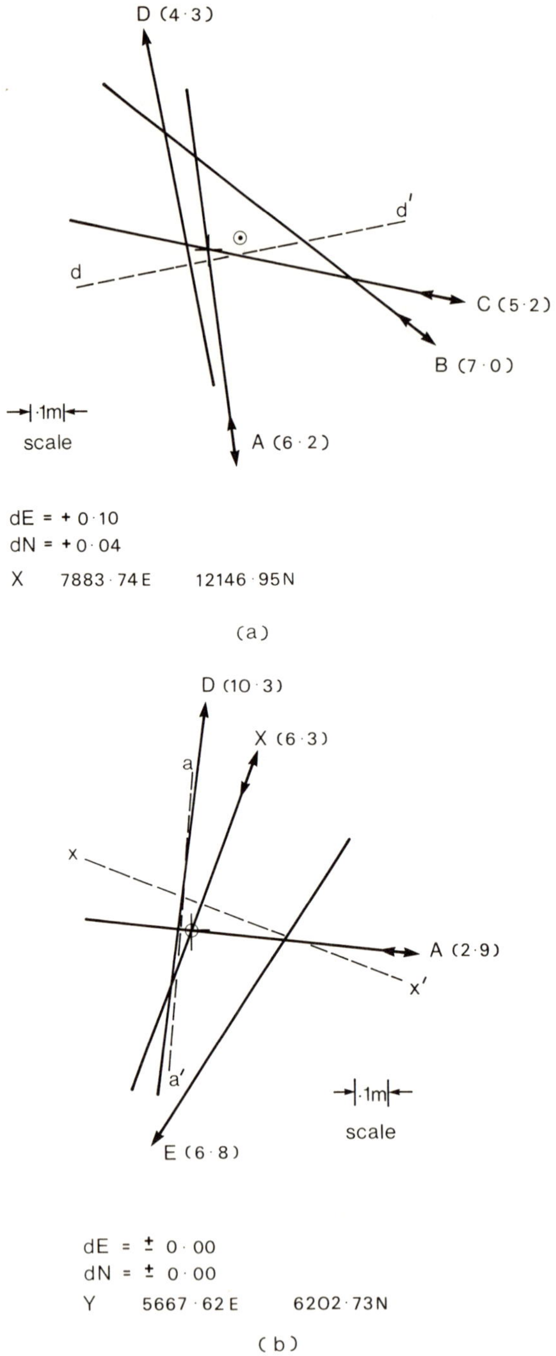

Figure 4.18 Error figure: *a* error figure for *X*; *b* error figure for *Y*.

Table 4.2 The direction sheet

(1) Stn	(2) Obs dist. (m)	(3) Obs dirn. ° ′ ″	(4) Approx. corn. ″	(5) Forw. dirn. ″	(6) Back dirn. ″	(7) Final corn. ″	(8) Final bearing ° ′ ″
At A							
X		353 48 45	+0	45		*+3*	*353 48 48*
C		40 33 37				+0	40 33 37
B		67 51 44				+1	67 51 45
Y		275 17 55	*+2*	*57*			
At B							
A		247 51 36				+9	247 51 45
X		307 04 46	+6	52		*+5*	*307 04 51*
F		172 49 35				+3	172 49 38
At C							
A		220 33 35				+2	22 33 37
X		281 04 31	+2	33			
F		172 35 40				+2	172 35 42
At X							
C		101 04 24	+9	28	33	*+8*	*101 04 32*
B		127 04 52	+0	56	52	*−1*	*127 04 51*
A		173 48 41	+4	45	45	*+7*	*173 48 48*
Y		200 26 44		*47*			
D	4339.90	349 15 09		13		*−4*	*349 15 05*
			(+4)			(*+3*)	
At Y							
E		213 41 11		*18*			*213 41 10*
D		7 50 41		*48*			*7 50 47*
X	6343.80	20 26 40	*+7*	*47*	*47*		*20 26 47*
A	2901.93	95 17 50	*+7*	*57*	*57*		*95 17 57*
			(+7)				

to surrounding control points, so that the process is essentially one of 'breaking down' from existing control rather than 'building up', the errors will not propagate very far. The advantages in terms of flexibility in planning the fieldwork make it a very useful method in practice. Further details and discussion on semigraphic methods based on directions are contained in [1], [5], [16], [20], [29], [30] and [41].

Circumstances in the field do not always permit the observation of so many control points. In fairly flat terrain, or when the density of control is limited, a traverse with additional observations is likely to be the pattern of the network. In these cases a modification of the procedure described above is used, to the

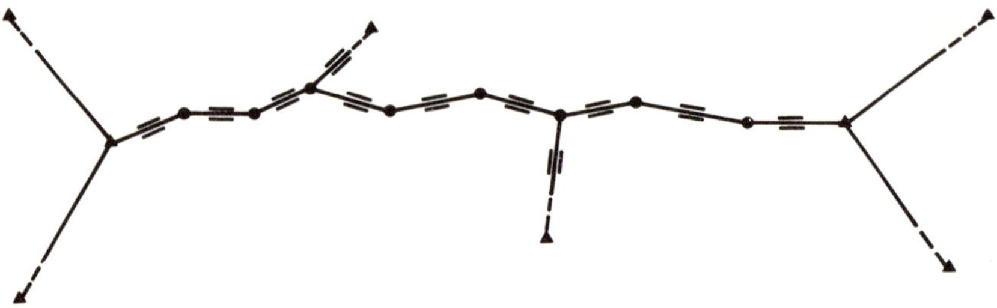

Figure 4.19 Traverse with additional observations—directions and distance. ——— observed direction—one way; ——— observed direction—both ways; ——— observed directions and distance.

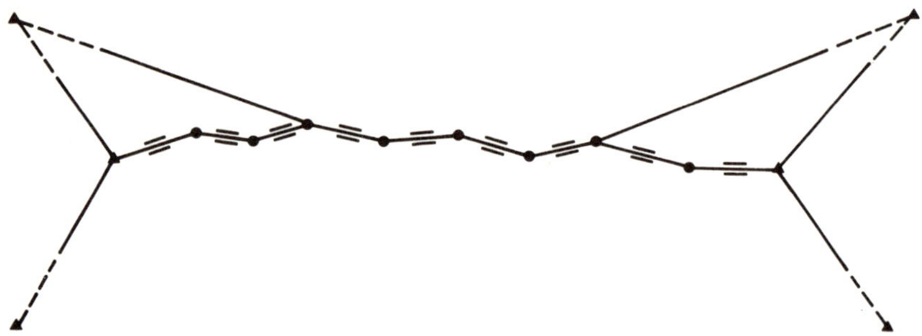

Figure 4.20 Traverse with additional observations—directions only. ——— observed direction—one way, ——— observed direction—both ways; ——— observed directions and distance.

extent that provisional coordinates might be obtained from a traverse calculation, and the additional observations used to improve the result—either via error figures, or by re-orientation at specific stations in the traverse and re-computation between these stations. These concepts would apply in situations such as those shown in Figures 4.19 and 4.20.

Clearly the variation in circumstances is endless. Each survey will merit its own particular solution, and the experienced surveyor will have an intuitive appreciation of the number and type of observations required to produce a specified accuracy. The guiding principle however, should be to use orientation checks as frequently as possible, and particularly after short traverse legs. Again, results will vary slightly depending on the sequence of computational steps taken, and the practitioner of many years standing will identify that which is more reliable. On the other hand, the one way to ensure a unique result is to use a least-squares solution described later in Chapter 10.

5 Mathematical models in photogrammetry

5.1 Introduction

Photogrammetry is, by definition, the solution of problems by making measurements on photographs. This normally implies measurements on a positive image of the scene, where the image base is paper, glass or film. Although it is possible to measure directly on the negative, this is seldom done in practice because of the possibility of damaging the emulsion. The measurements may be linear or angular, and may be between pairs of image points, or between an image point and some reference mark; or they may be coordinate values with respect to some imaginary reference point or line. In general rectangular cartesian coordinates are measured and the assumption is that the axes are at right angles. However, as a result of developments in computers generally, there are few disadvantages in measuring quantities related to other coordinate systems. Thus, by a process of triangulation or trilateration, measured angles or distances may be readily converted into rectangular coordinates. The main requirement in these cases is to ensure that the geometry of measurement is such that the necessary precision in the derived rectangular coordinates is achieved.

Another factor, related to improvements in computer technology, is that more sophisticated mathematical models may be used today for the processing of measured image coordinates. The less conventional imagery, such as that obtained from strip cameras, or cameras with focal plane shutters, may be put to greater use than in the past. What was once regarded as 'reconnaissance' photography might now be regarded as suitable for mapping purposes. Further detail on these matters falls outside the scope of this text, and attention is given solely to the more commonly used photogrammetric frame camera.

Other developments in recent years have been associated with the use of photogrammetry for 'non-topographic' applications—which encompass such diverse fields as underwater photogrammetry, and scanning electron microscope systems. Of great importance amongst the non-topographic applications are those which come under the general heading 'close-range photogrammetry', being that category where the distance from object to camera is no more than two or three hundred metres. Purpose-built metric camera systems have been designed by the leading manufacturers to cater for this category. More recently, with the increased availability of computers to process the measurements and compensate for distortions, conventional and inexpensive cameras have become viable tools in close-range photogrammetry. Thus the use of cameras generally is gaining in importance as a means of monitoring the size or change in shape of an object. Detailed accounts of the range of this equipment and the applications are given in [3], [6] and [24]. Although the sections to follow refer mainly to the traditional application to topographic mapping, the definitions and concepts involved apply also to close-range work. Very often the only difference is that of coordinate axis specification, although in some applications other variations are also present. For example, in high accuracy close-range engineering applications multi-station photogrammetry is recommended, and a review of the technique and its advantages is given in [19].

5.2 The frame camera

The frame camera with a single lens system is the most common camera in use for photogrammetric position fixing or mapping. The film is generally stationary at the time of exposure, and the shutter speed is normally fast enough that the change in position of the aircraft during exposure is considered negligible. Thus the incident light impinging on the film emulsion is

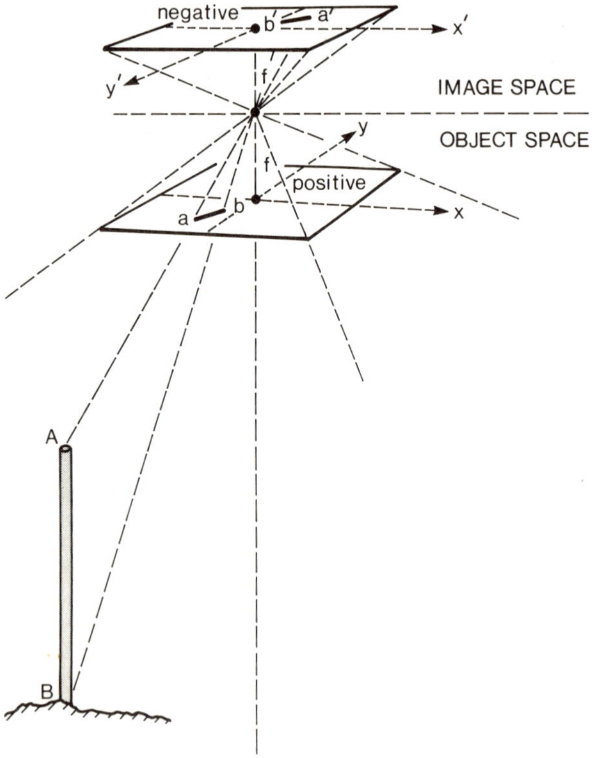

Figure 5.1 Projective transformation.

considered to come from a point source, and when the shutter is released the whole of the format area is exposed simultaneously. The geometry of the projective transformation that takes place for a vertical photograph is then as shown in Figure 5.1.

The camera lens, or more correctly the rear nodal point of the lens system, acts as the geometrical point source, and it defines the boundary between the *image space* and the *object space*. A negative image is produced on the emulsion of the film, and the equivalent positive image may be regarded as being in a parallel position and at the same distance from the lens in the object space. The *focal length* (f) of aerial cameras is set to focus objects which are at infinity, and these are brought to focus in the *focal plane*—the plane of the emulsion. The focal plane contains *fiducial marks* which also appear on each exposed photograph. Lines joining opposite fiducial marks intersect at the fiducial centre, or *collimation centre*, and this is meant to coincide with the *principal point*, which is defined as the foot of the perpendicular from the rear nodal point onto the focal plane. Lines joining fiducial marks serve also to define two reference axes which form the basis of a *rectangular coordinate system*. Image positions in the plane of the photograph may be established in this xy-coordinate system.

Although fiducial marks are located very precisely in cameras by manufacturers, there are unavoidable small discrepancies between the fiducial centre and the principal point. *Camera calibration* is the procedure used to define these discrepancies, as well as establish the *nominal focal length* of the camera and the distortion characteristics produced by the lens system. Other parameters such as distances between fiducial marks, angles between fiducial lines, flatness of the focal plane and resolving power, etc., are also included in the calibration process. The fiducial marks provide a means of assessing the amount of distortion that has taken place in the photographic material during processing, or over a period of time. The parameters relating to a specific camera, and determined through the process of camera calibration, are referred to as the *elements of interior orientation*. The most important of these are the location of the principal point, the nominal focal length, and the lens distortion characteristics, and they are vital to the photogrammetric processes that follow the taking of the photograph.

5.3 Coordinate systems

Coordinate systems in photogrammetry may be associated with the photographic image, the stereomodel, or the object (normally the terrain).

The *image system*, which is essentially a two-dimensional system, may be associated with either the positive image or the negative image, as shown in Figure 5.1. It has its origin at the fiducial centre and x-axis aligned along the fiducial line which is in the general direction of the camera base or flight line. Figure 5.2 shows two possible arrangements for fiducial marks; (a) represents the system described above, but it is clear that coordinates measured in the

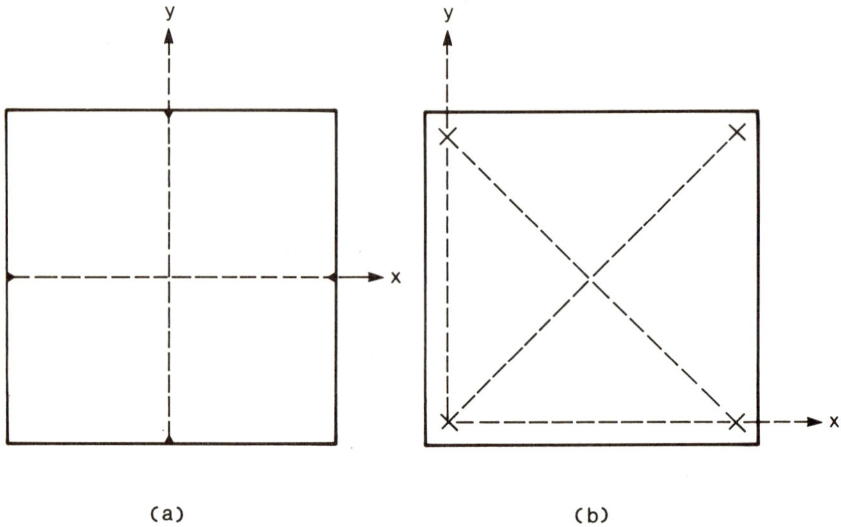

Figure 5.2 Two-dimensional image coordinate systems: *a*, side fiducial system; *b*, corner fiducial system.

system shown on (*b*) may be readily converted to the system defined in (*a*).

Modern cameras include all of the eight fiducial marks, and this provides a means of obtaining a more precise location of the fiducial centre, particularly when some fiducials lack definition due to poor lighting conditions. They also serve to provide more accurate corrections for deformations that have taken place in the photographic base material, or in the measuring instrument. Rectangular xy-coordinates are particularly useful as a means of representing position on a photograph, and they may be used to derive angular and linear relationships between points using the equations given in section 2.5. If rectangular coordinates have not been measured they may be derived using the equations given in Chapters 2 and 9.

Photographs which are introduced into a coordinate measuring instrument, such as a comparator, are not normally aligned so that the fiducial axes are exactly parallel to the instrument's axes, and it is not necessary that they should be. Measurements to the fiducials provide a means of transforming the values to such a system, and eqns (2.86) or (2.91) are generally used for this purpose. In some applications the rotation of the coordinate system is not necessary, and a simple translation of origin is all that is required.

In order to make the image coordinate system more general in application the perspective centre C is introduced, and a third axis z defined. Figure 5.3*a* shows the positive image coordinate system with origin at the fiducial centre, and z-axis perpendicular to the xy-plane.

The direction of z is such that the three axes define a three-dimensional right-handed cartesian system. The fiducial centre, principal point (p), and perspective centre (C) now have three-dimensional coordinates $(0,0,0)$, $(x_p, y_p, 0)$, (x_p, y_p, f) respectively. In an ideal camera the fiducial centre and the

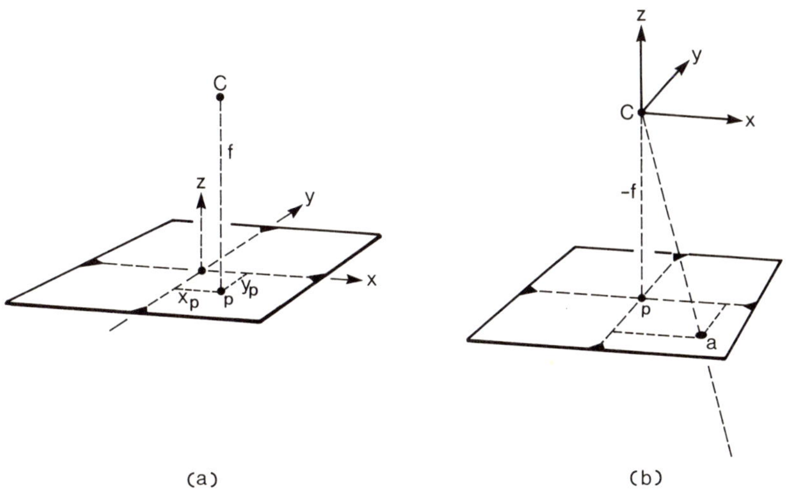

Figure 5.3 Three-dimensional image coordinate systems.

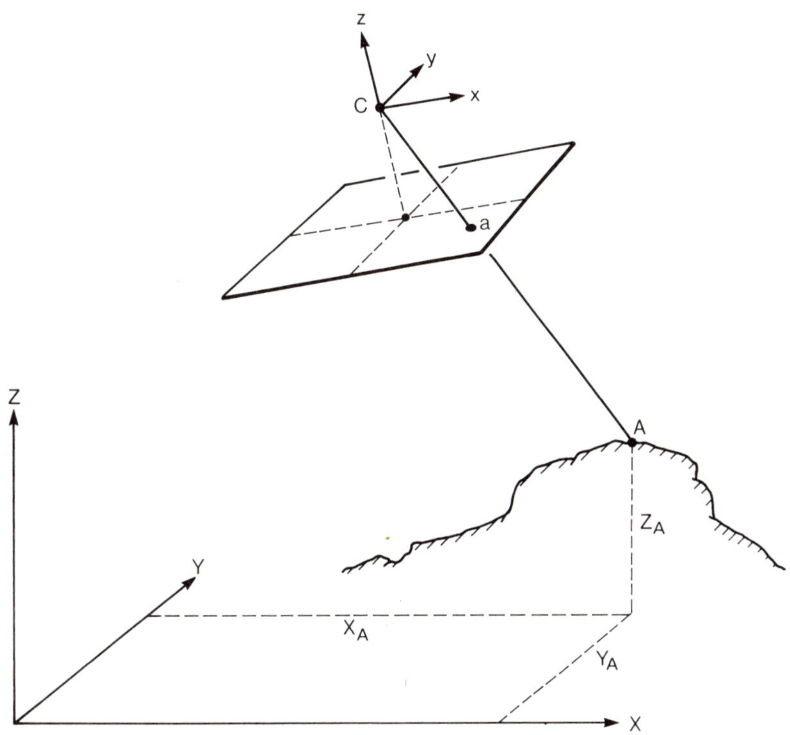

Figure 5.4 Object coordinate system: three-dimensional right-handed cartesian.

principal point coincide, and in many applications the difference between the two is small enough to be ignored. In such a case C has coordinates $(0, 0, f)$ and, if the origin is moved to the projection centre, p will have coordinates $(0, 0, -f)$, as indicated in Figure 5.3*b*. In order to simplify the mathematical derivations to follow, this assumption of an ideal camera will be adopted. A similar assumption is made regarding lens distortion, which will be considered to be negligible. If the amounts involved are not negligible then the assumption is that the measurements have been corrected for these distortions—i.e., so that values are distortion-free and refer to the principal point as origin.

Many applications in photogrammetry involve establishing the position and orientation of the camera at the time of exposure, and as the perspective centre is the point common to all projected lines it is the natural choice to act as origin. Therefore in Figure 5.3*b* the origin is at C and the right-handed system is parallel to that defined in Figure 5.3*a*. Therefore point a in the positive image plane will have coordinates $(x_a, y_a, -f)$, and C will have coordinates $(0, 0, 0)$.

The *object system* of coordinates is normally a three-dimensional right-handed cartesian system related to the object being measured or mapped. Figure 5.4 illustrates such a system (the XYZ-system), and in the case of aerial photogrammetry Z_c would be the height of the camera above the datum. A point A in the terrain would have coordinates X_A, Y_A, Z_A, and appear on the photograph as image point a.

The XYZ-system may be a geocentric cartesian system as defined in section 2.2. This system is useful if a particularly large portion of the earth's sufrace is being considered. Alternatively, and preferably, a local three-dimensional cartesian system may be used, whose relationship to the earth as a whole is shown in Figure 2.2. Another common system is the geodetic system, shown in Figure 2.1*b*. However, it is not used directly since it is not an orthogonal system, and two of the coordinates are in non-linear units. If the coordinates are quoted in such a geodetic system it is a simple matter to transform them to either of the cartesian systems. The transformation to geocentric coordinates would be accomplished by the following equations:

$$\left.\begin{aligned} X &= (v + h)\cos\phi\cos\lambda \\ Y &= (v + h)\cos\phi\sin\lambda \\ Z &= (v(1 - e^2) + h)\sin\phi \end{aligned}\right\} \qquad (5.1)$$

where v is the radius of curvature of the spheroid surface at the point, and in the plane of the prime vertical; and e is the eccentricity of the spheroid $(= ((a^2 - b^2)/a^2)^{1/2})$; where a and b are the values for the semimajor and semiminor axes of the spheroidal ellipse.

A transformation from a geocentric cartesian system to a local cartesian system may be achieved by using a *three-dimensional similarity transformation*, which is described in greater detail later (section 5.6).

Map projection coordinates (or grid coordinates, such as E, N, h) are often used in photogrammetry, particularly for the medium and larger scales of photography. However, since they are derived from geodetic coordinates (ϕ, λ, h), and relate to the spheroid surface, they are also non-orthogonal; therefore earth curvature must be accounted for separately. For the larger scales this will often be negligible; however, for many other applications it must be considered.

An intermediate system of coordinates, used frequently in practice, is referred to as the *model coordinate* system. These are three-dimensional cartesian coordinates, produced by the intersection of rays when two or more overlapping photographs are in their correct positions *relative to each other*. The system may be at any arbitrary scale and orientation, though usually the scale is close to the photographic scale and the orientation is defined by the orientation of the first photograph. Figure 5.5 shows a model coordinate system with the origin in the left-hand projection centre, the scale defined by the base-length Bx, and the orientation defined by the orientation of the left-hand camera.

Thus the model system may be seen to be an extension of the image system shown in Figure 5.3*b*, and if a number of models in a strip or block have been joined together the points have coordinates which are in one homogeneous system, defined by the position and orientation of the first photograph.

The relationships between the various coordinate systems—image/object, image/model, model/object—are established when the *elements of exterior*, *relative*, and *absolute orientation* respectively are known. These are dealt with in more detail in the sections to follow.

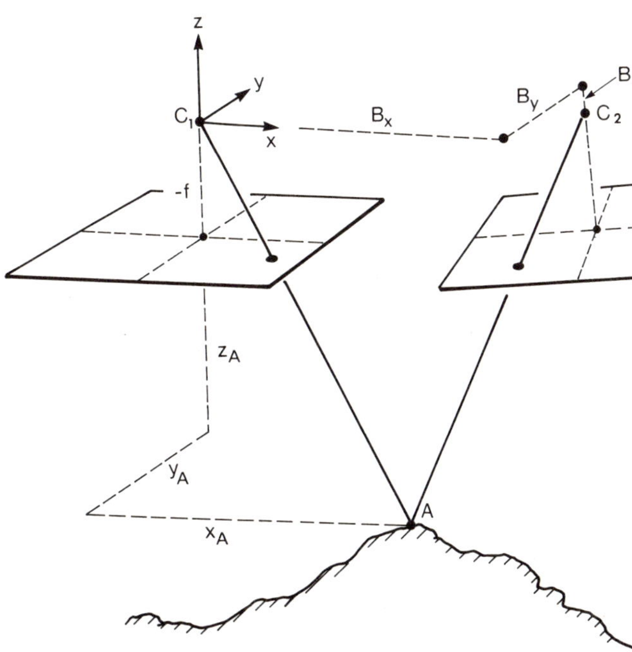

Figure 5.5 Model coordinate system.

5.4 The geometry of the vertical photograph

The expression 'vertical photography' refers to aerial photography taken by a camera with its optical axis aligned vertically. In practice this is difficult to achieve and small tilts, which are rarely greater than 3°, are usually present. If the assumption is that the axis is truly vertical then the following relationships may be established.

The *scale* (S) of a photograph is the ratio of a distance on the photograph to the corresponding distance on the ground, or alternatively the ratio of the focal length to the *flying height* (H) above the ground. Figures 5.6*a* and 5.6*b* are sections illustrating how scale may be determined for the cases of 'flat' terrain and 'variable' terrain respectively.

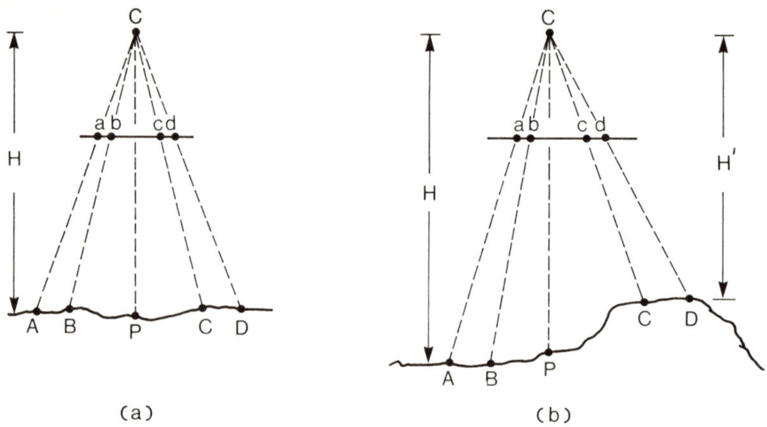

Figure 5.6 Scale determination: *a*, 'flat' terrain; *b*, 'variable' terrain.

In Figure 5.6*a* scale may be found from

$$S = \frac{ab}{AB} = \frac{cd}{CD} = \frac{f}{H} \tag{5.2}$$

In Figure 5.6*b* scale is not the same throughout the photograph: in the lower area

$$S = \frac{ab}{AB} = \frac{f}{H}$$

and in the higher area

$$S' = \frac{cd}{CD} = \frac{f}{H'}$$

The terrain is usually more variable than indicated in the diagrams, therefore H varies and so scale will vary continuously throughout a photograph. The scale for a small line segment on the photograph may be determined if the flying height above the corresponding line on the ground is known. Furthermore, the scale will vary depending on the orientation of the line segment. The *average* or *approximate scale* of a photograph implies the value obtained for the average height of the terrain. It may be derived from a pair of points known to be at the average height, or from a number of points distributed above and below the average height.

Relief displacement on a photograph is a characteristic of the central projection produced by the camera lens. It is radial from the nadir point which is also the principal point in truly vertical photography. Figure 5.7 shows the displacement (ab) on the photograph of the top of a vertical object of height h, relative to its base. The direction of the displacement is defined by the intersection of two planes—the photograph on the one hand and the vertical plane CAD on the other.

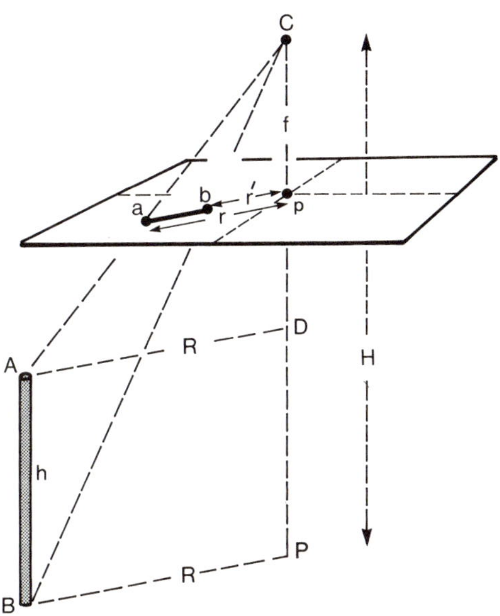

Figure 5.7 Relief displacement.

If R is the radial distance of the object from the nadir point on the ground, and r and r' are the radial distances of the images of the top and bottom respectively on the photograph, then the displacement is found as follows:

In the similar triangles ACD and aCp, $r/R = f/(H-h)$ therefore

$$f \cdot R = r(H-h) \tag{5.3}$$

and, in the similar triangles BCP and bCp, $r'/R = f/H$ therefore

$$f \cdot R = r'H \tag{5.4}$$

Equating the right-hand sides of (5.3) and (5.4)

$$r(H-h)=r'H$$

therefore

$$(r-r')=r\cdot\frac{h}{H}$$

or, if the outward displacement is designated Δ, then

$$\Delta=r\cdot\frac{h}{H} \tag{5.5}$$

The amount of displacement increases with increasing height of object and radial distance, and decreases with increased flying height. If h and H are heights above a specific datum then h will vary for different objects in the terrain and so the amount of displacement will vary.

Equation (5.5) may be reversed in order to determine the value of h for objects appearing on the photograph; thus

$$h=H\cdot\frac{\Delta}{r} \tag{5.6}$$

However, it should be made clear that if h is to be the height of an object above ground then both the top and bottom of the object must be visible on the photograph, and the value of H must refer to the bottom.

A value of *flying height* is necessary in many photogrammetric operations, and this could represent a value above the mean ground level appearing on the photograph, or it could be a value above a specific datum such as mean sea level. Equation (5.2) may be used to determine flying height above points in the terrain if the value of S relating to these points is known. Flying height is best determined from points which are at approximately the same height in the terrain, and which appear widely separated on the photograph, preferably equidistant from and on opposite sides of the principal point. An accurate value above a specific datum may be found using identifiable points whose heights above the same datum are known. Figure (5.8) shows points A and B in the terrain which have known heights h_A and h_B above datum.

If the XY-coordinate system shown in the diagram is a rectangular coordinate system in the horizontal datum plane, with axes parallel to the xy-axes of the image system, then the flying height (H) above datum may be derived as follows.

The horizontal distance D between the points A and B is calculated from

$$D^2=(X_B-X_A)^2+(Y_B-Y_A)^2 \tag{5.7}$$

but, from similar triangles $CA'A''$ and $Ca'p$

$$X_A=x_a\cdot\frac{(H-h_A)}{f} \tag{5.8a}$$

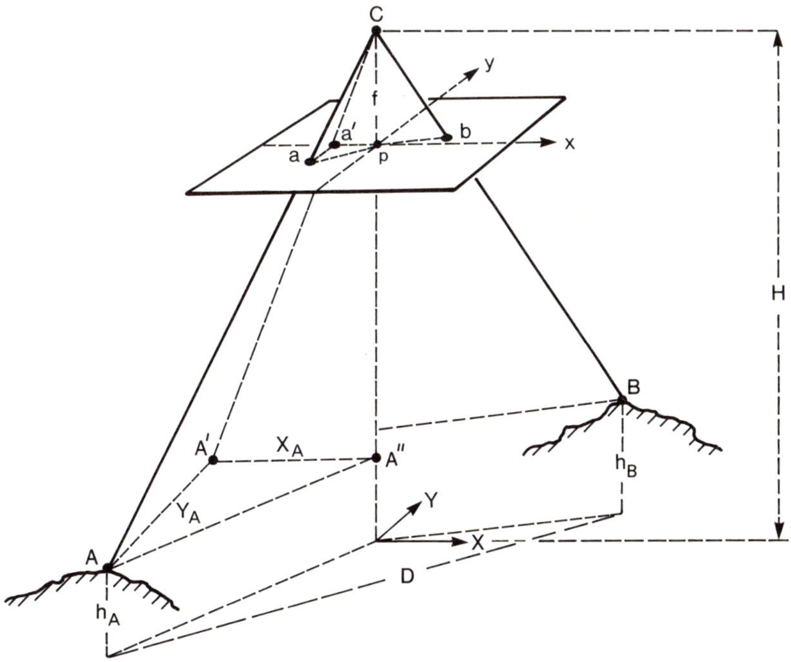

Figure 5.8 Determination of flying height.

and similarly

$$Y_A = y_a \cdot \frac{(H - h_A)}{f} \tag{5.8b}$$

Similar results may be obtained for point B.

Substituting in eqn (5.7) gives

$$D^2 = \left(x_b \frac{(H - h_B)}{f} - x_a \frac{(H - h_A)}{f} \right)^2 + \left(y_b \frac{(H - h_B)}{f} - y_a \frac{(H - h_A)}{f} \right)^2 \tag{5.9}$$

When known values for x_a, y_a, x_b, y_b, f, h_A, h_B, and D are introduced the equation becomes a quadratic equation in the unknown H,

$$aH^2 + bH + c = 0 \tag{5.10}$$

where a, b, c are known constants.

The solution is obtained by calculating the positive root, i.e.

$$H = \frac{-b + (b^2 - 4ac)^{1/2}}{2a} \tag{5.11}$$

The value of flying height found in this way is reliable within the limitations of the accuracy of measurement of the image coordinates x, y. However, it is based on the assumption that the optical axis of the camera is truly vertical, which is unrealistic in practice.

Equations (5.8) may be used to derive *ground coordinates* of points from measurements on a vertical photograph, provided the flying height above datum is known and the heights of the points above the same datum are also known. The ground coordinates will be in an arbitrary rectangular coordinate system and must be transformed if national grid coordinates (or values in some other system) are required. Two or more points with known coordinates in both systems are required for this purpose. The methods are described in more detail in sections 2.11 and 10.3.

5.5 The tilted photograph

In practice it is impossible to obtain truly vertical photographs, but by taking suitable precautions the tilts may be reduced to one or two degrees. However, even these small amounts have a significant effect on the positions of points, and they must be accounted for in any calculations for the accurate determination of position.

Figure 5.9 illustrates a tilted photograph taken from exposure station C. The principal point on the photograph is at p and the nadir point at n: the

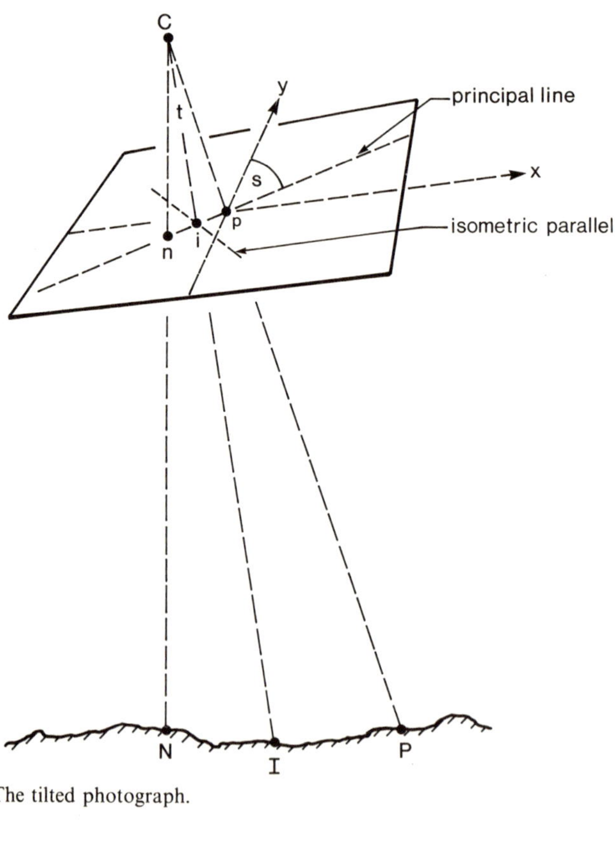

Figure 5.9 The tilted photograph.

corresponding points in the terrain are P and N respectively. The *principal plane* is that defined by Cnp, and the *principal line* np is the line of intersection between the principal plane and the photograph. The angle nCp, between the optical axis and the vertical, is the angle of tilt (t) of the photograph. The bisector of the angle nCp meets the principal line at i, the *isocentre* of the photograph. The angle s is the *swing* angle and is measured clockwise from the positive y-axis to the principal line. The line passing through i which is perpendicular to the principal line is the *isometric parallel*, and it is a horizontal line.

If the azimuth or bearing of the principal line in the ground coordinate system is known, then the three angles—tilt (t), swing (s), and azimuth (α)—provide a complete definition of the orientation of the photograph.

The determination of scale in the vicinity of a particular point on a tilted photograph depends on the height of the ground point, and its position on the photograph relative to the principal line and the isometric parallel. The point n on the principal line is defined by the angle t and the focal length f, and it is adopted as the origin of an auxiliary coordinate system (x' y') on the photograph. The system is shown in Figure 5.10*a*.

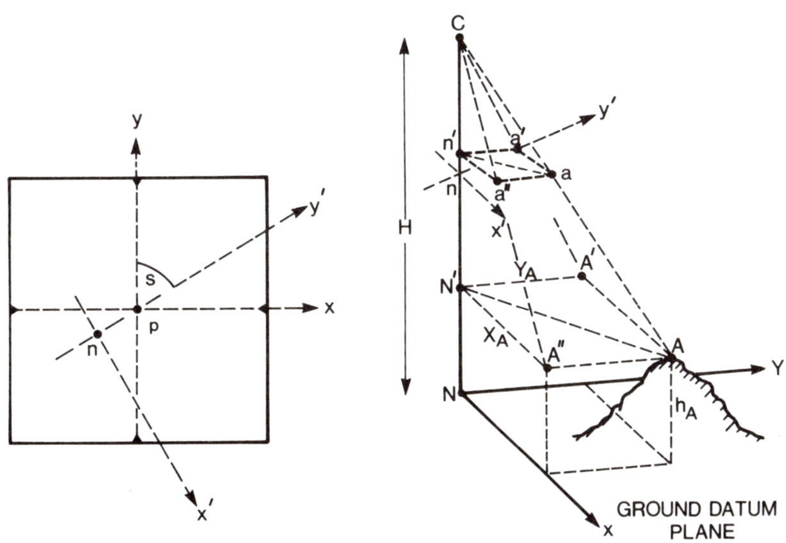

Figure 5.10 *a*, Auxiliary coordinate system; *b*, ground coordinates.

The distance np is $f \tan t$, and therefore the coordinates of points measured in the fiducial (or image) system may be transformed to the auxiliary system using eqns (2.87), as follows

$$\left.\begin{aligned} x' &= x \cos s - y \sin s \\ y' &= x \sin s + y \cos s + f \tan t \end{aligned}\right\} \tag{5.12}$$

Figure 5.10*b* shows a terrain point A which appears on the photograph at a. The orthogonal projection of a onto the principal line and the vertical produces points a' and n' respectively. Thus $a\,a'\,n'$ defines a horizontal plane passing through a. Similarly the points $AA'N'$ define a corresponding horizontal plane passing through the terrain point A, which is at height h_A above the datum. Therefore the scale (S_A) in the vicinity of a is the same as the scale ratio between the two horizontal planes, which may be determined from similar triangles $Cn'a$ and $CN'A$.
Thus

$$S_A = \frac{Cn'}{CN'} = \frac{Cn'}{H - h_A}$$

but

$$Cn' = Cn - n'n = f\sec t - y'\sin t$$

Therefore

$$S_A = \frac{f\sec t - y'\sin t}{H - h_A} \tag{5.13}$$

Figure 5.10*b* also shows a rectangular ground coordinate system (XY), where both axes lie in the datum plane and where the Y-axis also lies in the principal plane. The X-axis passes through the nadir point N in the datum plane. Again, from similar triangles it is clear that

$$\frac{A''N'}{a''n'} = \frac{CN'}{Cn'}, \quad \text{or} \quad A''N' = \frac{a''n' \cdot CN'}{Cn'}$$

i.e.

$$X_A = \frac{x' \cdot (H - h_A)}{f\sec t - y'\sin t} \tag{5.14a}$$

and similarly

$$Y_A = \frac{y'\cos t(H - h_A)}{f\sec t - y'\sin t} \tag{5.14b}$$

Once the coordinates X, Y of points in the terrain have been calculated, the angles, distances, areas, etc. may be derived. However, coordinates in an absolute system (such as the national grid system) may only be derived by applying suitable transformation formulae, such as eqns (2.87). This process requires that at least two points have known coordinates in both systems.

Tilt displacement may be derived by referring to Figure 5.11, which shows a number of points in the principal plane, both on the tilted photograph and on the equivalent vertical photograph. The equivalent vertical photograph intersects the tilted photograph along the isometric parallel passing through i.

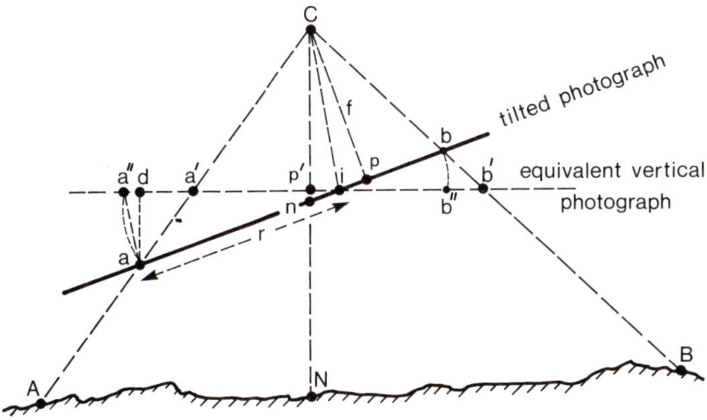

Figure 5.11 Tilt displacement on the principal line.

All images on the isometric parallel have the same positions in both photographs. Points off the parallel are displaced relative to their positions on the equivalent vertical photographs, and the amount of displacement may be interpreted as the difference in distance from the isocentre for the two photographs. In Figure 5.11, which refers specifically to the principal line, the terrain points A and B produce images at a and b on the tilted photograph, and at a' and b' on the equivalent vertical photograph. If point a'' is chosen such that $ia'' = ia = r$, then the displacement (Δ) is given as

$$\Delta = ia' - r = ia' - ia''$$

Now the triangles Cia' and $aa''a'$ are similar, therefore

$$\frac{a'a''}{ad} = \frac{ia'}{Cp'}, \quad \text{or} \quad \frac{\Delta}{r \sin t} = \frac{r + \Delta}{f}$$

Therefore, for a point on the principal line,

$$\Delta = \frac{r^2 \sin t}{f - r \sin t} \tag{5.15}$$

Figure 5.12 shows the more general situation on a tilted photograph, when the image points are in arbitrary positions, and r and Δ are radial distances and radial displacements respectively.

If β is the clockwise angle at the isocentre formed by the principal line (on that part of the photograph tilted upwards) and the radial line to the image point, then eqn (5.15) is made more general by replacing Δ and r by $\Delta \cos \beta$ and $r \cos \beta$ respectively.

Therefore the equation becomes

$$\Delta \cos \beta = \frac{(r \cos \beta)^2 \sin t}{f - r \cos \beta \sin t}$$

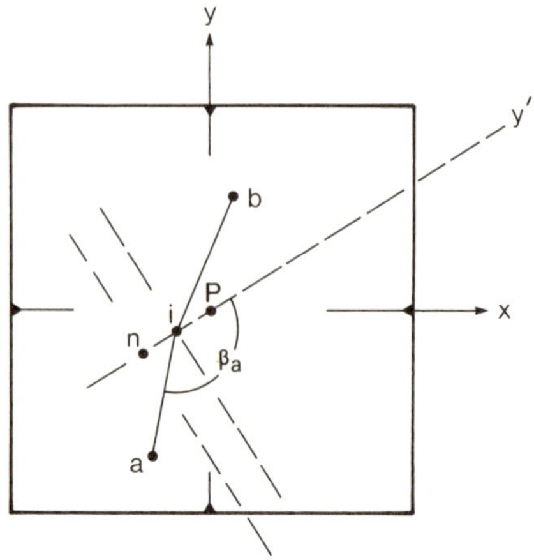

Figure 5.12 Radial tilt displacement.

or

$$\Delta = \frac{r^2 \cos^2 \beta \sin t}{f - r \cos \beta \sin t} \tag{5.16}$$

The angle β will have values from 0° to 360° and, since r and t are regarded as having positive values, the sign of Δ is not obtained automatically. It is assigned as negative for $90° < \beta < 270°$, and positive for other values of β. In other words displacement is towards the isocentre for points above the isocentre and away from it when below.

The calculation of tilt displacement requires a knowledge of f, t, r, and β; and before r and β may be determined the location of i and the principal line must be established.

Relief displacement on a tilted photograph occurs in much the same way as on a vertical photograph—that is, radially from the nadir point. Equation (5.5) is generally used to calculate relief displacement on near-vertical photography, and because the nadir point and principal point are in close proximity the latter is normally used to establish the radial distance r.

Most aerial photographs taken for mapping purposes contain small tilts, and the terrain photographed generally contains a certain amount of relief. Therefore a proper solution to the problem of establishing position can only be obtained if the position and tilt characteristics of the camera are known. Section 5.7 deals with this problems, which is known as *space resection*. However, before it can be considered the principles involved in three-dimensional conformal transformations must be developed.

5.6 Three-dimensional conformal transformation

The three-dimensional conformal (or similarity) transformation is one of the most commonly used transformations in analytical photogrammetry. It is a process which preserves three-dimensional shape and therefore it is extremely useful in applications such as

(i) converting three-dimensional image coordinates measured on a tilted photograph to values which would have been obtained on the equivalent vertical photograph;
(ii) connecting adjacent three-dimensional models, each with its own independent coordinate system, so that finally they form a block of models in one homogeneous coordinate system; and
(iii) transforming a three-dimensional model or block of models from its own independent coordinate system into the terrain system, or some other object system.

Details of these and other specific applications are given later whereas this section develops the transformation equations in fairly general terms.

Consider the right-handed systems shown in the diagrams of Figure 5.13.

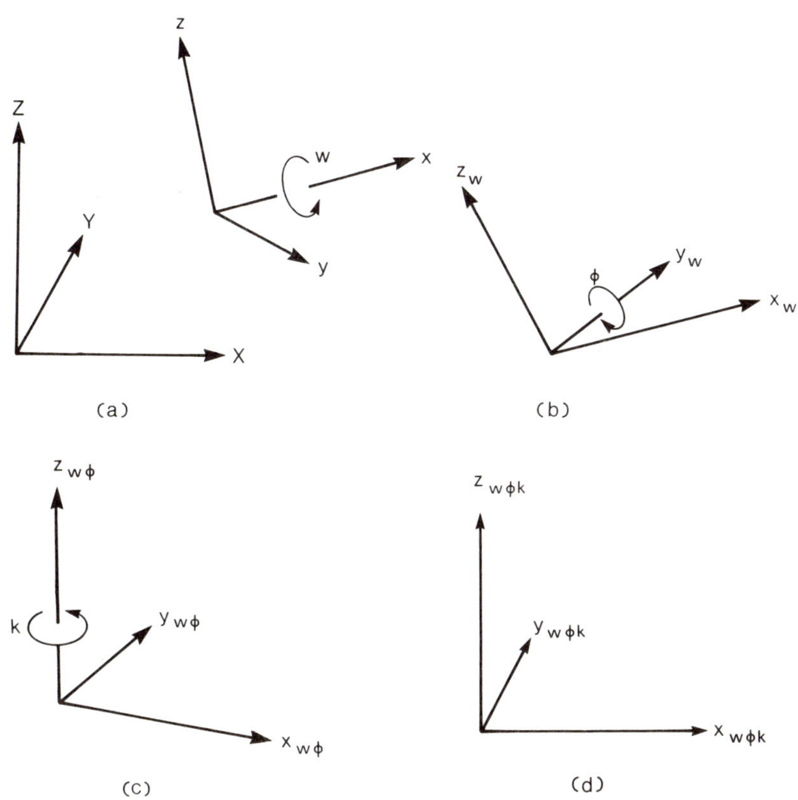

Figure 5.13 Sequence of rotations w, ϕ, k: a, initial position; b, after w-rotation; c, after ϕ-rotation; d, after k-rotation.

An XYZ-system is shown, and in (*a*) points with values in another xyz-system are to be transformed to it. The transformation involves the use of some or all of seven independent transformation parameters: three rotations (w, ϕ, k), three translations (X_C, Y_C, Z_C), and a scale-change (λ). The number used will depend on the application. For example, if the origins of the two systems coincide the translations will be 0, and if there is no scale-change the scale-factor λ will equal unity. In Figure 5.13*a* the system defined by the xyz-axes is to be rotated so that it is aligned parallel to the XYZ-system, and the sequence of diagrams (*a*)–(*d*) illustrates now this is achieved.

The diagrams show the effect of positive rotations ($w, \phi\, k$) taken in that order, and about the xyz-axes respectively. A positive rotation is defined as a clockwise rotation when viewed from the origin in the direction of the positive axis. A distinction should be made however, between a rotation of axes as shown in the diagrams of Figure 5.13 and a rotation of the line vector to a point when the axes are held fixed. In the latter case, the rotation will be in the opposite sense if the same transformation of coordinates is to be carried out. The final composite rotation matrix (***R***) is derived by considering the individual rotations in sequence. If the sequence is w, ϕ, k, then the xyz-axes are referred to as the primary, secondary, and tertiary axes respectively, and the secondary and tertiary axes are the once-removed and twice-removed axes respectively. Each individual rotation matrix represents a transformation in two dimensions and is assembled according to procedures described in section 2.11. The details with the appropriate diagrams are given below, and in all cases the positive direction of the axis of rotation is downwards into the paper. The diagrams illustrate a *rotation of axes* rather than a rotation of the line vector from the origin to the point.

The transformation due to a positive rotation w is

$$x_w = x$$
$$y_w = y \cos w + z \sin w$$
$$z_w = -y \sin w + z \cos w$$

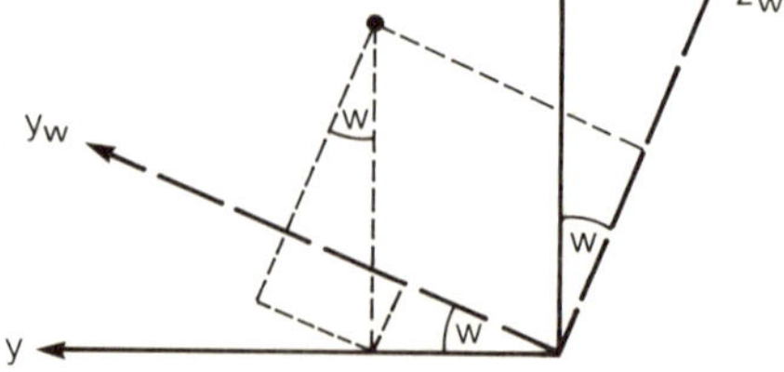

Figure 5.14

which may be expressed in matrix notation as

$$\begin{bmatrix} x_w \\ y_w \\ z_w \end{bmatrix} = \begin{bmatrix} 1 & 0 & 0 \\ 0 & \cos w & \sin w \\ 0 & -\sin w & \cos w \end{bmatrix} \begin{bmatrix} x \\ y \\ z \end{bmatrix}$$

or

$$\boldsymbol{x}_w = \boldsymbol{R}_x \cdot \boldsymbol{x} \qquad (5.17)$$

where

$$\boldsymbol{R}_x = \begin{bmatrix} 1 & 0 & 0 \\ 0 & \cos w & \sin w \\ 0 & -\sin w & \cos w \end{bmatrix}$$

At this stage the y_w-axis is parallel to the XY-plane (see Figure 5.13*b*).

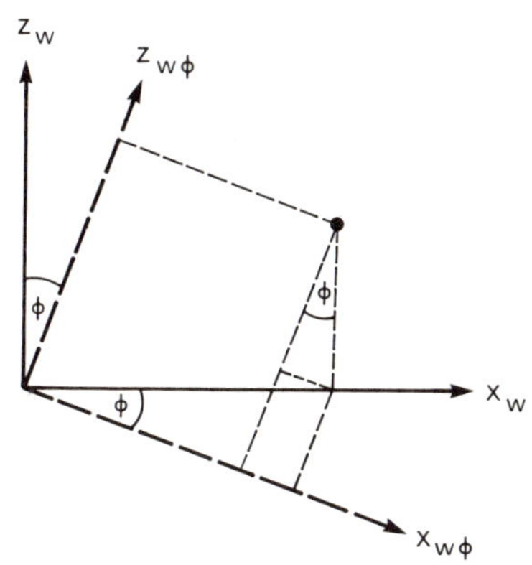

Figure 5.15

A positive ϕ-rotation gives

$$\begin{bmatrix} x_{w\phi} = x_w \cos\phi - z \sin\phi \\ y_{w\phi} = y_w \\ z_{w\phi} = x_w \sin\phi + z_w \cos\phi \end{bmatrix}$$

which in matrix notation is

$$\begin{bmatrix} x_{w\phi} \\ y_{w\phi} \\ z_{w\phi} \end{bmatrix} = \begin{bmatrix} \cos\phi & 0 & -\sin\phi \\ 0 & 1 & 0 \\ \sin\phi & 0 & \cos\phi \end{bmatrix} \begin{bmatrix} x_w \\ y_w \\ z_w \end{bmatrix}$$

or

$$\boldsymbol{x}_{w\phi} = \boldsymbol{R}_y \cdot \boldsymbol{x}_w$$

where

$$\boldsymbol{R}_y = \begin{bmatrix} \cos\phi & 0 & -\sin\phi \\ 0 & 1 & 0 \\ \sin\phi & 0 & \cos\phi \end{bmatrix} \tag{5.18}$$

At this stage the $\boldsymbol{x}_{w\phi}$-axis is also parallel to the XY-plane, and the $z_{w\phi}$-axis is parallel to the Z-axis (see Figure 5.13*c*).

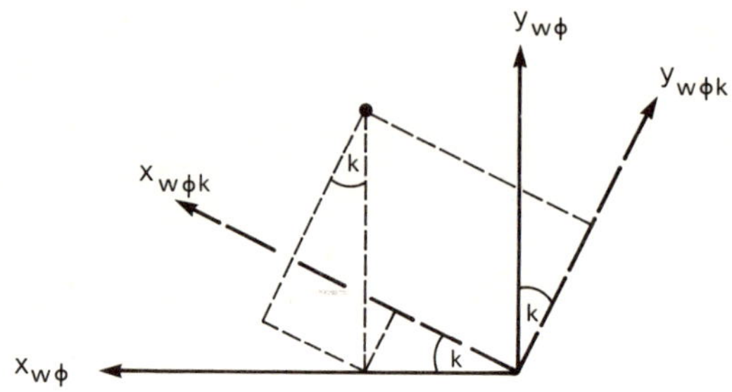

Figure 5.16

Finally, the transformation due to positive k has the following effect

$$\begin{aligned} x_{w\phi k} &= \quad x_{w\phi}\cos k + y_{w\phi}\sin k \\ y_{w\phi k} &= -x_{w\phi}\sin k + y_{w\phi}\cos k \\ z_{w\phi k} &= \quad z_{w\phi} \end{aligned}$$

which in matrix notation is

$$\begin{bmatrix} x_{w\phi k} \\ y_{w\phi k} \\ z_{w\phi k} \end{bmatrix} = \begin{bmatrix} \cos k & \sin k & 0 \\ -\sin k & \cos k & 0 \\ 0 & 0 & 1 \end{bmatrix} \begin{bmatrix} x_{w\phi} \\ y_{w\phi} \\ z_{w\phi} \end{bmatrix}$$

or

$$\boldsymbol{x}_{w\phi k} = \boldsymbol{R}_z \cdot \boldsymbol{x}_{w\phi} \tag{5.19}$$

where

$$\boldsymbol{R}_z = \begin{bmatrix} \cos k & \sin k & 0 \\ -\sin k & \cos k & 0 \\ 0 & 0 & 1 \end{bmatrix}$$

At this stage all three axes are parallel to the corresponding XYZ-axes.

Note that the pattern of the elements in the individual rotation matrices is such that, once $\boldsymbol{R}_x$ is formed, matrices $\boldsymbol{R}_y$ and $\boldsymbol{R}_z$ may be formed directly—simply by allocating the negative sign to the sine function in a cyclic fashion.

When the individual rotation matrices in eqns (5.17), (5.18) and (5.19) are combined together in the proper sequence to form one composite rotation matrix, the result is

$$\boldsymbol{R} = \boldsymbol{R}_z \cdot \boldsymbol{R}_y \cdot \boldsymbol{R}_x = \begin{bmatrix} r_{11} & r_{12} & r_{13} \\ r_{21} & r_{22} & r_{23} \\ r_{31} & r_{32} & r_{33} \end{bmatrix}$$

$$= \begin{bmatrix} \cos\phi\cos k & \sin w \sin\phi\cos k + \cos w \sin k & -\cos w \sin\phi\cos k + \sin w \sin k \\ -\cos\phi\sin k & -\sin w\sin\phi\sin k + \cos w\cos k & \cos w\sin\phi\sin k + \sin w\cos k \\ \sin\phi & -\sin w\cos\phi & \cos w\cos\phi \end{bmatrix} \tag{5.20}$$

and so

$$\begin{bmatrix} x_{w\phi k} \\ y_{w\phi k} \\ z_{w\phi k} \end{bmatrix} = \boldsymbol{R} \cdot \begin{bmatrix} x \\ y \\ z \end{bmatrix} \tag{5.21}$$

The formulae that represent the nine elements of $\boldsymbol{R}$ are applicable for all values of w, ϕ, k between $+180°$ and $-180°$; they relate to a specific sequence of rotations, and will be different for another sequence. For example the sequence ϕ, w, k (i.e. y primary, x secondary, and z tertiary) would produce the following rotation matrix:

$$\boldsymbol{R} = \boldsymbol{R}_z \cdot \boldsymbol{R}_x \cdot \boldsymbol{R}_y = \begin{bmatrix} \cos\phi\cos k + \sin\phi\sin w\sin k & \cos w\sin k & -\sin\phi\cos k + \cos\phi\sin w\sin k \\ -\cos\phi\sin k + \sin\phi\sin w\cos k & \cos w\cos k & \sin\phi\sin k + \cos\phi\sin w\cos k \\ \sin\phi\cos w & -\sin w & \cos\phi\cos w \end{bmatrix} \tag{5.22}$$

Although the formulae are different and the values of ϕ and w would be different, the numerical values of the nine elements would be the same, for the same two sets of axes, as those derived using eqn (5.20).

The orthogonal matrix may also be formed from another set of independent parameters, which are associated with the direction cosines of the *one* axis of rotation needed to effect the transformation. Such a matrix has been found to be useful because trigonometrical functions are not required to establish the matrix. However, it would appear that with developments in computer technology the advantages have largely disappeared. The matrix is known as Rodrigues matrix and more details may be found in [38].

Returning to eqns (5.20) and (5.22), if the numerical values of $\boldsymbol{R}$ are known, and there is a requirement to derive values for w, ϕ, k, then a decision must be made as to which are primary and secondary axes; only then can the appropriate formulae be used to solve for the tilts. This requirement could arise if values for the tilts, derived say from aerial triangulation, are to be set on a stereoplotting instrument. The choice of formulae ((5.20) or (5.22)) will depend on which axis has been made primary axis in the instrument.

$\boldsymbol{R}$ is compiled from three orthogonal matrices, therefore it is itself orthogonal; and because it is orthogonal the inverse matrix $\boldsymbol{M}$ will equal its transpose.

Thus

$$\boldsymbol{M} = \boldsymbol{R}^{-1} = \boldsymbol{R}^{\mathrm{T}}$$

and therefore

$$\begin{bmatrix} x \\ y \\ z \end{bmatrix} = \boldsymbol{R}^{\mathrm{T}} \cdot \begin{bmatrix} x_{w\phi k} \\ y_{w\phi k} \\ z_{w\phi k} \end{bmatrix} \tag{5.23}$$

An important characteristic of an orthogonal matrix is that the sum of the squares of the elements of any row or column is equal to 1, and this property should be used to verify that the matrix has been formed correctly. Alternatively, and preferably, the condition for an orthogonal matrix should be verified—namely that $\boldsymbol{R}^{\mathrm{T}}\boldsymbol{R} = \boldsymbol{I}$.

The remaining parameters of a three-dimensional conformal transformation—viz. scale-change (λ), and translation (X_C, Y_C, Z_C)—are applied to the coordinates derived in eqn (5.21) as follow

$$\begin{bmatrix} X \\ Y \\ Z \end{bmatrix} = \lambda \cdot \begin{bmatrix} x_{w\phi k} \\ y_{w\phi k} \\ z_{w\phi k} \end{bmatrix} + \begin{bmatrix} X_C \\ Y_C \\ Z_C \end{bmatrix} \tag{5.24}$$

Therefore the total transformation, expressed in terms of the original xyz-coordinates, may be written as

$$\left.\begin{aligned} \begin{bmatrix} X \\ Y \\ Z \end{bmatrix} &= \lambda \cdot \boldsymbol{R} \cdot \begin{bmatrix} x \\ y \\ z \end{bmatrix} + \begin{bmatrix} X_C \\ Y_C \\ Z_C \end{bmatrix} \\ \text{or} \qquad \mathbf{X} &= \lambda \cdot \boldsymbol{R} \cdot \boldsymbol{x} + \boldsymbol{X}_C \end{aligned}\right\} \tag{5.25}$$

There are a variety of different methods of obtaining values for the seven parameters, given a number of points with known coordinates in both systems. Some of these methods are considered later.

5.7 Collinearity equations and space resection (exterior orientation)

Space resection is a process which involves determining the position and orientation of a photograph, using measurements to images appearing on the photograph. The most common application is to aerial photogrammetry when the measured points are ground control points—i.e. points which have known values in the object system of coordinates. The process is analogous to resection discussed in section 2.8, except that in the case of photogrammetry it is a three-dimensional process, and therefore more demanding in computational terms. Many methods of computation have been devised in the past, the variations being due to the complexities presented by the problem, and the limitations of available computing facilities. However, with the development of powerful and readily available computers most of these limitations have disappeared, and this section deals solely with the method involving the principle of collinearity. The collinearity equations may be regarded as the most generally applicable set of equations in photogrammetry. They express the relationship between the coordinates of image points measured on the photograph and the coordinates of the same points in the object system. At least three points with known object system coordinates, and the nominal focal length, are required in order to solve the problem of space resection—or, in other words, determine the parameters of exterior orientation.

The collinearity condition states that, for any given photograph, the projection centre, image point, and object point must lie on the same straight line. Thus in Figure 5.17 points C, a, and A must be collinear. Consider a plane $CA'N$, which is parallel to the XZ-plane of the object system, and another plane $CA''N$ which is parallel to the YZ-plane. A' and A'' are formed by projecting A orthogonally onto these vertical planes; and similarly a' and a'' are formed by projecting the image point a onto the same planes. From similar triangles in the diagram it is clear that the scale-change λ_a is given by

$$\lambda_a = \frac{CA}{Ca} = \frac{CA'}{Ca'} = \frac{CA''}{Ca''} = \frac{CN}{Ca'''} = \frac{(Z_A - Z_C)}{(Z_a - Z_C)} \qquad (5.26a)$$

and similarly,

$$\lambda_a = \frac{NA'}{a'''a'} = \frac{(X_A - X_C)}{(X_a - X_C)} \quad \text{and} \quad \lambda_a = \frac{NA''}{a'''a''} = \frac{(Y_A - Y_C)}{(Y_a - Y_C)} \qquad (5.26b)$$

Therefore

$$\begin{bmatrix} X_A - X_C \\ Y_A - Y_C \\ Z_A - Z_C \end{bmatrix} = \lambda_a \cdot \begin{bmatrix} X_a - X_C \\ Y_a - Y_C \\ Z_a - Z_C \end{bmatrix} \qquad (5.27)$$

The coordinate components $(X_a - X_C)$, $(Y_a - Y_C)$, $(Z_a - Z_C)$ are the image coordinates (x_a, y_a, z_a), which have been transformed by a rotation matrix R. The matrix R is an orthogonal matrix formed from values for the tilt parameters w, ϕ, k (see section 5.6).

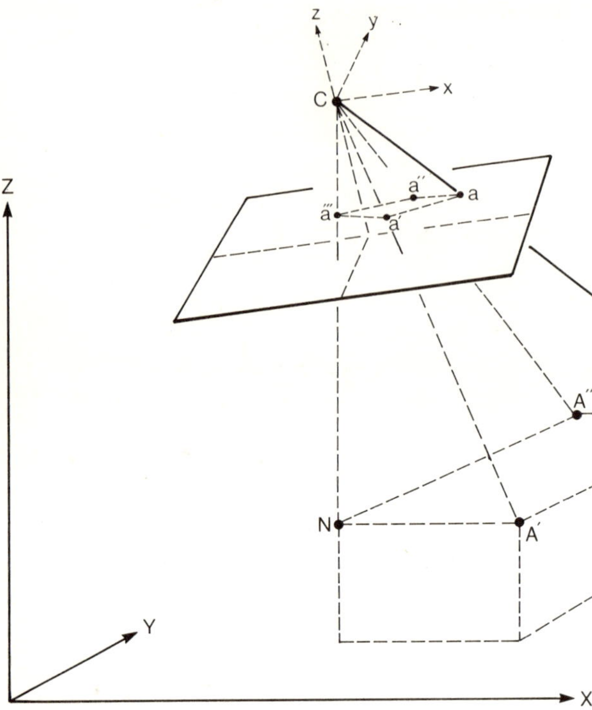

Figure 5.17 Collinearity.

Therefore eqn (5.27) becomes

$$\begin{bmatrix} X_A - X_C \\ Y_A - Y_C \\ Z_A - Z_C \end{bmatrix} = \lambda_a \cdot \boldsymbol{R} \cdot \begin{bmatrix} x_a \\ y_a \\ z_a \end{bmatrix} \tag{5.28}$$

which may also be written

$$\begin{bmatrix} X_A \\ Y_A \\ Z_A \end{bmatrix} = \lambda_a \cdot \boldsymbol{R} \cdot \begin{bmatrix} x_a \\ y_a \\ z_a \end{bmatrix} + \begin{bmatrix} X_C \\ Y_C \\ Z_C \end{bmatrix} \tag{5.29}$$

and is recognizable as being of the same form as eqn (5.25). The coordinate z_a is the value of the nominal focal length f, but with a negative sign, and it will be the same for all image points.

Therefore eqn (5.28) becomes

$$\begin{bmatrix} X_A - X_C \\ Y_A - Y_C \\ Z_A - Z_C \end{bmatrix} = \lambda_a \cdot \boldsymbol{R} \cdot \begin{bmatrix} x_a \\ y_a \\ -f \end{bmatrix} \tag{5.30}$$

Equations (5.30) may be rewritten as

$$\begin{bmatrix} x_a \\ y_a \\ -f \end{bmatrix} = \frac{1}{\lambda_a} \cdot \boldsymbol{M} \cdot \begin{bmatrix} X_A - X_C \\ Y_A - Y_C \\ Z_A - Z_C \end{bmatrix} \tag{5.31}$$

where

$$\boldsymbol{M} = \boldsymbol{R}^{-1} = \begin{bmatrix} m_{11} & m_{12} & m_{13} \\ m_{21} & m_{22} & m_{23} \\ m_{31} & m_{32} & m_{33} \end{bmatrix}$$

Written in full, eqn (5.31) becomes

$$\left.\begin{aligned} x_a &= \frac{1}{\lambda_a} \cdot (m_{11}(X_A - X_C) + m_{12}(Y_A - Y_C) + m_{13}(Z_A - Z_C)) \\ y_a &= \frac{1}{\lambda_a} \cdot (m_{21}(X_A - X_C) + m_{22}(Y_A - Y_C) + m_{23}(Z_A - Z_C)) \\ -f &= \frac{1}{\lambda_a} \cdot (m_{31}(X_A - X_C) + m_{32}(Y_A - Y_C) + m_{33}(Z_A - Z_C)) \end{aligned}\right\} \tag{5.32}$$

and if the first and second rows are divided by the third, and $-f$ transferred to the right-hand sides, the following *collinearity equations* are obtained:

$$\left.\begin{aligned} x_a &= -f \cdot \frac{(m_{11}(X_A - X_C) + m_{12}(Y_A - Y_C) + m_{13}(Z_A - Z_C))}{(m_{31}(X_A - X_C) + m_{32}(Y_A - Y_C) + m_{33}(Z_A - Z_C))} \\ y_a &= -f \cdot \frac{(m_{21}(X_A - X_C) + m_{22}(Y_A - Y_C) + m_{23}(Z_A - Z_C))}{(m_{31}(X_A - X_C) + m_{32}(Y_A - Y_C) + m_{33}(Z_A - Z_C))} \end{aligned}\right\} \tag{5.33}$$

The inverse form of these equations may be derived from eqns (5.30) and the result would be

$$\left.\begin{aligned} (X_A - X_C) &= (Z_A - Z_C) \cdot \frac{(r_{11}x_a + r_{12}y_a + r_{13}(-f))}{(r_{31}x_a + r_{32}y_a + r_{33}(-f))} \\ (Y_A - Y_C) &= (Z_A - Z_C) \cdot \frac{(r_{21}x_a + r_{22}y_a + r_{23}(-f))}{(r_{31}x_a + r_{32}y_a + r_{33}(-f))} \end{aligned}\right\} \tag{5.34}$$

Since $\boldsymbol{R}$ is an orthogonal matrix, and $\boldsymbol{M} = \boldsymbol{R}^{-1}$, therefore

$$\boldsymbol{R} = \boldsymbol{M}^{-1} = \boldsymbol{M}^{\mathrm{T}}$$

Therefore eqns (5.34) may be written

$$\left.\begin{aligned} (X_A - X_C) &= (Z_A - Z_C) \cdot \frac{(m_{11}x_a + m_{21}y_a + m_{31}(-f))}{(m_{13}x_a + m_{23}y_a + m_{33}(-f))} \\ (Y_A - Y_C) &= (Z_A - Z_C) \cdot \frac{(m_{12}x_a + m_{22}y_a + m_{32}(-f))}{(m_{13}x_a + m_{23}y_a + m_{33}(-f))} \end{aligned}\right\} \tag{5.35}$$

Eqns (5.33) are the most commonly used form of the collinearity equations. The equations contain twelve independent quantities: the image coordinates $(x_a, y_a, -f)$, the projection centre coordinates (X_C, Y_C, Z_C), the object point coordinates (X_A, Y_A, Z_A), and the elements of rotation w, ϕ, k which are implicit in the quantities $m_{11}, \ldots, m_{33}$. Notice that the scale-factor λ_a, which would have a different value for each image point, has been eliminated.

The equations may be expressed in a more compact matrix form as follows: Let

$$\boldsymbol{X} = \begin{bmatrix} X_A - X_C \\ Y_A - Y_C \\ Z_A - Z_C \end{bmatrix}$$

and

$$\boldsymbol{M}_1 = (m_{i1} \quad m_{i2} \quad m_{i3}), \quad \text{where } i = 1, 2, 3$$

then

$$\left.\begin{aligned} x_a &= -f \cdot \frac{\boldsymbol{M}_1 \boldsymbol{X}}{\boldsymbol{M}_3 \boldsymbol{X}} \\ y_a &= -f \cdot \frac{\boldsymbol{M}_2 \boldsymbol{X}}{\boldsymbol{M}_3 \boldsymbol{X}} \end{aligned}\right\} \tag{5.36}$$

The nominal focal length f is usually a known quantity, and x_a, y_a are generally observed quantities. If sufficient points with known object system coordinates are observed then the unknown tilts w, ϕ, k and the projection centre coordinates (X_C, Y_C, Z_C) may be derived. This then represents the solution to the space resection problem. Alternatively, if the tilts and projection centre coordinates of two overlapping photographs are known, and observations x_a, y_a are made on these photographs, then the object system coordinates (X_A, Y_A, Z_A) may be derived. Other problems in photogrammetry may also be solved using these equations. For example, they may be used to determine a more reliable value of f: or with known quantities on the right-hand sides they may be used to derive values for x_a and y_a.

A difficulty that arises when the equations are used to solve for the unknowns is the fact that they are non-linear as they stand. Therefore a process of linearization is necessary before a solution of simultaneous equations may take place. Eqns (5.33) may be written as functions F and G which are equal to 0

$$\left.\begin{aligned} F &= x_a(m_{31}(X_A - X_C) + m_{32}(Y_A - Y_C) + m_{33}(Z_A - Z_C)) \\ &\quad + f(m_{11}(X_A - X_C) + m_{12}(Y_A - Y_C) + m_{13}(Z_A - Z_C)) = 0 \\ G &= y_a(m_{31}(X_A - X_C) + m_{32}(Y_A - Y_C) + m_{33}(Z_A - Z_C)) \\ &\quad + f(m_{21}(X_A - X_C) + m_{22}(Y_A - Y_C) + m_{23}(Z_A - Z_C)) = 0 \end{aligned}\right\} \tag{5.37}$$

and if f is considered a known constant, then the linearized form using Taylor's theorem (see Appendix A) would be

$$\left.\begin{aligned} F_0 &+ \left(\frac{\partial F}{\partial x_a}\right)_o \cdot dx_a + \left(\frac{\partial F}{\partial w}\right)_o \cdot dw + \left(\frac{\partial F}{\partial \phi}\right)_o \cdot d\phi + \left(\frac{\partial F}{\partial k}\right)_o \cdot dk \\ &+ \left(\frac{\partial F}{\partial X_C}\right)_o \cdot dX_C + \left(\frac{\partial F}{\partial Y_C}\right)_o \cdot dY_C + \left(\frac{\partial F}{\partial Z_C}\right)_o \cdot dZ_C \\ &+ \left(\frac{\partial F}{\partial X_A}\right)_o \cdot dX_A + \left(\frac{\partial F}{\partial Y_A}\right)_o \cdot dY_A + \left(\frac{\partial F}{\partial Z_A}\right)_o \cdot dZ_A = 0 \end{aligned}\right\} \quad (5.38a)$$

and

$$\left.\begin{aligned} G_0 &+ \left(\frac{\partial G}{\partial y_a}\right)_o \cdot dy_a + \left(\frac{\partial G}{\partial w}\right)_o \cdot dw + \left(\frac{\partial G}{\partial \phi}\right)_o \cdot d\phi + \left(\frac{\partial G}{\partial k}\right)_o \cdot dk \\ &+ \left(\frac{\partial G}{\partial X_C}\right)_o \cdot dX_C + \left(\frac{\partial G}{\partial Y_C}\right)_o \cdot dY_C + \left(\frac{\partial G}{\partial Z_C}\right)_o \cdot dZ_C \\ &+ \left(\frac{\partial G}{\partial X_A}\right)_o \cdot dX_A + \left(\frac{\partial G}{\partial Y_A}\right)_o \cdot dY_A + \left(\frac{\partial G}{\partial Z_A}\right)_o \cdot dZ_A = 0 \end{aligned}\right\} \quad (5.38b)$$

In these equations the expressions $(\partial F/\partial x_a)_0$, $(\partial F/\partial w)_0, \ldots, (\partial F/\partial Z_A)_0$, and $(\partial G/\partial y_a)_0, \ldots, (\partial G/\partial Z_A)_0$ are the partial derivatives of the functions with respect to the unknown variables, and $(\partial F/\partial x_a)_0$ and $(\partial G/\partial y_a)_0$ are in fact equal to the denominator in eqns (5.33). The subscript o implies that the expression is evaluated using approximations for the unknowns, which therefore become known constants for a particular computational cycle. Details concerning the partial derivatives vary depending on the way in which the rotation matrix M is constructed—i.e. on the sequence of rotations. The derivation is lengthy, but an example is given in Appendix B. The unknowns dx_a, dw, $d\phi$, etc. are corrections to the initial approximations. If x_a and y_a represent observations the equations may be written as

$$\left.\begin{aligned} dx_a &= -\left(F_o + \left(\frac{\partial F}{\partial w}\right)_o \cdot dw + \cdots + \left(\frac{\partial F}{\partial Z_A}\right)_o \cdot dZ_A\right) \Big/ \left(\frac{\partial F}{\partial x_a}\right)_o \\ dy_a &= -\left(G_o + \left(\frac{\partial G}{\partial w}\right)_o \cdot dw + \cdots + \left(\frac{\partial G}{\partial Z_A}\right)_o \cdot dZ_A\right) \Big/ \frac{(\partial G)}{(\partial y_a)_o} \end{aligned}\right\} \quad (5.39)$$

and if observations (x_b, y_b), $(x_c, y_c), \ldots$, etc. have been made to other points then pairs of equations such as (5.39) may be established for each point.

If the number of equations is equal to the number of unknowns the values of dx and dy will be 0. If there are more observations ('*redundant*' observations) than the minimum required to solve the unknowns the dx and dy are residual discrepancies, denoted by v_x and v_y. These are the amounts by which the equations are not satisfied when the derived values of the unknowns are substituted into the equations. In photogrammetric practice it is invariably the case that redundant observations exist, and therefore a least-squares determination of unknowns becomes desirable. Details of least-squares procedures are given in Chapter 9.

In order to simplify the notation, and considering measurements to any point, eqns (5.39) may be written as

$$\left.\begin{aligned} v_x &= c_{11}\,\mathrm{d}w + c_{12}\,\mathrm{d}\phi + c_{13}\,\mathrm{d}k + c_{14}\,\mathrm{d}X_C + c_{15}\,\mathrm{d}Y_C + c_{16}\,\mathrm{d}Z_C \\ &\quad + c_{14}\,\mathrm{d}X + c_{15}\,\mathrm{d}Y + c_{16}\,\mathrm{d}Z + C \\ v_y &= c_{21}\,\mathrm{d}w + c_{22}\,\mathrm{d}\phi + c_{23}\,\mathrm{d}k + c_{24}\,\mathrm{d}X_C + c_{25}\,\mathrm{d}Y_C + c_{26}\,\mathrm{d}Z_C \\ &\quad + c_{24}\,\mathrm{d}X + c_{25}\,\mathrm{d}Y + c_{26}\,\mathrm{d}Z + D \end{aligned}\right\} \quad (5.40)$$

where v_x, v_y are unknown residuals (corrections to observations), which will be 0 if the minimum number of observations are used; $\mathrm{d}w \ldots \mathrm{d}Z$ are unknown corrections to approximate values for the nine parameters; $c_{11} \ldots c_{26}$ are known coefficients, evaluated using approximate values for the nine parameters (details are given in Appendix A); and C, D are the values of the functions in (5.37), which are obtained when approximate values for the parameters are used.

Equations (5.40) are applicable to virtually all problems in photogrammetric practice—including space resection, intersection, relative orientation, aerial triangulation, etc. The equations are modified to suit the particular application. In space resection the ground coordinates of control points will be known and so the terms involving $\mathrm{d}X$, $\mathrm{d}Y$, $\mathrm{d}Z$ will not appear. In space intersection the position and angles of tilt of the photograph will be known and therefore the terms in $\mathrm{d}w$, $\mathrm{d}\phi$, $\mathrm{d}k$, $\mathrm{d}X_C$, $\mathrm{d}Y_C$, $\mathrm{d}Z_C$ will not appear. Because the process starts with linearization of a non-linear mathematical model, it is usually necessary to repeat the computations using improved values for the initial approximations of the unknowns. Very often several of these 'iteration cycles' are necessary, the number depending on the accuracy with which the initial values were determined in the first place.

The equations which are used for space resection are:

$$\left.\begin{aligned} v_x &= c_{11}\,\mathrm{d}w + c_{12}\,\mathrm{d}\phi + c_{13}\,\mathrm{d}k + c_{14}\,\mathrm{d}X_C + c_{15}\,\mathrm{d}Y_C + c_{16}\,\mathrm{d}Z_C + C \\ v_y &= c_{21}\,\mathrm{d}w + c_{22}\,\mathrm{d}\phi + c_{23}\,\mathrm{d}k + c_{24}\,\mathrm{d}X_C + c_{25}\,\mathrm{d}Y_C + c_{26}\,\mathrm{d}Z_C + D \end{aligned}\right\} \quad (5.41)$$

If three ground control points with known XYZ-coordinates are observed, then six equations may be formed, and the six unknown corrections solved simultaneously. These values are added to the initial approximations to provide a new set of approximate values. The coefficients $c_{11} - c_{26}$, and constants C, D, are re-evaluated and the cycle is repeated. The size of the corrections should diminish rapidly with each iteration, and the iterations continue until the corrections are negligible. If more than three control points are observed a least-squares solution of the unknowns is used.

Establishing initial approximate values for the unknowns should not be difficult, although it could be inconvenient. With near-vertical photography the values for w and ϕ will be 0; k may be estimated by identifying the direction of north on the photograph; X_C and Y_C may be obtained by scaling distances on the photograph from the principal point to control points; and Z_C may be calculated on the basis of a vertical photograph, the focal length

f, and ground and image distances for a line between two control points.

5.8 Analytical rectification

If the XYZ ground coordinates of a number of control points are known then a process known as *analytical rectification* may be carried out, based on parameters derived from these coordinates. The rectification is achieved by the application of equations referred to as *two-dimensional projective transformation equations*. When these are applied to measured xy-image coordinates of a tilted photograph, the corresponding XY-coordinates in the ground system are obtained. The projective transformation equations are derived as follows.

Consider Figure 5.17 and eqn (5.30), which is repeated here for convenience:

$$\begin{bmatrix} X_A - X_C \\ Y_A - Y_C \\ Z_A - Z_C \end{bmatrix} = \lambda_a \cdot \boldsymbol{R} \cdot \begin{bmatrix} x_a \\ y_a \\ -f \end{bmatrix}$$

Written in full the equations become

$$\left.\begin{aligned} (X_A - X_C) &= \lambda_a(r_{11}x_a + r_{12}y_a + r_{13}(-f)) \\ (Y_A - Y_C) &= \lambda_a(r_{21}x_a + r_{22}y_a + r_{23}(-f)) \\ (Z_A - Z_C) &= \lambda_a(r_{31}x_a + r_{32}y_a + r_{33}(-f)) \end{aligned}\right\} \quad (5.42)$$

Dividing the first two equations in (5.42) by the third, and transferring $(Z_A - Z_C)$ to the right-hand sides, produces

$$\left.\begin{aligned} (X_A - X_C) &= (Z_A - Z_C) \cdot \frac{(r_{11}x_a + r_{12}y_a + r_{13}(-f))}{(r_{31}x_a + r_{32}y_a + r_{33}(-f))} \\ (Y_A - Y_C) &= (Z_A - Z_C) \cdot \frac{(r_{21}x_a + r_{22}y_a + r_{23}(-f))}{(r_{31}x_a + r_{32}y_a + r_{33}(-f))} \end{aligned}\right\} \quad (5.43)$$

which are the same as eqns (5.34).

By extracting $(-r_{33}f)$ in the denominator these may be converted to

$$\left.\begin{aligned} (X_A - X_C) &= \frac{(Z_A - Z_C)}{(-r_{33}f)} \cdot \frac{(r_{11}x_a + r_{12}y_a + r_{13}(-f)}{(-(r_{31}/r_{33}f)x_a - (r_{32}/r_{33}f)y_a + 1)} \\ (Y_A - Y_C) &= \frac{(Z_A - Z_C)}{(-r_{33}f)} \cdot \frac{(r_{21}x_a + r_{22}y_a + r_{23}(-f)}{(-(r_{31}/r_{33}f)x_a - (r_{32}/r_{33}f)y_a + 1)} \end{aligned}\right\} \quad (5.44)$$

Now f, X_C, Y_C, Z_C, and all the r's are constants; and if the Z_A coordinate is constant for all points (i.e. flat terrain), then the equations may be written

$$\left.\begin{aligned} X &= \frac{a_1x + b_1y + c_1}{a_3x + b_3y + 1} \\ Y &= \frac{a_2x + b_2y + c_2}{a_3x + b_3y + 1} \end{aligned}\right\} \quad (5.45)$$

where X, Y are ground coordinates; x, y are image coordinates; and $a_1, \ldots, c_2$ are the transformation parameters.

Equations (5.44) are rigorous and may be used to determine XY-coordinates of any point appearing on the image, provided the Z-coordinate for the point is known. If the Z value is not known, and an assumed value is used instead, then the X and Y values will be in error by amounts which correspond with the error in Z. If, as might often be the case in practice, the variation in relief is small and $(Z_A - Z_C)$ is assumed to be a constant (the flying height above the terrain) for all points, then eqns (5.45) may be used for rectification purposes. They are general equations expressing the projective transformation between two planes, and they produce a linear transformation of image coordinates in any direction in the plane of the image. The projective transformation of a square grid in the image plane onto the object plane is illustrated in Figure 5.18.

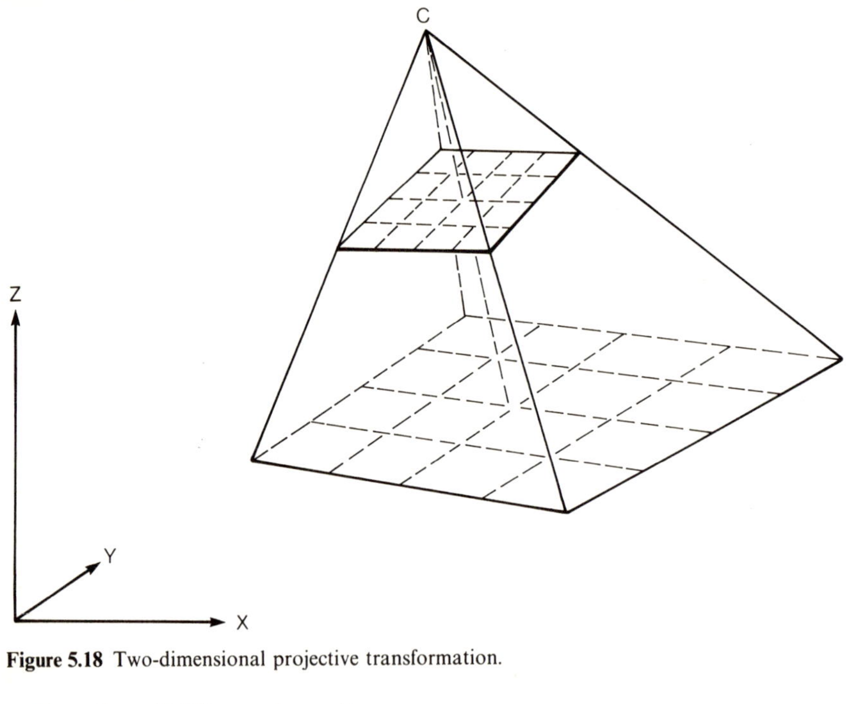

Figure 5.18 Two-dimensional projective transformation.

Equations (5.45) may be written

$$X(a_3x + b_3y + 1) = a_1x + b_1y + c_1$$
$$Y(a_3x + b_3y + 1) = a_2x + b_2y + c_2$$

or, by re-arrangement,

$$\left.\begin{aligned} X &= a_1x - a_3xX + b_1y - b_3yX + c_1 \\ Y &= a_2x - a_3xY + b_2y - b_3yY + c_2 \end{aligned}\right\} \qquad (5.46)$$

and, in matrix notation,

$$\boldsymbol{X} = \boldsymbol{A}\boldsymbol{p} \tag{5.47}$$

where

$$\boldsymbol{X} = \begin{pmatrix} X \\ Y \end{pmatrix}, \quad \boldsymbol{A} = \begin{pmatrix} x & 0 & -xX & y & 0 & -yX & 1 & 0 \\ 0 & x & -xY & 0 & y & -yY & 0 & 1 \end{pmatrix}, \quad \boldsymbol{p} = \begin{bmatrix} a_1 \\ a_2 \\ a_3 \\ b_1 \\ b_2 \\ b_3 \\ c_1 \\ c_2 \end{bmatrix}$$

If the values for exterior orientation of the camera are not known the eight parameters $a_1, \ldots, c_2$ may be found from ground control information. At least four suitably distributed (non-collinear) points are required to produce eight equations for a unique solution. If more ground control points are available a least-squares determination of parameters may be used, according to procedures described in Chapters 9 and 10.

5.9 The geometry of overlapping photographs and space intersection

A single photograph is a two-dimensional representation of a three-dimensional scene, and therefore it is not sufficient in itself to locate the three-dimensional positions of objects. Although rectification may determine the planimetric position (X, Y) of an object, the process relies on a known value for the third dimension (Z). If the Z-coordinate is not known, and all three object system coordinates are required, the object must be photographed from at least two different exposure stations. The overlapping photographs then define two intersecting rays to each object point, as shown in Figure 5.19, and the intersection of these *conjugate* rays will determine the object system coordinates. The computational procedure that achieves this is termed *space intersection.*

The conjugate rays may be represented mathematically by collinearity equations such as eqns (5.33). Thus a single object point appearing on two photographs will have XYZ-coordinates contained in four equations. If the positions and tilts of the exposures are known, in addition to the elements of interior orientation and the image coordinates of the point, then the equations may be solved to produce values for the unknowns X, Y, Z. Therefore the linearized eqns (5.40) reduce to

$$\left.\begin{aligned} v_x &= c_{14}\mathrm{d}X + c_{15}\mathrm{d}Y + c_{16}\mathrm{d}Z + C \\ v_y &= c_{24}\mathrm{d}X + c_{25}\mathrm{d}Y + c_{26}\mathrm{d}Z + D \end{aligned}\right\} \tag{5.48}$$

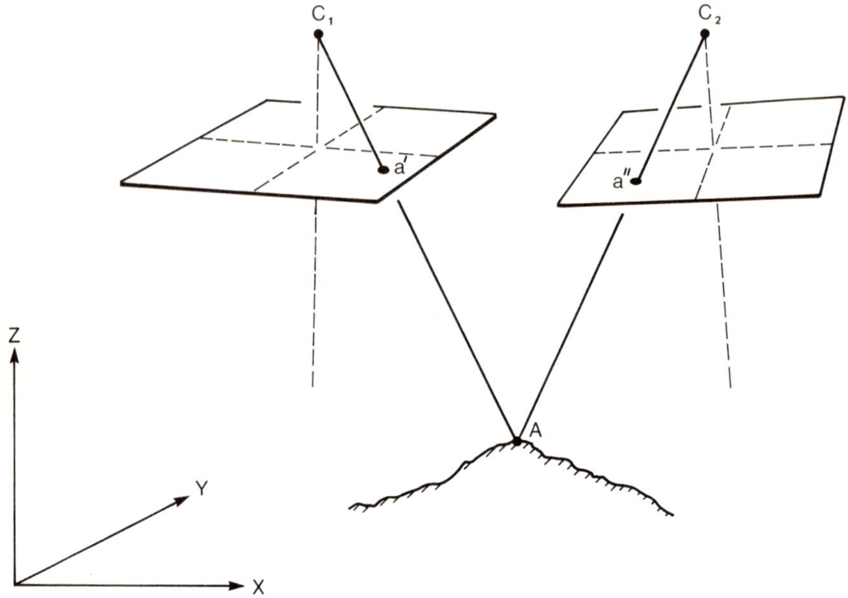

Figure 5.19 Space intersection.

and there will be two such equations for each image point. When both photographs are considered there will be four equations and three unknowns (dX, dY, dZ), and a least-squares solution may be used. These values are added to the initial approximations of X, Y, Z to obtain improved values, and the solution is repeated. The iterations continue until the corrections become negligible.

It is useful to note that the original collinearity equations (5.33), when applied to conjugate rays, contain a total of 25 variables. These are the nine elements of interior and exterior orientation for each exposure, two measured coordinates on each image, and the three object system coordinates for the point. If the equations are expressed in the form of eqns (5.32) then two additional variables are included—the scale factors for the two rays—making a total of 27. Thus it becomes clear how flexible the collinearity equations are, and each application will have its own particular set of known and unknown variables. Other photogrammetric conditions may be derived from the collinearity equations, and one in particular is the *coplanarity condition* used for relative orientation, and discussed in the next section.

The use of linearized equations such as eqns (5.48) requires that initial approximate values for X, Y, Z are specified, and it may not be convenient to derive these values. Therefore another method of space intersection is described below, which is not a least-squares solution but gives acceptable esults for most purposes. Consider the diagram shown in Figure 5.20. The ordinates x', y', z' and x'', y'', z'' are the image coordinates for points on the

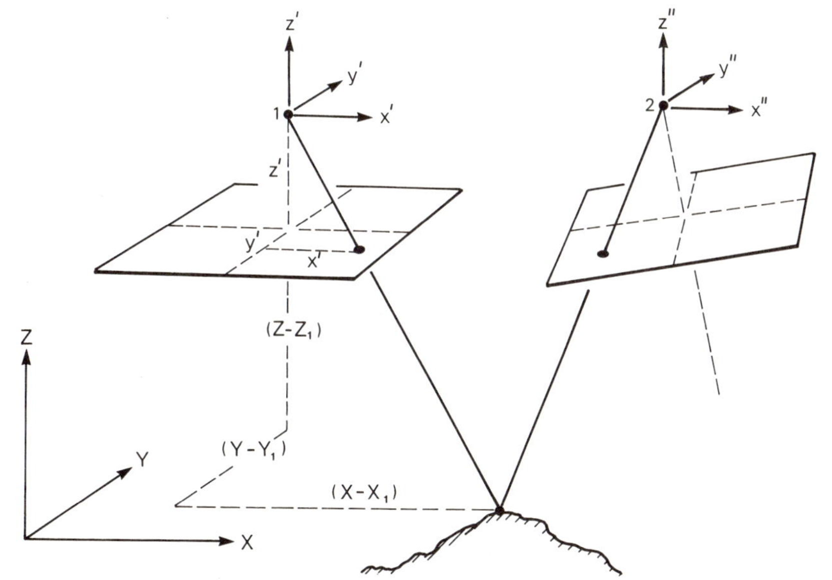

Figure 5.20 Model system of coordinates.

first and second exposures respectively, but rotated into systems which are parallel to the object system of coordinates. If the object system is aligned approximately with the direction of flight, as shown in the diagram, then it is referred to as the *model system* of coordinates.

From similar triangles the following relationships may be established. For the first photograph

$$\frac{X - X_1}{x'} = \frac{Y - Y_1}{y'} = \frac{Z - Z_1}{z'} \tag{5.49}$$

and similarly for the second photograph

$$\frac{X - X_2}{x''} = \frac{Y - Y_2}{y''} = \frac{Z - Z_2}{z''} \tag{5.50}$$

Therefore from (5.49) and (5.50)

$$\left.\begin{aligned} X &= X_1 + x' \cdot \frac{(Z - Z_1)}{z'} \\ X &= X_2 + x'' \cdot \frac{(Z - Z_2)}{z''} \end{aligned}\right\} \tag{5.51}$$

Also

$$\left.\begin{aligned} Y &= Y_1 + y' \cdot \frac{(Z - Z_1)}{z'} \\ Y &= Y_2 + y'' \cdot \frac{(Z - Z_2)}{z''} \end{aligned}\right\} \tag{5.52}$$

Equations (5.51) contain the two unknowns, X and Z, therefore equating the right-hand sides and solving for Z,

$$Z = Z_1 + z' \cdot \frac{(z''(X_2 - X_1) - x''(Z_2 - Z_1))}{(x'z'' - x''z')} \tag{5.53}$$

which may be written

$$Z = Z_1 + \lambda \cdot z' \tag{5.54}$$

where

$$\lambda = \frac{z''(X_2 - X_1) - x''(Z_2 - Z_1)}{(x'z'' - x''z')} \tag{5.54a}$$

is the scale factor for the first ray.

Substitution of Z in the first of eqns (5.51) gives

$$X = X_1 + \lambda \cdot x' \tag{5.55}$$

Finally two values for Y may be obtained by substituting for Z in eqns (5.52). The two values will differ and the mean is usually adopted. The difference, referred to as *residual Y-parallax*, is caused by a combination of observational and other errors and the fact that there is one redundant observation. Notice that the equation for scale factor (eqn (5.54*a*)) contains the base components $(X_2 - X_1)$ and $(Z_2 - Z_1)$. It is these components that determine the scale of the model produced by intersected points and, with the alignment of the object system as specified, $(X_2 - X_1)$ becomes the dominant component. Therefore if $(X_2 - X_1)$ is approximately equal to the base length on the photographs the Y-parallax will also be a representative value for y-parallax in the scale of the imagery.

Although the method outlined above is useful because it avoids the matter of initial approximate values for X, Y, Z, it is not the best solution, as it allows the observational errors to produce discrepancies in the Y-direction alone. A better solution would be one which distributes the errors correctly amongst the observations—that is, the xy image coordinates; the least-squares solution based on eqns (5.48) does this.

5.10 Relative orientation by collinearity and coplanarity conditions

Relative orientation is the process of establishing the correct position and orientation of one photograph relative to the adjacent overlapping photograph. This is achieved by enforcing the condition that corresponding or conjugate rays intersect. When relative orientation is complete the so-called photogrammetric model is established; that is the set of points appearing in the common overlap, and whose three-dimensional model (object system) coordinates may be computed, using equations developed in section 5.9. Two methods of relative orientation are described in this section, the first using the collinearity equations, and the second based on the coplanarity condition.

If one photograph is considered fixed then six parameters (three translations and three rotations) determine the position and orientation of the second photograph relative to it. Consider Figure 5.21, which shows a pair of overlapping photographs taken from positions 1 and 2. It is usually convenient to specify values (X_1, Y_1, Z_1) for the first projection centre and align the model system of coordinates with the photograph, as shown in the diagram. Thus w', ϕ', k' would be 0 and X_1, Y_1, Z_1 would be known quantities: very often the latter are also made equal to 0—i.e. the origin for the model system would be made to coincide with the first projection centre.

In the case of the second photograph one of the translations (usually the X-component) is assigned a value which determines the scale of the model, therefore the remaining five parameters associated with the second photograph must be determined by the relative orientation process. If the collinearity equations (5.40) are used it is clear that there are also three additional unknowns (dX, dY, dZ) associated with the object space coordinates for each observed point. Therefore, with reference to Figure 5.21, the following four equations may be written for each set of observed image coordinates $(x', y'$ and $x'', y'')$:

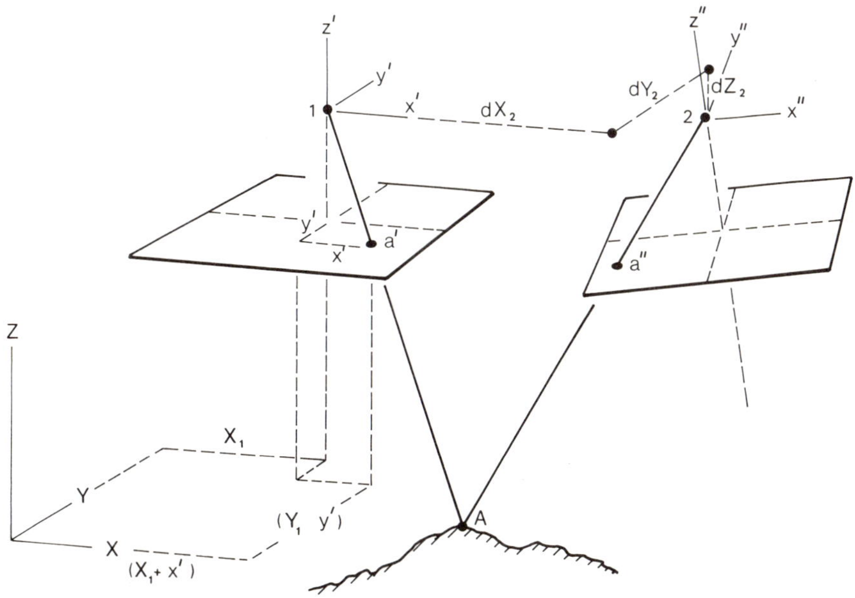

Figure 5.21 Relative orientation.

$$\left.\begin{aligned}
v_{x'} &= c_{14}\mathrm{d}X + c_{15}\mathrm{d}Y + c_{16}\mathrm{d}Z + C' \\
v_{y'} &= c_{24}\mathrm{d}X + c_{25}\mathrm{d}Y + c_{26}\mathrm{d}Z + D' \\
v_{x''} &= c_{11}\,\mathrm{d}w'' + c_{12}\,\mathrm{d}\phi'' + c_{13}\,\mathrm{d}k'' + c_{15}\,\mathrm{d}Y_2 + c_{16}\,\mathrm{d}Z_2 \\
&\quad + c_{14}\,\mathrm{d}X + c_{15}\,\mathrm{d}Y + c_{16}\,\mathrm{d}Z + C'' \\
v_{y''} &= c_{21}\,\mathrm{d}w'' + c_{22}\,\mathrm{d}\phi'' + c_{23}\,\mathrm{d}k'' + c_{25}\,\mathrm{d}Y_2 + c_{26}\,\mathrm{d}Z_2 \\
&\quad + c_{24}\,\mathrm{d}X + c_{25}\,\mathrm{d}Y + c_{26}\,\mathrm{d}Z + D''
\end{aligned}\right\} \quad (5.56)$$

where $v_{x'}, v_{y'}, v_{x''}, v_{y''}$ are residual discrepancies, which will be 0 if the minimum number of observations have been made to solve for the unknowns; $\mathrm{d}Y_2, \mathrm{d}Z_2, \mathrm{d}w'', \mathrm{d}\phi'', \mathrm{d}k''$ are the five unknown parameters of relative orientation for the second photograph—i.e. corrections to approximate values for the parameters; $\mathrm{d}X, \mathrm{d}Y, \mathrm{d}Z$ are the unknown corrections to approximate values for the model coordinates of the point; $c_{14}, \ldots, c_{26}$ are constant coefficients, which relate to the values specified for the parameters of orientation of the first photograph, and the approximate values for the model coordinates of the point; $c_{11}, \ldots, c_{26}$ are constant coefficients which relate to approximate values assigned to the parameters of the second photograph and the model coordinates of the point; and C', D', C'', D'' are constants evaluated by using assigned or approximate values for parameters and model coordinates.

The four equations in (5.56) contain eight unknowns, and for each additional point observed there will be four more equations and three more unknowns. Therefore the number of equations will equal the number of unknowns when five points have been observed, giving a total of 20 equations in 20 unknowns. Five points distributed evenly across the overlap area produce fairly reliable values for the parameters, but more satisfactory answers are obtained by using six or more points and a least-squares solution to the problem. It is worth noting that the redundancy increases rather slowly with the increase in the number of points observed: six points produces 24 equations with 23 unknowns (i.e. one redundancy), and 12 points produces 48 equations with 41 unknowns (seven redundancies). In other words, although each extra point above five adds four observations to the data it adds only one redundant observation to the solution. If matrix partitioning is used in the solution, the number of unknowns may be reduced to only the five parameters relating to the second photograph—i.e. the elements of relative orientation. Another method that achieves the same economy in computational effort is a relative orientation solution based on the coplanarity condition. The derivation given below is based on a detailed description by Schut in [35].

The *coplanarity condition* states that the conjugate rays from two exposure stations, to the same point in the object space, must lie in the same plane—the epipolar plane. Thus in Fig. 5.21 points $1, 2, a', a''$ must all lie in the same plane. If x', y' and x'', y'' are the measured image coordinates for point A then the coordinates of the point in the second photograph, and in a system parallel to the model system, are given by

$$\begin{bmatrix} x''' \\ y''' \\ z''' \end{bmatrix} = \boldsymbol{R} \cdot \begin{bmatrix} x'' \\ y'' \\ z'' \end{bmatrix} \tag{5.57}$$

where the nine elements of the orthogonal matrix $\boldsymbol{R}$ are functions of the three independent parameters—in this case (ϕ, w, k).

When the photographs are in the correct relative orientation the coplanarity condition may be expressed in the form of a determinant as follows:

$$\begin{vmatrix} X_1 & Y_1 & Z_1 & 1 \\ X_2 & Y_2 & Z_2 & 1 \\ (X_1 + x') & (Y_1 + y') & (Z_1 + z') & 1 \\ (X_2 + x''') & (Y_2 + y''') & (Z_2 + z''') & 1 \end{vmatrix} = 0 \tag{5.58}$$

If the second row is subtracted from the fourth, and the first subtracted from the second and third, the equation becomes

$$\begin{vmatrix} X_1 & Y_1 & Z_1 & 1 \\ \mathrm{d}X_2 & \mathrm{d}Y_2 & \mathrm{d}Z_2 & 0 \\ x' & y' & z' & 0 \\ x''' & y''' & z''' & 0 \end{vmatrix} = 0 \tag{5.59}$$

where $\mathrm{d}X_2 = X_2 - X_1, \quad \mathrm{d}Y_2 = Y_2 - Y_1, \quad \mathrm{d}Z_2 = Z_2 - Z_1.$

To simplify the notation let $\Delta X = \mathrm{d}X_2, \Delta Y = \mathrm{d}Y_2, \Delta Z = \mathrm{d}Z_2$. Therefore the determinant may be expressed as

$$\begin{vmatrix} \Delta X & \Delta Y & \Delta Z \\ x' & y' & z' \\ x''' & y''' & z''' \end{vmatrix} = 0 \tag{5.60}$$

or, in the expanded form,

$$\Delta X(y'z''' - y'''z') - \Delta Y(x'z''' - x'''z') + \Delta Z(x'y''' - x'''y') = 0 \tag{5.61}$$

$\Delta X, \Delta Y, \Delta Z$ are the base components and, as in the case of collinearity equations, one of these (usually ΔX) is assigned a value which determines the model scale.

Each object point (or pair of corresponding image points) produces an equation of the type (5.60), and it is worth noting that object system coordinates do not appear in the equation. This can be a distinct advantage since it becomes unnecessary to derive initial approximate values for the model coordinates.

With ΔX assigned a specific value the elements of relative orientation in (5.60) are

$\Delta Y, \Delta Z$ in row 1, and

ϕ, w, k implicit in row 3, via eqn (5.57).

Eqns (5.57) (and therefore (5.60)) are not linear with respect to the unknown parameters ϕ, w, k. Therefore linearization of eqn (5.60) is necessary before a solution of unknowns may take place.

The linearized form of eqn (5.60) is

$$\begin{vmatrix} \Delta X & \Delta Y_o & \Delta Z_o \\ x' & y' & z' \\ x_o''' & y_o''' & z_o''' \end{vmatrix} + \begin{vmatrix} 0 & \mathrm{d}Y & \mathrm{d}Z \\ x' & y' & z' \\ x_o''' & y_o''' & z_o''' \end{vmatrix} + \begin{vmatrix} \Delta X & \Delta Y_o & \Delta Z_o \\ x' & y' & z' \\ \mathrm{d}x''' & \mathrm{d}y''' & \mathrm{d}z''' \end{vmatrix} = 0 \quad (5.62)$$

where the subscript o indicates that the quantity is evaluated using approximate values for the unknown parameters. The unknowns $\mathrm{d}x''$, $\mathrm{d}y''$, $\mathrm{d}z''$ must be expressed as functions of corrections to the elements ϕ, w, k, and these functions are found by linearizing eqns (5.57). The details are not given here, but typically, if ϕ, w, k are primary, secondary, and tertiary rotations respectively, then

$$\left.\begin{aligned} \mathrm{d}x''' &= -y_o'''\,\mathrm{d}k + z_o'''\,\mathrm{d}\phi \\ \mathrm{d}y''' &= -z_o'''\,\mathrm{d}w + x_o'''\,\mathrm{d}k \\ \mathrm{d}z''' &= -x_o'''\,\mathrm{d}\phi + y_o'''\,\mathrm{d}w \end{aligned}\right\} \quad (5.63)$$

If the first two determinants in eqn (5.62) are added together and developed in terms of the elements of the first row, which contain the unknowns, the result is

$$\begin{vmatrix} \Delta X & \Delta Y_{o+1} & \Delta Z_{o+1} \\ x' & y' & z' \\ x_o''' & y_o''' & z_o''' \end{vmatrix}$$

$$= \Delta X \cdot \begin{vmatrix} y' & z' \\ y_o''' & z_o''' \end{vmatrix} - \Delta Y_{o+1} \cdot \begin{vmatrix} x' & z' \\ x_o''' & z_o''' \end{vmatrix} + \Delta Z_{o+1} \cdot \begin{vmatrix} x' & y' \\ x_o''' & z_o''' \end{vmatrix} \quad (5.64)$$

which is of the same form as (5.60), except that ΔY_{o+1} and ΔZ_{o+1} represent *improved* values for the base components.

The third determinant in eqn (5.62) is developed by substituting the functions of $\mathrm{d}x'''$, $\mathrm{d}y'''$, $\mathrm{d}z'''$ contained in (5.63), and writing it as the sum of three determinants with unknown coefficients $\mathrm{d}\phi, \mathrm{d}w, \mathrm{d}k$.

Thus

$$\begin{vmatrix} \Delta X & \Delta Y_o & \Delta Z_o \\ x' & y' & z' \\ (-y_o'''\,\mathrm{d}k + z_o'''\,\mathrm{d}\phi) & (-z_o'''\,\mathrm{d}w + x_o'''\,\mathrm{d}k) & (-x_o'''\,\mathrm{d}\phi + y_o'''\,\mathrm{d}w) \end{vmatrix}$$

$$= \mathrm{d}\phi \begin{vmatrix} \Delta X & \Delta Y_o & \Delta Z_o \\ x' & y' & z' \\ z_o''' & 0 & -x_o''' \end{vmatrix} + \mathrm{d}w \begin{vmatrix} \Delta X & \Delta Y_o & \Delta Z_o \\ x' & y' & z' \\ 0 & -z_o''' & y_o''' \end{vmatrix}$$

$$+ \mathrm{d}k \begin{vmatrix} \Delta X & \Delta Y_o & \Delta Z_o \\ x' & y' & z' \\ -y_o''' & x_o''' & 0 \end{vmatrix} \tag{5.65}$$

Combining eqns (5.64) and (5.65) the linearized coplanarity condition is obtained in its final form:

$$\Delta X \begin{vmatrix} y' & z' \\ y_o''' & z_o''' \end{vmatrix} - \Delta Y_{o+1} \begin{vmatrix} x' & z' \\ x_o''' & z_o''' \end{vmatrix} + \Delta Z_{o+1} \begin{vmatrix} x' & y' \\ x_o''' & z_o''' \end{vmatrix}$$

$$+ \mathrm{d}\phi \begin{vmatrix} \Delta X & \Delta Y_o & \Delta Z_o \\ x' & y' & z' \\ z_o''' & 0 & -x_o''' \end{vmatrix} + \mathrm{d}w \begin{vmatrix} \Delta X & \Delta Y_o & \Delta Z_o \\ x' & y' & z' \\ 0 & -z_o''' & y_o''' \end{vmatrix}$$

$$+ \mathrm{d}k \begin{vmatrix} \Delta X & \Delta Y_o & \Delta Z_o \\ x' & y' & z \\ -y_o''' & x_o''' & 0 \end{vmatrix} = 0 \tag{5.66}$$

where ΔX is an assumed value; $\Delta Y_{o+1}, \Delta Z_{o+1}, \mathrm{d}\phi, \mathrm{d}w, \mathrm{d}k$ are the unknowns; and the determinants represent constants which are evaluated using approximate values for the unknowns; these are either assigned values or values determined from the previous iteration. Note that no assumptions have been made regarding the positions of the points in the model—i.e. incomplete models and mountainous terrain are also catered for by eqn (5.66). The equation may be written in a simpler form as follows:

$$c_1\Delta Y + c_2\Delta Z + c_3\mathrm{d}\phi + c_4\mathrm{d}w + c_5\mathrm{d}k + c_6 = 0 \tag{5.67}$$

where the c's are known constants; in particular c_6 is the y-parallax, which is found from the expression

$$\Delta X \begin{vmatrix} y' & z' \\ y_o''' & z_o''' \end{vmatrix}$$

Five of these equations for five suitably located points will provide a unique solution for the five unknowns. If more equations are used, then a least-squares procedure will give the best values for the unknowns, and the equations will not be satisfied exactly. The equations then have the form

$$v = c_1 \Delta Y + c_2 \Delta Z + c_3 \mathrm{d}\phi + c_4 \mathrm{d}w + c_5 \mathrm{d}k + c_6 \tag{5.68}$$

and Σvv is made a minimum by the least-squares procedure.

The values $\mathrm{d}\phi, \mathrm{d}w, \mathrm{d}k$ derived above may be used to form an orthogonal rotation matrix $\boldsymbol{R}'$, and new coordinates for the second photograph are computed using eqn (5.57). Thus

$$\begin{bmatrix} x^{iv} \\ y^{iv} \\ z^{iv} \end{bmatrix} = \boldsymbol{R}' \cdot \begin{bmatrix} x''' \\ y''' \\ z''' \end{bmatrix} \tag{5.69}$$

These new coordinates are used as a starting point for the next iteration. The cycles are repeated until the required accuracy for relative orientation is achieved—i.e. so that changes in $\Delta Y, \Delta Z, \phi, w, k$ are negligible.

The methods described above are two of many which are possible for the solution of relative orientation. For example, eqn (5.58) may be modified to express the condition that the minimum distance between rays must equal 0, or other combinations of five unknowns (out of ten) may be chosen to solve the problem. Note that the solution by coplanarity condition does not involve ground or model coordinates. Therefore for problems such as aerial triangulation and absolute orientation, which involve fitting to ground control as well, the coplanarity condition alone is not enough.

Examples illustrating relative orientation by the coplanarity condition are given in Chapters 7 and 10.

5.11 Absolute orientation and model connection

Once relative orientation for a pair of overlapping photographs has been established, it should be possible to determine the three-dimensional coordinates for all points appearing in the overlap area. If relative orientation was achieved in a stereoplotting instrument the three coordinates would be measured, whereas if analytical relative orientation was used, based on the measurement of image coordinates, then the three-dimensional coordinates would be calculated. With either method the result is a set of independent *model* coordinates, which has an arbitrary origin and orientation. In practice the origin is often the first projection centre, and the orientation linked with that of the first bundle of rays. The problem that arises frequently in photogrammetry is to carry out a three-dimensional conformal (similarity) transformation to convert the coordinates into the ground system of coordinates, or those of the adjacent model, or indeed some other three-dimensional system. The transformation is the familiar seven-parameter transformation which was developed in some detail in section 5.6, and then used in subsequent sections. However, before transformation is possible the seven parameters (three rotations, three translations, and a scale change) must be determined, and common points with known coordinates in both systems are used to accomplish this.

Figure 5.22 illustrates the problem of *absolute orientation.* Control points *A*, *B*, *C* have known coordinates in both the ground system (*X*, *Y*, *Z*) and the model system (*x*, *y*, *z*).

Model coordinates are regarded as the observed quantities, and they may be expressed as functions of ground coordinates using eqn (5.25).

Therefore

$$\left.\begin{aligned} \begin{bmatrix} x \\ y \\ z \end{bmatrix} &= \lambda\cdot \boldsymbol{R}\cdot \begin{bmatrix} X \\ Y \\ Z \end{bmatrix} + \begin{bmatrix} x_T \\ y_T \\ z_T \end{bmatrix} \\ \text{or} \quad \boldsymbol{x} &= \lambda\cdot \boldsymbol{R}\cdot \boldsymbol{X} + \boldsymbol{x}_T \end{aligned}\right\} \tag{5.70}$$

where *x*, *y*, *z* is the model system of coordinates; *X*, *Y*, *Z* is the ground system of coordinates; ***R*** is the orthogonal rotation matrix formed from the rotations Ω, Φ, *K*, which are primary, secondary, and tertiary respectively; λ is the scale factor; and x_T, y_T, z_T are the translations.

The seven unknown parameters may be determined from the control points *A*, *B*, *C*. If all three coordinates for each point are known in both systems there will be nine equations to solve for the seven unknowns. On the other hand, if *A* and *B* have known three-dimensional coordinates and *C* has known *z* and *Z*

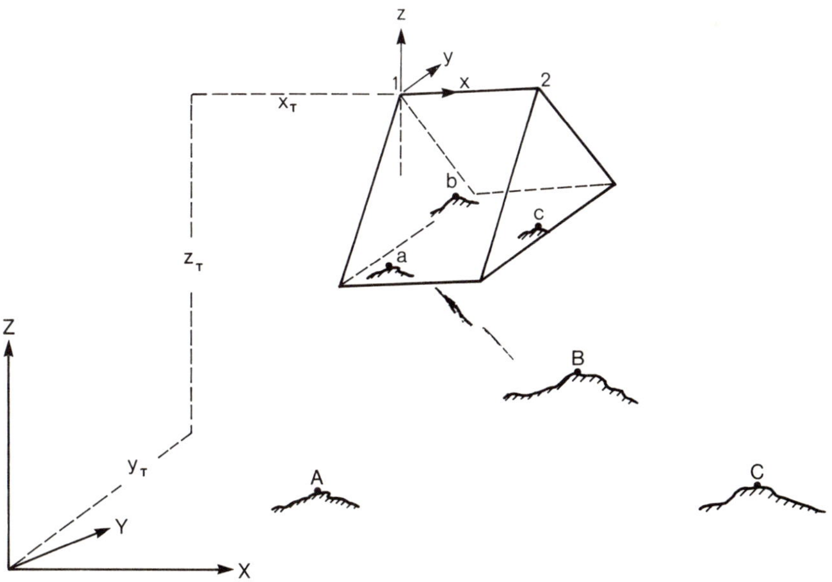

Figure 5.22 Absolute orientation.

values, then seven equations may be formed, the last being established from the third equation in (5.70). With seven equations in seven unknowns a unique solution is obtained. Thus it is not essential that every point has known three-dimensional coordinates in both systems, and in fact five separate control points could be used—two with known XY-coordinates and three with known Z coordinates. In practice more equations than unknowns is usual, and so a least-squares determination of parameters is preferred. Regardless of the number of equations, however, they are non-linear in the form presented by eqn (5.70), and linearization is necessary before a solution of simultaneous equations may be carried out.

Linearization is achieved using the principles outlined in section 5.7 and Appendix A, with the difference that in this case all three equations are retained.

Therefore, if the three functions are

$$\left.\begin{aligned} F &= x - \lambda(r_{11}X + r_{12}Y + r_{13}Z) - x_T = 0 \\ G &= y - \lambda(r_{21}X + r_{22}Y + r_{23}Z) - y_T = 0 \\ H &= z - \lambda(r_{31}X + r_{32}Y + r_{33}Z) - z_T = 0 \end{aligned}\right\} \tag{5.71}$$

the linearized form of these equations would be

$$\left.\begin{aligned} F_o &+ \left(\frac{\partial F}{\partial x}\right)_o \cdot dx + \left(\frac{\partial F}{\partial \lambda}\right)_o \cdot d\lambda + \left(\frac{\partial F}{\partial \Omega}\right)_o \cdot d\Omega + \left(\frac{\partial F}{\partial \Phi}\right)_o \cdot d\Phi + \left(\frac{\partial F}{\partial K}\right)_o \cdot dK \\ &+ \left(\frac{\partial F}{\partial X}\right)_o \cdot dX + \left(\frac{\partial F}{\partial Y}\right)_o \cdot dY + \left(\frac{\partial F}{\partial Z}\right)_o \cdot dZ + \left(\frac{\partial F}{\partial x_T}\right)_o \cdot dx_T = 0 \end{aligned}\right\} \tag{5.72a}$$

and similarly for the functions G and H,

$$\left.\begin{aligned} G_o + \left(\frac{\partial G}{\partial y}\right)_o \cdot dy + \cdots + \left(\frac{\partial G}{\partial y_T}\right)_o \cdot dy_T = 0 \\ H_o + \left(\frac{\partial H}{\partial z}\right)_o \cdot dz + \cdots + \left(\frac{\partial H}{\partial z_T}\right)_o \cdot dz_T = 0 \end{aligned}\right\} \tag{5.27b}$$

where $(\partial F/\partial x)_o$, $(\partial F/\partial \lambda)_o$, etc. are partial derivatives; the subscripts o implying that the expression is evaluated using approximate values for the unknown parameters; dx, $d\lambda$, etc. are unknown corrections to approximate values for the parameters; F_o, G_o, H_o are values for the functions computed using approximate values for the parameters.

In absolute orientation the XYZ-coordinates are constant, therefore dX, dY, dZ become 0, and terms involving these disappear from the equations. If the minimum amount of control is provided then dx, dy, dz will also be 0, and a unique solution for the remaining unknowns ($d\lambda, d\Omega$, $d\Phi$, dK, dx_T, dy_T, dz_T) may be found. If more than the minimum is provided a least-squares determination is used, and the residuals dx, dy, dz at x, y, z are minimized. These residuals may be expressed as functions of the other terms as follows

$$\left.\begin{aligned}
\mathrm{d}x &= \left(F_{\mathrm{o}} + \left(\frac{\partial F}{\partial \lambda}\right)_{\mathrm{o}} \cdot \mathrm{d}\lambda + \left(\frac{\partial F}{\partial \Omega}\right)_{\mathrm{o}} \cdot \mathrm{d}\Omega + \left(\frac{\partial F}{\partial \Phi}\right)_{\mathrm{o}} \cdot \mathrm{d}\Phi \right.\\
&\qquad \left. + \left(\frac{\partial F}{\partial K}\right)_{\mathrm{o}} \cdot \mathrm{d}K + \left(\frac{\partial F}{\partial x_{\mathrm{T}}}\right)_{\mathrm{o}} \cdot \mathrm{d}x_{\mathrm{T}}\right) = 0 \\
\mathrm{d}y &= -\left(G_{\mathrm{o}} + \left(\frac{\partial G}{\partial \lambda}\right)_{\mathrm{o}} \cdot \mathrm{d}\lambda + \cdots + \left(\frac{\partial G}{\partial y_{\mathrm{T}}}\right)_{\mathrm{o}} \cdot \mathrm{d}y_{\mathrm{T}}\right) = 0 \\
\mathrm{d}z &= -\left(H_{\mathrm{o}} + \left(\frac{\partial H}{\partial \lambda}\right)_{\mathrm{o}} \cdot \mathrm{d}\lambda + \cdots + \left(\frac{\partial H}{\partial z_{\mathrm{T}}}\right)_{\mathrm{o}} \cdot \mathrm{d}z_{\mathrm{T}}\right) = 0
\end{aligned}\right\} \quad (5.73)$$

where $(\partial F/\partial x)_{\mathrm{o}}$, $(\partial G/\partial y)_{\mathrm{o}}$, $(\partial H/\partial z)_{\mathrm{o}}$ in eqns (5.72) are all equal to unity.

The residuals are usually denoted v_x, v_y, v_z, and simplifying the notation for the other terms, the equations may be written

$$\left.\begin{aligned}
v_x &= c_{11}\,\mathrm{d}\lambda + c_{12}\,\mathrm{d}\Omega + c_{13}\,\mathrm{d}\Phi + c_{14}\,\mathrm{d}K + c_{15}\,\mathrm{d}x_{\mathrm{T}} + C \\
v_y &= c_{21}\,\mathrm{d}\lambda + c_{22}\,\mathrm{d}\Omega + c_{23}\,\mathrm{d}\Phi + c_{24}\,\mathrm{d}K + c_{25}\,\mathrm{d}y_{\mathrm{T}} + D \\
v_z &= c_{31}\,\mathrm{d}\lambda + c_{32}\,\mathrm{d}\Omega + c_{33}\,\mathrm{d}\Phi + c_{34}\,\mathrm{d}K + c_{35}\,\mathrm{d}z_{\mathrm{T}} + E
\end{aligned}\right\} \quad (5.74)$$

where v_x, v_y, v_z are unknown residuals (corrections to observations), which will be 0 if the minimum number of observations are used; $\mathrm{d}\lambda, \ldots, \mathrm{d}z_{\mathrm{T}}$ are unknown corrections to approximate values for the seven parameters; $c_{11}, \ldots, c_{35}$ are known coefficients, evaluated using approximate values for the seven parameters: details are given in Appendix B; and C, D, E are the values of the functions in (5.71), which are obtained when approximate values for the parameters are used.

When the unknowns $(\mathrm{d}\lambda, \ldots, \mathrm{d}z_{\mathrm{T}})$ have been solved they are added to the initial approximations to provide a new set of approximate values. The coefficients $(c_{11}, \ldots, c_{35})$ and constants (C, D, E) are re-evaluated and the cycle is repeated. The number of iteration cycles depends on the accuracy of the initial values for the parameters, but it is not difficult to determine these such that two or three iterations suffice. The assignment of initial values is made that much simpler if the origin of the model system is chosen to be the first projection centre, and the z-axis is made to coincide with the optical axis of the camera. In such a case the initial values for Ω and Φ would be 0; K and λ could be calculated using xy- and XY-coordinates for two control points; x_{T} and y_{T} could be established by measuring distances on the photograph from the origin to a control point; and z_{T} could be established from a value of flying height.

It is not essential that all seven parameters are determined simultaneously, and some may be conveniently separated from the others without affecting the result. The unknown translations may be removed temporarily by reducing the coordinates in both systems to the common centroid. Further-

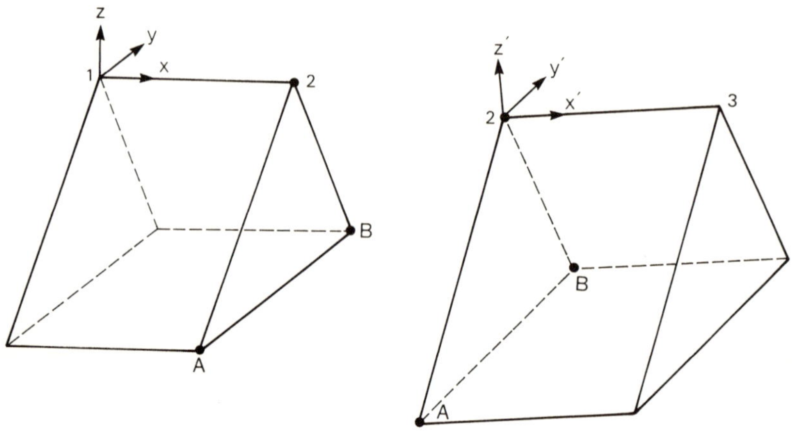

Figure 5.23 Model connection.

more, a value for λ may be determined independently, thus leaving only three unknowns $(d\Omega, d\Phi, dK)$ to be solved simultaneously. This reduces the computer storage requirements considerably, and allows the implementation of a least-squares solution on computers with a much smaller capacity. Further details are given in sections 7.4 and 10.3. The same result is achieved by a process of partitioning of matrices.

Equations (5.70) express the transformation from the ground system to the model system, whereas in practice it is more usual that the reverse transformation is required. Since $\boldsymbol{R}$ is an orthogonal matrix, the reverse transformation may be found from

$$\begin{bmatrix} X \\ Y \\ Z \end{bmatrix} = \frac{1}{\lambda} \cdot \boldsymbol{R}^{\mathrm{T}} \cdot \begin{bmatrix} x - x_{\mathrm{T}} \\ y - y_{\mathrm{T}} \\ z - z_{\mathrm{T}} \end{bmatrix} \tag{5.75}$$

Equally satisfactory results may be obtained however, by using the relationship

$$\begin{bmatrix} X \\ Y \\ Z \end{bmatrix} = \lambda \cdot \boldsymbol{R} \cdot \begin{bmatrix} x \\ y \\ z \end{bmatrix} + \begin{bmatrix} X_{\mathrm{T}} \\ Y_{\mathrm{T}} \\ Z_{\mathrm{T}} \end{bmatrix} \tag{5.76}$$

as a basis for the solution, and regarding the XYZ-coordinates as observations; the objective after all is a least-squares *fit* between the two systems. Examples on absolute orientation are given in Chapters 7 and 10.

Model connection may be regarded as a slight variation of the problem of absolute orientation. A model with its own independent coordinate system must be transformed to fit an adjacent model in the same strip, and one of the vital points used in the process is the common projection centre. The other points are located in the common overlap between the models, and preferably as far apart as possible, so that the three groups of common points define a large triangle. Figure 5.23 illustrates two independent models formed by three bundles with projection centres 1, 2 and 3. The model 2–3 is to be transformed into the coordinate system specified by model 1–2.

Points A and B are the common pass points, and the three points 2, A, B serve to establish equations for the determination of the transformation parameters. The computational procedure is the same as for absolute orientation, though the rotation parameters have a different meaning. Also, it is easier to specify initial values for the parameters, as all the rotation parameters will be made 0, and λ made equal to unity. Computer storage may be reduced, as in absolute orientation, by reducing the coordinates to a common centroid or, as in some algorithms, to the common projection centre. A reduction may also be achieved by calculating λ independently.

Examples on model connection are given in Chapters 7 and 10.

6 Image coordinate refinement

6.1 Introduction

The two-dimensional photographic image represents a three-dimensional scene in an analogue way. The ideal image is one which is formed when the geometry of a central projection is followed exactly, and when the object is defined by a three-dimensional cartesian system in a homogeneous medium. The ideal numerical model of such an image is obtained if the measuring instrument is error-free, and if perfect observations are made. These ideal conditions are never realized, and the result is the presence of systematic errors in the observations. Mathematical models, which may be used to compensate for these systematic errors, are discussed in this chapter.

It is the relative positions of points on the photograph that determine their three-dimensional location in space, and the image positions are best represented by rectangular xy-coordinates, established from measurements of either angle or distance in the plane of the photograph. Convenient rectangular systems have been defined in section 5.3. The coordinates shown in Figure 5.2 are not measured directly, but are derived from measurements related to an arbitrary origin, often well outside the format area of the photograph. Figure 6.1 shows a photograph with the image system of coordinates (x', y'). The origin is the fiducial centre, and the axes are aligned at an angle θ to the comparator (measuring) system (x, y) with origin elsewhere.

The image coordinates for a point may be derived using the two-dimensional similarity transformation, given in section 2.11. Thus the observed coordinates x, y of a point may be transformed into image coordinates x', y', with origin at the fiducial centre, using an equation such as (2.81): thus

$$\begin{pmatrix} x' \\ y' \end{pmatrix} = \begin{pmatrix} \cos\theta & -\sin\theta \\ \sin\theta & \cos\theta \end{pmatrix} \begin{pmatrix} x - x_p \\ y - y_p \end{pmatrix} \tag{6.1}$$

where x_p, y_p are the coordinates of the fiducial centre, derived from observations to fiducial marks 1, 2, 3 and 4; and θ is the angle of rotation, also derived from observations to the fiducial marks.

Equation (6.1) expresses the most fundamental transformation that must be applied to the observed coordinates, and the corrections involved are likely to be the largest discussed in this chapter. Small additional translations may be necessary if the fiducial centre does not coincide with the principal

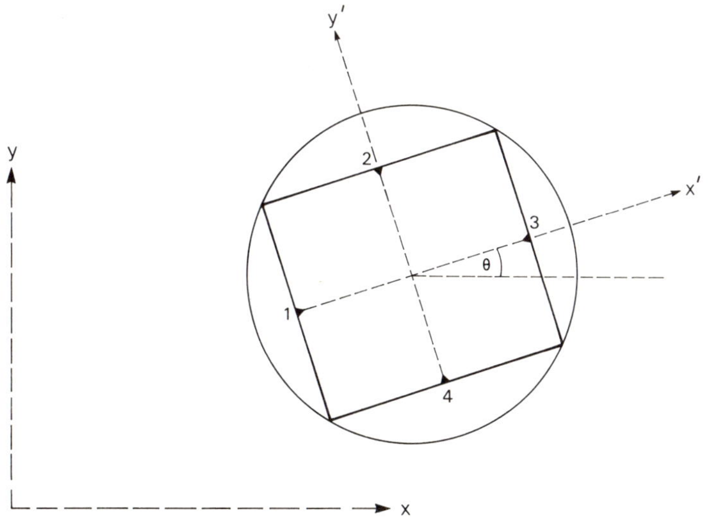

Figure 6.1 Comparator and image coordinate systems.

point, and the camera calibration process establishes these translations. Other more refined corrections are described in the following sections; the corrections are related to either the measuring instrument, the image base, the camera lens, the atmosphere, or the object coordinate system.

6.2 The measuring instrument

The measuring instrument is normally a comparator or an analytical plotter. The comparator may be either a monocomparator or a stereocomparator. Furthermore, the observations may be either xy rectangular coordinates, or angular values which are transformed to xy-coordinates; or distances which are used to derive xy-coordinates by a trilateration intersection calculation. The most common are those which measure xy-coordinates directly. The calibration of comparators is carried out using precision grid plates. By comparing the known positions of grid intersections with measured coordinates of these points it is possible to establish values for parameters which model the systematic errors in measurement. The corrections are usually small, and are often not considered worthwhile except for the most demanding applications. However, with the appearance of inexpensive comparators, some of the more common deformations, such as non-orthogonality of axes and scale-errors, have become significant. Corrections for these errors may be accomplished using affine transformation equations such as (2.91). Therefore, the corrected coordinates x', y' may be derived from

$$\left.\begin{aligned} x' &= ax + by + c \\ y' &= dx + ey + f \end{aligned}\right\} \qquad (6.2)$$

where x, y are observed coordinates; and $a,\ldots,f$ are the six transformation parameters which account for the deformations mentioned above, and which may be derived from grid plate measurements.

6.3 The image base

The emulsion and the base material of either the film negative or the positive may be subject to changes in geometry over a period of time. These distortions may be caused by changes in the environment, or by a simple ageing process. The changes may be of either a random or systematic nature, and observations to fiducial marks, or preferably a denser grid of reseau marks, will indicate the type of distortion present. Observed distances between marks are compared with known distances, obtained from camera calibration procedures, and the relevant correction parameters are derived.

Most cameras are constructed so that at least four fiducial marks are visible on a photograph. Therefore sufficient points are available to determine any linear changes in dimensions, whether conformal, affine, or projective in nature. Non-linear distortions may only be modelled if more than four calibrated marks are observed, as with a full reseau grid. If at least three fiducial marks are observed either a conformal or an affine transformation may be used, such as eqns (2.87) and (2.91) respectively. Applied to the current situation the equations are as follows:

$$\left.\begin{aligned} x' &= ax - by + c \\ y' &= bx + ay + d \end{aligned}\right\} \qquad (6.3)$$

$$\left.\begin{aligned} x' &= ax + by + c \\ y' &= dx + ey + f \end{aligned}\right\} \qquad (6.4)$$

where $a,\ldots,f$ are the parameters derived from observations to fiducials.

In practice at least four fiducials should be observed, thus providing some redundancy and allowing a least-squares determination of the parameters. Observations to four fiducial marks also allow the determination of eight correction parameters, as specified by analytical rectification, and given by eqns (5.45). In the present context the equations would be

$$\left.\begin{aligned} x' &= \frac{ax + by + c}{gx + hy + 1} \\[6pt] y' &= \frac{dx + ey + f}{gx + hy + 1} \end{aligned}\right\} \qquad (6.5)$$

where $a,\ldots,h$ are the eight correction parameters.

However, no redundancy is involved when four fiducials are observed, and a transformation using these equations would be inadvisable unless more fiducial marks (say all eight) were used to determine the parameters. Such a transformation corrects for linear deformations in any direction on the image.

6.4 Lens distortion

A lens system is never perfect, and the bundle of rays striking the object face of a lens assembly will not be reproduced exactly in the image space. A ray will be deflected from its original direction in a manner specified by the design of the lens system. However, any deviation from the specified direction produces a distortion of the image in the image plane, and what is important is to establish the amount of this distortion. Camera calibration establishes the distortions for different parts of the image, and the components of distortion may be classified as symmetric radial distortion, asymmetric radial distortion, and tangential distortion. Mathematical models may be established to cater for all three types; however, the last two are usually very small and ignored, and are not considered further. Radial lens distortion, on the other hand, is often significant, and it may be represented in several ways. A convenient form in the context of this chapter is to express it as a distortion in linear units in the plane of the image, and as a function of distance from the principal point, as shown by Figure 6.2.

The radial distance (r) for a particular point is derived from the xy-coordinates of the point

$$r = (x^2 + y^2)^{1/2} \tag{6.6}$$

and the amount of distortion may be determined from the graph in Figure 6.2, or interpolated from an equivalent table of values. Alternatively, the distortion may be represented as a polynomial, such as

$$\Delta = c_1 r + c_2 r^3 + c_3 r^5 + \cdots \tag{6.7}$$

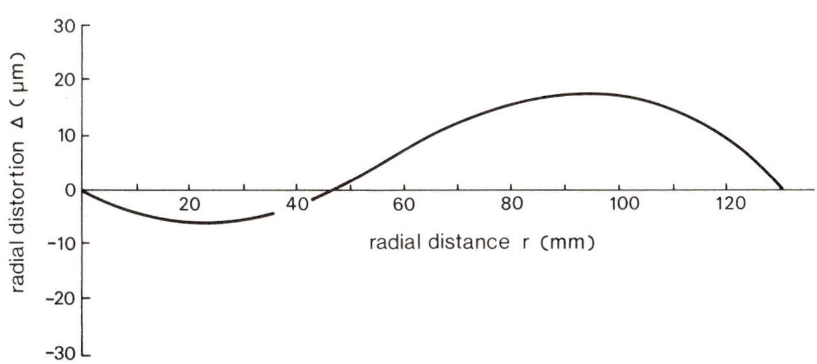

Figure 6.2 Radial lens distortion.

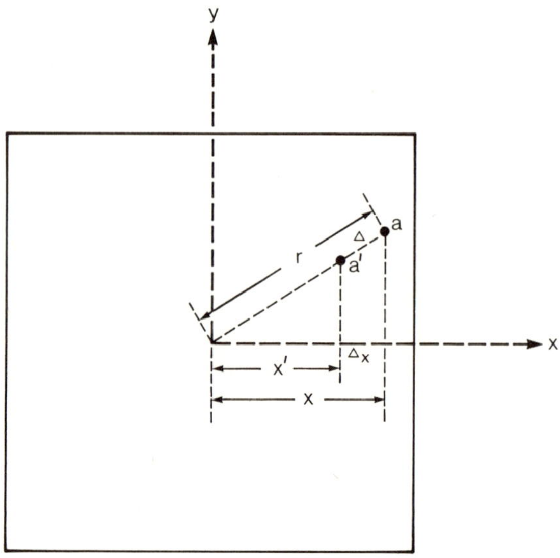

Figure 6.3 Correction for radial lens distortion.

where the coefficients $c_1, c_2, c_3, \ldots$ have been determined by a least-squares fitting routine, and using the camera calibration results. A form such as eqn (6.7) is useful since the coefficients may be stored in the computer, and used as necessary when values for the xy-coordinates of the point are available.

Corrections to the image coordinates are obtained by applying Δ in the opposite sense; and since Δ is the radial distortion the components in the x- and y-directions will be proportional to it, as shown in Figure 6.3, and given by the following equations:

$$x' = x - \Delta_x = x\left(1 - \frac{\Delta_x}{x}\right)$$

and

$$\frac{\Delta_x}{x} = \frac{\Delta}{r}$$

Therefore

$$x' = x\left(1 - \frac{\Delta}{r}\right) \tag{6.8a}$$

and similarly

$$y' = y\left(1 - \frac{\Delta}{r}\right) \tag{6.8b}$$

Therefore, when a value for Δ from eqn (6.7) is substituted, the final equations become

$$\left.\begin{aligned} x' &= x(1 - c_1 - c_2 r^2 - c_3 r^4 - \cdots) \\ y' &= y(1 - c_1 - c_2 r^2 - c_3 r^4 - \cdots) \end{aligned}\right\} \tag{6.9}$$

6.5 Atmospheric refraction

The atmosphere is a medium of decreasing density with increase in altitude. Therefore a ray of light from an object on the earth's surface, passing through the atmosphere to the camera lens in an aircraft, will be refracted away from the vertical. The amount of refraction follows Snell's law for deviation at the interface between two surfaces of different density. As the density changes continuously the ray bends as indicated in Figure 6.4, and after passing through the camera lens it forms an image at a instead of a'. The ray from N is not refracted since it is normal to successive homogeneous layers of the

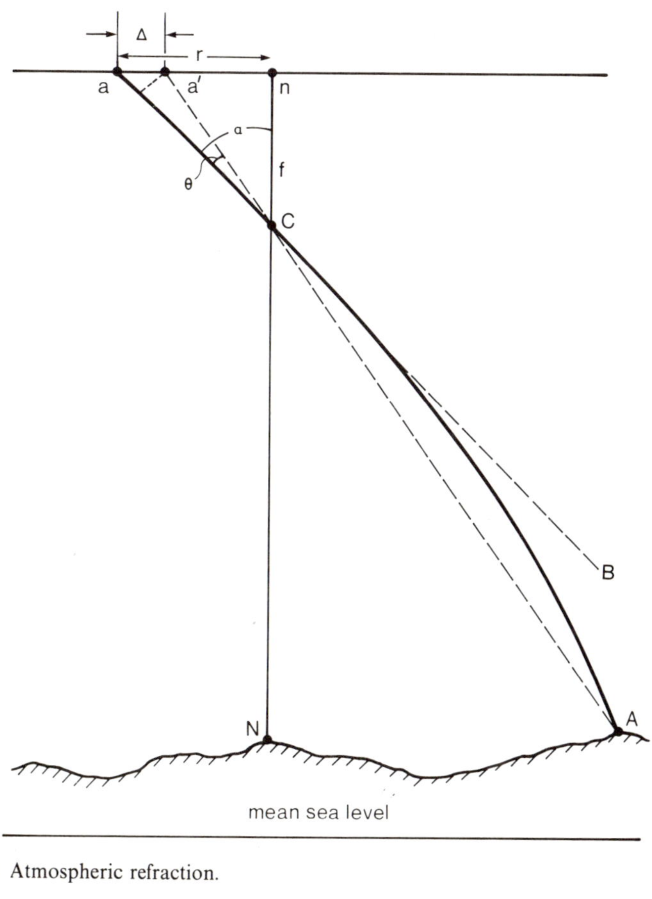

Figure 6.4 Atmospheric refraction.

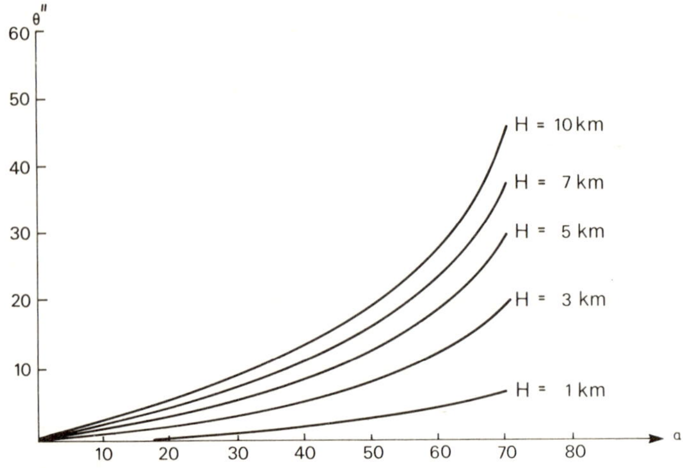

Figure 6.5 Variation of θ with α and flying height.

atmosphere. Let α be the angle made by the ray with the vertical, θ the angle of refraction generated at the lens C, and r the radial distance from the image point to the nadir point—which will be the principal point on vertical photography.

The amount of image displacement Δ, caused by refraction, is then found from

$$\Delta = \frac{aC \cdot \theta}{\cos \alpha} \tag{6.10}$$

where θ is in radians.

Now $\cos \alpha = f/aC$, and $aC = (r^2 + f^2)^{1/2}$, therefore, by substituting in eqn (6.10),

$$\Delta = \left(\frac{r^2 + f^2}{f}\right) \cdot \theta \tag{6.11}$$

Values for θ may be derived for a standard atmosphere, and for a particular flight altitude, average terrain height, and angle α. References [3] and [36] give details of methods used to obtain values of θ, and Figure 6.5 illustrates how θ varies with α and flying height. The quantity Δ is the radial displacement, and once it is established the xy-image coordinates may be corrected using eqns (6.8).

6.6 Earth curvature

If the terrain or object system of coordinates is a three-dimensional orthogonal system, the relationship between the object system and the image system is simplified, and a correction for earth curvature becomes unnecessary. However, terrain coordinate systems are generally not orthogonal; heights are values which are normal to a curved surface, whereas planimetric coordinates are quoted in a plane rectangular system. Therefore, if such a system is used to define the points in the object space, then a correction for earth curvature must be applied to the image coordinates. Figure 6.6 shows the effect of earth curvature on the image of a point A. The point A' represents the map position for point A, and the corresponding image point is a'.

If H is the flying height above average ground level, and r is the radial distance, then Δ the displacement may be calculated from

$$\Delta = \frac{Hr^3}{2Rf^2} \tag{6.12}$$

where R is the radius of the earth, and may be taken to be 6372 km.

Note that the correction for earth curvature is in the opposite sense to atmospheric refraction. The correction Δ is radial from the nadir point, therefore corrections to the xy-coordinates may also be derived using eqns (6.8).

Earth curvature correction is propotional to r^3 and inversely proportional

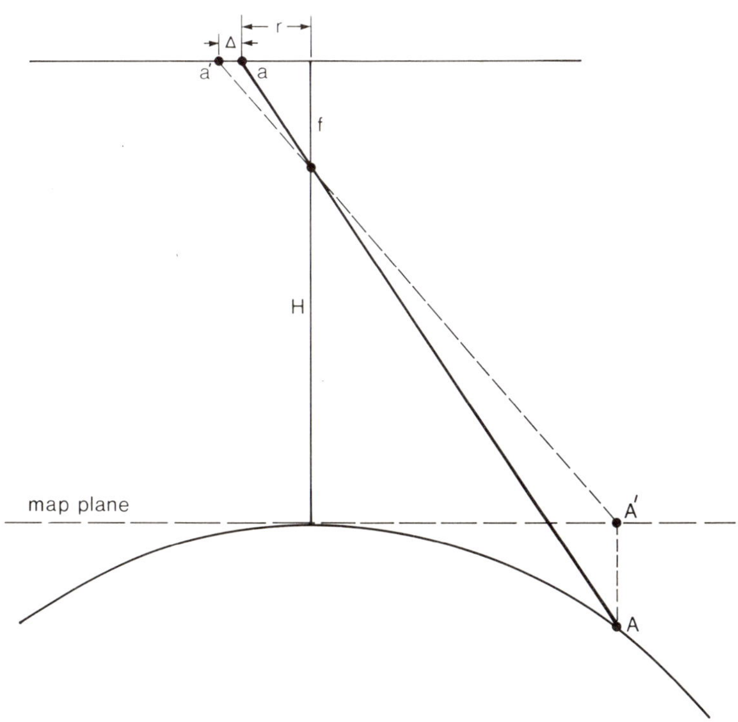

Figure 6.6 Earth curvature correction.

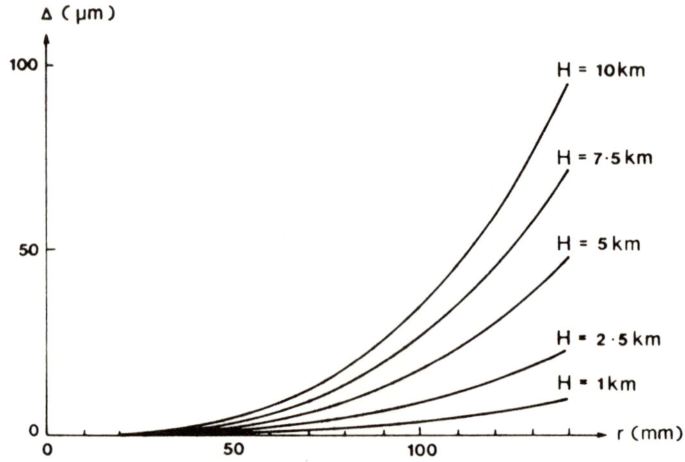

Figure 6.7 Variation in earth curvature correction.

to f^2, therefore it will be large for large radial distances and short focal lengths. It will also increase directly with an increase in flying height, therefore it becomes particularly important when high-altitude photography, taken with wide- and superwide-angle cameras, is used for mapping purposes. Figure 6.7 illustrates how the correction varies with radial distance and flying height, when a camera of focal length 150 mm is used.

7 Computation with redundant data

7.1 Introduction

Analytical photogrammetry is the process of obtaining positional information by computation alone. The procedures involved may be approximate or rigorous, depending on the mathematical models used and the attention given to the relative precision of the observed data. Other distinguishing features are the assumptions and simplifications introduced in order to reduce the problem to one which can be handled by the computational facilities available.

Chapter 5 dealt with the fundamental mathematical models in photogrammetry, and concentrated on developing equations for specific operations when the minimum number of observations are recorded. This chapter takes matters further, and considers solutions to problems when more than the minimum amount of data is available. More specifically, it considers the use of *enough* data to give the solution geometric stability, without necessarily using *all* the data in a manner which gives the *best* answer. For example six points (instead of five) are used for relative orientation in order to improve the coverage throughout the overlap. The same comments apply to absolute orientation, where four control points—one in each corner of the overlap—are preferable to three, in order to avoid the inaccuracies resulting from an extrapolation from control. The matter of making the best use of all the observed data is considered in the section on least-squares applications (section 10.3).

7.2 Relative orientation

The problem is the determination of the five parameters of relative orientation when image coordinates have been measured. Equation (5.66) expresses the general differential equation for relative orientation, based on the coplanarity condition, where the first photograph is fixed and the unknowns are the elements of orientation of the second photograph relative to the first. The notation is associated with Figure 5.21, and the equation is restated here for convenience:

$$\Delta X \begin{vmatrix} y' & z' \\ y_o''' & z_o''' \end{vmatrix} - \Delta Y_{o+1} \begin{vmatrix} x' & z' \\ x_o''' & z_o''' \end{vmatrix} + \Delta Z_{o+1} \begin{vmatrix} x' & y' \\ x_o''' & y_o''' \end{vmatrix} +$$

$$+\mathrm{d}\phi\begin{vmatrix}\Delta X & \Delta Y_o & \Delta Z_o\\ x' & y' & z'\\ z_o''' & 0 & -x_o'''\end{vmatrix}+\mathrm{d}w\begin{vmatrix}\Delta X & \Delta Y_o & \Delta Z_o\\ x' & y' & z'\\ 0 & -z_o''' & y_o'''\end{vmatrix}$$

$$+\mathrm{d}k\begin{vmatrix}\Delta X & \Delta Y_o & \Delta Z_o\\ x' & y' & z'\\ -y_o''' & x_o''' & 0\end{vmatrix}=0 \qquad (7.1)$$

where x', y', z' are the image coordinates for the first photograph; x_o''', y_o''', z_o''' are rotated image coordinates for the second photograph, obtained from the previous iteration; the rotation is based on values for the parameters determined during that iteration, and initially the values are specified; ΔX is an assumed value for the base-component in the X-direction—normally the base length on the photograph; ΔY_o, ΔZ_o are approximate values for the base-components in the Y- and Z-directions; they are assigned values initially (usually 0), but become improved values after each iteration cycle; ΔY_{o+1}, ΔZ_{o+1} are two unknowns—the improved values for the base components in the Y- and Z-directions; $\mathrm{d}\phi$, $\mathrm{d}w, \mathrm{d}k$ are the unknown corrections to the rotation parameters ϕ, w, k for the second photograph; values for ϕ, w, k are specified in the first instance; and the determinants are constants evaluated using approximate values for the unknowns; these are either assigned values or values determined from the previous iteration.

The equation as it stands is general in application, because no assumption has been made regarding the positions of points in the object space—i.e. incomplete models and mountainous terrain are catered for equally well. However, the computation is complex and lengthy, and requires considerable computer storage. Economy in this respect may be achieved by introducing one or two simplifications and assumptions, and this is done without too great a loss in accuracy.

The first simplification is one which makes the numbers involved more manageable, particularly for small computers. Equation (7.1) above is divided throughout by $f^2 \cdot \Delta X$, where f is the focal length of the camera, and βY and βZ are defined as follows:

$$\beta Y=\frac{\Delta Y}{\Delta X}; \qquad \beta Z=\frac{\Delta Z}{\Delta X} \qquad (7.2)$$

If the initial values for the elements of orientation of the second photograph are $\beta Y, \beta Z, \phi, w, k$, and are assumed to be 0, the new unknowns are corrections to these values. Let these be $\delta\beta Y$, $\delta\beta Z$, $\delta\phi$, δw, δk.

The first term, associated with y-parallax, is

$$\frac{1}{f^2\cdot\Delta X}\cdot\Delta X\begin{vmatrix}y' & f\\ y'' & f\end{vmatrix}=\frac{1}{f^2\cdot\Delta X}\cdot\Delta X(y'f-y''f)$$

$$=\frac{y'}{f}-\frac{y''}{f} \tag{7.3}$$

The coefficients are obtained as follows

For $\delta\beta Y$: The coefficient for ΔY is

$$-\frac{1}{f^2\cdot\Delta X}\begin{vmatrix} x' & f \\ x'' & f \end{vmatrix}=\frac{1}{f^2\cdot\Delta X}(x''f-x'f)$$

$$=\frac{x''}{f\cdot\Delta X}-\frac{x'}{f\cdot\Delta X}$$

Therefore the coefficient for $\delta\beta Y$ becomes

$$=\frac{x''}{f}-\frac{x'}{f} \tag{7.4}$$

For $\delta\beta Z$: The coefficient for ΔZ is

$$\frac{1}{f^2\cdot\Delta X}\begin{vmatrix} x' & y' \\ x'' & y'' \end{vmatrix}=\frac{1}{f^2\cdot\Delta X}(x'y''-x''y')$$

$$=\frac{x'y''}{f^2\Delta X}-\frac{x''y'}{f^2\Delta X}$$

Therefore the coefficient for $\delta\beta Z$ becomes

$$=\frac{x'y''}{f^2}-\frac{x''y'}{f^2} \tag{7.5}$$

For $\delta\phi$:

$$\frac{1}{f^2\cdot\Delta X}\begin{vmatrix} \Delta X & 0 & 0 \\ x' & y' & f \\ f & 0 & -x'' \end{vmatrix}=\frac{1}{f^2\cdot\Delta X}(-x''y'\Delta X)$$

$$=-\frac{-x''y'}{f^2} \tag{7.6}$$

For δw:

$$\frac{1}{f^2\cdot\Delta X}\begin{vmatrix} \Delta X & 0 & 0 \\ x' & y' & f \\ 0 & -f & y' \end{vmatrix}=\frac{1}{f^2\cdot\Delta X}(\Delta Xy'y''+\Delta Xf^2)$$

$$=1+\frac{y'y''}{f^2} \tag{7.7}$$

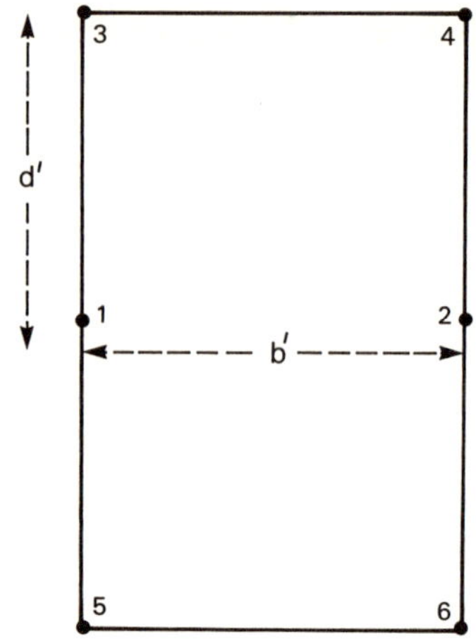

Figure 7.1 Relative orientation points.

For δk:

$$\frac{1}{f^2 \cdot \Delta X}\begin{vmatrix} \Delta X & 0 & 0 \\ x' & y' & f \\ -y'' & x'' & 0 \end{vmatrix} = \frac{1}{f^2 \cdot \Delta X}(-x'' f \Delta X)$$

$$= -\frac{x''}{f} \qquad (7.8)$$

An assumption is made that six points are used for relative orientation, and that they are located close to the ideal positions in the model, as shown in Figure 7.1. Under this assumption two dimensions b and d are calculated,

Table 7.1 Coordinates of the six points used for relative orientation

Points	$\frac{x'}{f}$	$\frac{y'}{f}$	$\frac{x''}{f}$	$\frac{y''}{f}$
1	0	0	$-b$	0
2	b	0	0	0
3	0	d	$-b$	d
4	b	d	0	d
5	0	$-d$	$-b$	$-d$
6	b	$-d$	0	$-d$

Table 7.2 Coefficients of the unknowns

Point	Coefficients of $\delta\beta Y$	$\delta\beta Z$	$\delta\phi$	δw	δk
1	$-b$	0	0	1	b
2	$-b$	0	0	1	0
3	$-b$	bd	bd	$(1+d^2)$	b
4	$-b$	bd	0	$(1+d^2)$	0
5	$-b$	$-bd$	$-bd$	$(1+d^2)$	b
6	$-b$	$-bd$	0	$(1+d^2)$	0

which are associated with b' and d' shown in the diagram:

$$b = \frac{\sum\left(\frac{x'}{f} - \frac{x''}{f}\right)}{6}; \qquad d = \frac{\sum\left|\frac{y}{f}\right|}{8} \tag{7.9}$$

where | | indicates modulus.

The new coordinates for the six points are as indicated in Table 7.1 and the coefficients for the five unknowns are as listed in Table 7.2.

These coefficients may be regarded as the elements of a matrix A, in a set of equations

$$\boldsymbol{v} = \boldsymbol{A} \cdot \boldsymbol{\Delta} - \boldsymbol{l} \tag{7.10}$$

where

$$\boldsymbol{v} = \begin{bmatrix} v_1 \\ v_2 \\ v_3 \\ v_4 \\ v_5 \\ v_6 \end{bmatrix}, \quad \boldsymbol{\Delta} = \begin{bmatrix} \delta\beta Y \\ \delta\beta Z \\ \delta\phi \\ \delta w \\ \delta k \end{bmatrix}, \quad \boldsymbol{l} = \begin{bmatrix} p_1 \\ p_2 \\ p_3 \\ p_4 \\ p_5 \\ p_6 \end{bmatrix}$$

and $\boldsymbol{v}$ is the vector of residuals at the six points; $\boldsymbol{\Delta}$ is the vector of unknowns; **1** is the vector of y-parallaxes at the six points, derived using eqn (7.3) for each point.

A least-squares determination of the parameters is obtained by applying the matrix procedures described in section 9.5, i.e.

$$\boldsymbol{\Delta} = \boldsymbol{N}^{-1}\boldsymbol{F} \quad \text{where} \quad \boldsymbol{N} = \boldsymbol{A}^{\mathrm{T}}\boldsymbol{A} \quad \text{and} \quad \boldsymbol{F} = \boldsymbol{A}^{\mathrm{T}}\boldsymbol{l} \tag{7.11}$$

and, when these procedures are used the solution for the unknowns is found to be

$$\left.\begin{aligned}
\delta w &= \frac{1}{4d^2}(2(p_1 + p_2) - (p_3 + p_4 + p_5 + p_6)) \\
\delta \phi &= \frac{1}{2bd}((p_4 - p_6) - (p_3 - p_5)) \\
\delta k &= \frac{1}{3b}((p_2 + p_4 + p_6) - (p_1 + p_3 + p_5)) \\
\delta \beta Y &= \frac{1}{3b}(\mathrm{d}w(3 + 2d^2) + (p_2 + p_4 + p_6)) \\
\delta \beta Z &= \frac{1}{2bd}(p_6 - p_4)
\end{aligned}\right\} \tag{7.12}$$

These values are added to the previous values for the parameters of relative orientation.

$$\left.\begin{aligned}
w_{i+1} &= w_i + \delta w \\
\phi_{i+1} &= \phi_i + \delta \phi \\
k_{i+1} &= k_i + \delta k \\
\beta Y_{i+1} &= \beta Y_i + \delta \beta Y \\
\beta Z_{i+1} &= \beta Z_i + \delta \beta Z
\end{aligned}\right\} \tag{7.13}$$

and the new values of w, ϕ, k are used to form an orthogonal matrix R. New residual y-parallaxes may then be found directly from,

$$\left(\frac{x''}{f} \quad \frac{y''}{f} \quad 1\right) \cdot \boldsymbol{R}^{\mathrm{T}} \boldsymbol{\beta} \cdot \begin{bmatrix} x'/f \\ y'/f \\ 1 \end{bmatrix} = p \tag{7.14}$$

where

$$\boldsymbol{\beta} = \begin{bmatrix} 0 & -\beta Z & \beta Y \\ \beta Z & 0 & -1 \\ -\beta Y & 1 & 0 \end{bmatrix} \tag{7.15}$$

Equation (7.14) is another way of expressing eqn (5.60), bearing in mind the simplification introduced by eqn (7.2), and the division throughout by $f^2 \cdot \Delta x$.

The new parallaxes are the starting point for the next iteration, and the process is repeated until residual parallaxes are negligible. Once final values for the parameters have been obtained, and the image coordinates for the second photograph transformed, the independent three-dimensional model coordinates may be calculated using eqns (5.51) to (5.55).

A numerical example of relative orientation by this method is given below, and the results at intermediate stages are quoted.

Example 7.1

Rectangular coordinates of six points appearing on two overlapping photographs have been observed in a comparator. The points are located approximately in the ideal positions according to Figure 7.1. After reduction to the respective principal points the plate coordinates (in mm) are:

Plate coordinates:

Point	x'	y'	x''	y''
1	0.35	3.31	−83.61	4.85
2	82.51	0.37	−0.70	2.48
3	3.66	100.50	−87.46	103.67
4	95.55	99.41	10.93	105.26
5	0.40	−100.94	−80.47	−96.46
6	105.31	−96.78	23.20	−93.75

The focal length of the camera is 152.07 mm.
Derive values for the five elements of relative orientation of the second photograph.

Solution

The plate coordinates are divided by f and the quantities $\left(\frac{x'}{f}-\frac{x''}{f}\right)$ and $\left(\frac{y'}{f}-\frac{y''}{f}\right)$ are calculated:

Point	$\frac{x'}{f}$	$\frac{y'}{f}$	$\frac{x''}{f}$	$\frac{y''}{f}$	$\left(\frac{x'}{f}-\frac{x''}{f}\right)$	$p=\left(\frac{x'}{f}-\frac{y''}{f}\right)$
1	0.002302	0.021766	−0.549813	0.031893	0.552114	−0.010127
2	0.542579	0.002433	−0.004603	0.016308	0.547182	−0.013875
3	0.024068	0.660880	−0.575130	0.681726	0.599198	−0.020846
4	0.628329	0.653712	0.071875	0.692181	0.556454	−0.038469
5	0.002630	−0.663773	−0.529164	−0.634313	0.531795	−0.029460
6	0.692510	−0.636417	0.152561	−0.616492	0.539949	−0.019925

Values for b and d are obtained as follows:

$$b=\frac{\sum\left(\frac{x'}{f}-\frac{x''}{f}\right)}{6}; \quad \text{and} \quad d=\frac{\sum|y/f|}{8} \quad \text{for the points 3 to 6.}$$

Thus values for b and d, and the coefficients, are found to be

$$b = 0.554449; \qquad d = 0.654937$$

$$\frac{1}{4d^2} = 0.582829; \quad \frac{1}{2bd} = 1.376922; \quad \frac{1}{3b} = 0.601198; \quad 3 + 2d^2 = 3.857885$$

Equations (7.12) are used to derive values for the five parameters:

$$\delta w = 0.035375; \quad \delta\phi = -0.037395; \quad \delta k = -0.007116; \quad \delta\beta Y = 0.038599;$$
$$\delta\beta Z = 0.025534$$

where δw, $\delta\phi$, δk are in radians and

$\delta\beta Y$, $\delta\beta Z$ are the ratios $\delta\Delta Y/f$ and $\delta\Delta Z/f$ respectively.

These values are added to the previous values (see eqns (7.13)—in this case the previous values are zero- and the values for the rotation elements are used to form the rotation matrix ***R***.

Thus

$$\boldsymbol{R} = \begin{bmatrix} 0.999276 & 0.006450 & -0.037496 \\ -0.007772 & 0.999349 & -0.035218 \\ 0.037244 & 0.035484 & 0.998676 \end{bmatrix}$$

R is used in eqn (7.14) to obtain new residual parallaxes $p\left(=\frac{y'}{f}-\frac{y''}{f}\right)$, and a new iteration cycle commences.

Successive values for the residual parallaxes, given in the form $(y' - y'')$ and in units of millimetres at plate scale are listed below. The associated values for the parameters of relative orientation are also given, with rotation elements in radians and translations in millimetres.

	Iteration Number					
	0	1	2	3	4	5
y-parallax:						
Point 1	−1.54	−0.307	0.075	−0.019	−0.002	−0.005
2	−2.11	−0.226	0.084	−0.011	0.007	0.004
3	−3.17	−0.370	0.067	−0.011	0.005	0.002
4	−5.85	0.181	−0.021	−0.003	−0.002	−0.002
5	−4.48	−0.257	−0.098	−0.015	0.005	0.002
6	−3.03	−0.682	0.179	−0.036	0.004	−0.004
R.O. element:						
w	0	0.03538	03561	0.03558	0.03561	0.03560
ϕ	0	−0.03740	−0.02856	−0.03008	−0.02981	−0.02986
k	0	−0.00712	−0.00630	−0.00629	−0.00632	−0.00631
ΔY	0	5.870	5.516	5.652	5.630	5.633
ΔZ	0	3.883	2.694	2.969	2.923	2.931

Comment

Five iterations are required to reduce the residual parallaxes to a minimum. If the assumptions regarding the ideal distribution of the points had not been made then fewer iterations would have been required, but only at the cost of an increase in the computer storage requirements.

7.3 Model connection

The process of model connection involves applying a three-dimensional similarity transformation to the coordinates of one model, so that it fits the adjacent model as closely as possible. The transformation is accomplished using equations derived in section 5.11, but in order to establish the equations the seven parameters of transformation must be determined in the first place. Therefore, when the two models with independent coordinate systems have points in common, these points may be used to derive values for the parameters. At least three common points defining a plane are required to establish the seven parameters. In the case of model connection one of these points should be the common projection centre.

In Figure 7.2, point 2 is the common projection centre, and points A and B are common pass points; the latter are located symmetrically about the x-axis, and are chosen so that together with the projection centre they form as large a triangle as possible. The point C is defined as being midway between A and B.

Therefore,

$$\left.\begin{aligned} x_C &= \frac{x_A + x_B}{2}; & y_C &= \frac{y_A + y_B}{2}; & z_C &= \frac{z_A + z_B}{2} \\ x'_C &= \frac{x'_A + x'_B}{2}; & y'_C &= \frac{y'_A + y'_B}{2}; & z'_C &= \frac{z'_A + z'_B}{2} \end{aligned}\right\} \qquad (7.16)$$

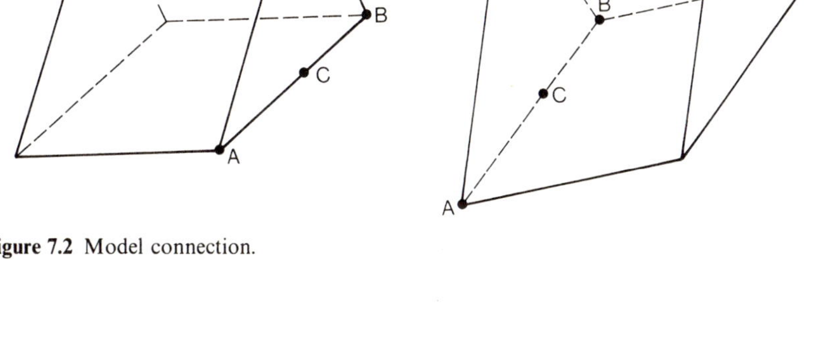

Figure 7.2 Model connection.

The differences in scale and orientation between the models are likely to be small, therefore it is a relatively simple matter to obtain estimates for the parameters. The procedure is as follows:

(i) Reduce the coordinates of the common points in both systems to the common projection centre.

(ii) Calculate a value for the increment $d\lambda$ to the scale-factor λ, and apply it to the $x'\ y'\ z'$ coordinates of A and B obtained in (i), or from the previous iteration. The increment may be determined by least squares (see section 10.3), or in some other way. The least-squares value is found from

$$d\lambda = \frac{[xx' + yy' + zz']}{[x'^2 + y'^2 + z'^2]} \tag{7.17}$$

where x, y are the coordinates obtained in (i); x', y' are the coordinates obtained in (i), or as a result of a previous iteration; and [ ] indicates summation.

(iii) Calculate a new value for scale-factor (λ') from

$$\lambda' = d\lambda \cdot \lambda \tag{7.18}$$

where λ is the value from the previous iteration: initially it would be unity.

(iv) Approximate values for the increments to the rotation parameters may be found from

$$\left.\begin{aligned} dk &= \frac{(x'_B - x'_A) - (x_B - x_A)}{y_B - y_A} \\ d\phi &= -\frac{(x'_C - x_C)}{z_C} \\ dw &= \frac{y'_C - y_C}{z_C} \end{aligned}\right\} \tag{7.19}$$

where the coordinates are those derived in stage (ii).

(v) Add these values to the values obtained in the previous iteration, i.e.

$$\left.\begin{aligned} k' &= k + dk \\ \phi' &= \phi + d\phi \\ w' &= w + dw \end{aligned}\right\} \tag{7.20}$$

Initially the values would be 0.

(vi) Use the latest values of k, ϕ, w to form the rotation matrix $\boldsymbol{R}$ from eqn (5.20) or (5.22); and use the latest values of λ and $\boldsymbol{R}$ to transform the coordinates derived in stage (i).

i.e.

$$\begin{bmatrix} x'' \\ y'' \\ z'' \end{bmatrix} = \lambda \boldsymbol{R} \begin{bmatrix} x' \\ y' \\ z' \end{bmatrix} \tag{7.21}$$

(vii) Repeat the cycle (ii)–(vi) until the differences between the coordinates in both systems are small enough to be ignored. The final rotation matrix and scale-factor are adopted and used to transform all points in the second model: if the coordinates of these points have been reduced to the common projection centre the transformation will be as follows:

$$\begin{bmatrix} x \\ y \\ z \end{bmatrix} = \lambda \boldsymbol{R} \begin{bmatrix} x' & x_2 \\ y' + y_2 \\ z' & z_2 \end{bmatrix} \tag{7.22}$$

where $(x_2 y_2 z_2)^{\mathrm{T}}$ are the coordinates of the common projection centre in the system of the first model.

A numerical example which illustrates this approximate method is given below.

Example 7.2

The three-dimensional coordinates of adjacent models have been observed, and the values for the common projection centre (2) and common pass points (A and B) are given below. Derive suitable values for the seven parameters of transformation.

Observed model coordinates (mm):

Point	x	y	z	x'	y'	z'
2	90.00	2.90	169.95	0	0	170.00
A	87.69	− 81.38	25.54	− 1.32	− 88.91	9.82
B	83.66	75.10	29.62	− 5.41	82.44	17.88

Solution:

Let the initial values of k, ϕ, and w be 0, and let the initial value of λ be unity. After reduction to the common projection centre, and the calculation of coordinates for C, the values become

Point	x	y	z	x'	y'	z'
2	0	0	0	0	0	0
A	− 2.310	− 84.280	− 144.410	− 1.320	− 88.910	− 160.180
B	− 6.340	72.200	− 140.330	− 5.410	82.440	− 152.120
C	− 4.325	− 6.040	− 142.370	− 3.365	− 3.235	− 156.150

The scale-factor increment is derived from points A and B, using eqn (7.17):

$$\mathrm{d}\lambda = \frac{37.3486 + 13445.5028 + 44478.5934}{31.0105 + 14701.3417 + 48798.1268} = \frac{57961.4448}{63530.4790} = 0.912341$$

Hence $\lambda' = \mathrm{d}\lambda \cdot \lambda = 0.912341 \times 1 = 0.912341$

which is applied to the $x'y'z'$-coordinates of A, B and C, to give the following values:

Point	x'	y'	z'
2	0	0	0
A	−1.204	−81.116	−146.139
B	−4.936	75.213	−138.785
C	−3.070	−2.951	−142.462

Equations (7.19) are used to derive values for the rotation increments. Thus

$$dk = \frac{-3.731-(-4.03)}{156.48} = \frac{0.299}{156.48} = 0.001908$$

$$d\phi = \frac{-1.255}{-142.37} = 0.008815$$

$$dw = \frac{3.089}{-142.37} = -0.021694$$

The rotation matrix $\boldsymbol{R}$ is formed:

$$\boldsymbol{R} = \begin{bmatrix} 0.999959 & -0.001908 & 0.008815 \\ 0.001716 & 0.999763 & 0.021691 \\ -0.008854 & -0.021675 & 0.999726 \end{bmatrix}$$

and the reduced coordinates of A and B are transformed using eqn (7.21).

The computation of new increments to the parameters is repeated, and the iterations continue until discrepancies between the models at A and B are minimum. Successive values for the increments and the discrepancies in coordinates are given below.

	Iteration number 0	1	2	3	Parameter	Final Value
Parameter increment:						
$d\lambda$	1	0.912341	1.000261	1.000000	λ	0.912579
dk	0	0.001908	0.000410	0.000002	k	0.002320
$d\phi$	0	0.008815	0.000027	0.000009	ϕ	0.008851
dw	0	−0.021694	0.000055	0.000009	w	−0.021630
Coordinate discrepancies (mm):						
A dx		0.028	−0.001	0.000		
dy		−0.011	0.004	0.002		
dz		−0.080	−0.038	−0.037		

B dx	−0.038	−0.001	0.000
dy	0.023	−0.001	−0.002
dz	0.004	0.036	0.035

Comment

The number of iterations is not large, and the effect of the last iteration is minimal. This is due to the pass points being located fairly close to the ideal positions—thus giving suitable values for C, and allowing the calculation of good approximations for the rotation parameters. With a less favourable location of pass points the number of iterations would be greater.

7.4 Absolute orientation

Absolute orientation is, in principle, the same as model connection—that is, the determination of seven parameters of transformation and the use of these to convert independent model coordinates into the ground system. The distinguishing features however, are that usually two parameters (scale-factor (λ) and azimuth change (K)) will have large values. Therefore, in the first instance, approximate values for these two parameters must be determined and used (eqn 5.76) to transform the model coordinates into a system which is close to the ground system. It is then possible to use the linearized equations in section 5.11 to derive values for all seven parameters.

Independent models normally cover an area which is roughly rectangular in shape. Therefore four control points—one in each corner—is the minimum required in order to avoid errors due to extrapolation. Only three are required to establish the seven parameters, and the fourth serves to improve the accuracy, or expose a gross error if it exists. Many approximate methods are used for calculating the seven parameters, and the sequence of steps required

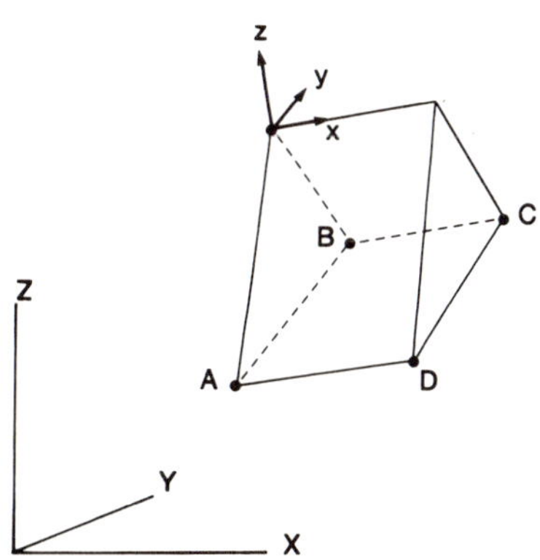

Figure 7.3.

for one of these is given below. The assumption is that four control points (A, B, C, D) exist, located in the vicinity of the four corners, and that the x and y axes indicate the alignment of the model, as shown in Figure 7.3.
The steps are:

(i) Reduce the coordinates of the four points in both systems to their respective centroids.

(ii) Use the coordinates derived in (i), or those obtained from a previous iteration, to calculate the scale-factor increment

$$d\lambda = \left(\frac{\Delta X^2 + \Delta Y^2 + \Delta Z^2}{\Delta x^2 + \Delta y^2 + \Delta z^2}\right)^{1/2} \tag{7.23}$$

where $\Delta x, \Delta y, \Delta z$ are coordinate components for a particular line in the model system, and derived from current values for the model coordinates; and $\Delta X, \Delta Y, \Delta Z$ are coordinate components for the same line in the ground system.

The longest lines in the model should be used for this calculation—i.e. either AC or BD, or both, and the mean value derived.

(iii) Apply the scale-factor increment to the current values of the model coordinates, and then derive increments to the rotation elements as follows.

$$dK = \alpha_g - \alpha_m \tag{7.24}$$

where

$$\alpha_g = \tan^{-1}\left(\frac{\Delta X}{\Delta Y}\right), \quad \text{and} \quad \alpha_m = \tan^{-1}\left(\frac{\Delta x}{\Delta y}\right) \tag{7.25}$$

Also

$$\left.\begin{aligned} d\Phi &= \frac{\Delta Z - \Delta z}{\delta x} \\ d\Omega &= \frac{-(\Delta Z - \Delta z)}{\delta y} \end{aligned}\right\} \tag{7.26}$$

where δx and δy are the original x and y components for an appropriate line, but converted into the scale of the ground system.

(iv) Calculate a new value for scale-factor (λ') from

$$\lambda' = d\lambda \cdot \lambda \tag{7.27}$$

where λ is the value from the previous iteration; initially it would be unity. Calculate new values for the rotations from

$$\left.\begin{aligned} K' &= K + dK \\ \Phi' &= \Phi + d\Phi \\ \Omega' &= \Omega + d\Omega \end{aligned}\right\} \tag{7.28}$$

where, in the first cycle, K, Φ, Ω will be zero.

(v) Use the latest values of k, Φ, Ω to form the rotation matrix $\boldsymbol{R}$ from eqn (5.20) or (5.22), and use the latest values of λ and R to transform the coordinates derived in (i), i.e.

$$\begin{bmatrix} x' \\ y' \\ z' \end{bmatrix} = \lambda R \begin{bmatrix} x \\ y \\ z \end{bmatrix} \qquad (7.29)$$

(vi) Repeat the cycle (ii)–(v) until the differences between the coordinates in both systems are small enough to be ignored. The final rotation matrix and scale-factor are adopted to transform all points in the model. If the coordinates of these points have been reduced to the centroid calculated in (i), the transformation will be

$$\begin{bmatrix} X \\ Y \\ Z \end{bmatrix} = \lambda \boldsymbol{R} \begin{bmatrix} x \\ y \\ z \end{bmatrix} + \begin{bmatrix} X_T \\ Y_T \\ Z_T \end{bmatrix} \qquad (7.30)$$

where $(X_T\ Y_T\ Z_T)^T$ are the ground coordinates of the centroid.
If the coordinates have not been reduced to the centroid, the translations (X_T, Y_T, Z_T) may be found from eqn (7.30), by substituting values for any point which are known in both systems.

A numerical example illustrating the method is given below.

Example 7.3

An independent three-dimensional model has been formed, and four points in the model have known ground coordinates. Derive suitable values for the seven parameters which may be used to transform model coordinates into the ground system.

Model coordinates (mm) and ground coordinates (m) are:

	Model coordinates (mm)			Ground coordinates (m)		
Point	x	y	z	X	Y	Z
A	29.791	−33.564	−145.923	503.30	6077.64	0.67
B	1.690	92.850	−142.345	3942.55	6638.43	52.64
C	80.629	80.660	−146.171	3491.81	4534.98	42.44
D	85.675	−33.135	−148.164	427.51	4574.83	4.77
$\sum$	197.785	106.811	−582.603	8365.17	21825.88	100.52
$\sum/4$	49.446	26.703	−145.651	2091.29	5456.47	25.13

Solution

Let the initial values of K, Φ, Ω be zero, and let the initial value of λ be unity. After reduction to the common centroid the coordinate values are

Point	Model coordinates (mm) x	y	z	Ground coordinates (m) X	Y	Z
A	−19.655	−60.267	−0.272	−1587.99	621.17	−24.46
B	−47.756	66.147	3.306	1851.26	1181.96	27.51
C	31.183	53.957	−0.520	1400.52	−921.49	17.31
D	36.229	−59.838	−2.513	−1663.78	−881.64	−20.36
$\sum$	.001	−.001	.001	.01	·00	·00

A value for $d\lambda$ is calculated for the lines AC and BD as follows.
From eqn 7.23,

$$d\lambda_{AC}=\left(\frac{(2988.51)^2+(-1542.66)^2+(41.77)^2}{(50.838)^2+(114.224)^2+(-0.248)^2}\right)^{1/2}=26.9017745$$

$$d\lambda_{BD}=\left(\frac{(-3515.04)^2+(-2063.60)^2+(-47.87)^2}{(83.985)^2+(-125.985)^2+(5.819)^2}\right)^{1/2}=26.9020033$$

Therefore the mean value is

$$d\lambda=26.901889$$

and, when it is applied to the model coordinates, the coordinates and coordinate differences become

Point	x	y	z	Line	Δx	Δy	Δz
A	−528.76	−1621.30	−7.32	AD	1503.39	11.54	−60.28
B	−1284.73	1779.48	88.94	BC	2123.61	−327.94	−102.93
C	838.88	1451.54	−13.99	AB	−755.97	3400.78	96.26
D	974.63	−1609.76	−67.60	DC	−135.75	3061.30	53.61

Increments to the rotation elements are found from eqns (7.24), (7.25) and (7.26). For lines AC and BD respectively, the values of α_g, α_m, and dK, in radians, are

	α_g	α_m	$dK=\alpha_g-\alpha_m$
AC	2.047313	0.418750	1.628563
BD	4.181526	2.553617	1.627909
		Mean $dK=$	1.628236

Values for $d\Phi$, and $d\Omega$, in radians, and derived from eqns (7.26), are

	AD	BC	AB	DC	Mean
$d\Phi$	0.042823	0.043666	———	———	0.043244
$d\Omega$	———	———	0.000130	0.005207	0.002668

Values for λ', K', Φ' Ω', are obtained from eqns (7.27) and (7.28).
The rotation matrix $\boldsymbol{R}$ is formed:

$$\boldsymbol{R} = \begin{bmatrix} -0.057356 & 0.998340 & 0.005145 \\ -0.997417 & -0.057525 & 0.043006 \\ 0.043230 & -0.002666 & 0.999062 \end{bmatrix}$$

and the reduced coordinates obtained in step (i) are transformed using eqn (7.29), to give the following:

Point	x'	y'	z'
A	−1588.32	620.34	−24.85
B	1850.67	1182.87	28.57
C	1400.95	−920.82	18.42
D	−1663.33	−882.42	−21.12

Steps (ii)–(iv) are repeated for the required number of iterations, with the following results.

	Iteration number 1	2	3	Parameter	Final Value
Parameter Increment					
$d\lambda$	26.901889	0.999999	1.000000	λ	26.901860
dK	1.628236	0.000494	−0.000030	K	1.628700
$d\Phi$	0.043244	0.000198	−0.000468	Φ	0.042974
$d\Omega$	0.002668	0.000680	−0.000041	Ω	0.003307

Coordinate Discrepancies (m)		1	2	3
A	dX	−0.33	−0.02	−0.03
	dY	−0.83	0.00	−0.04
	dZ	−1.39	−0.39	−0.21
B	dX	−0.59	0.05	0.01
	dY	0.91	−0.06	−0.01
	dZ	1.06	−0.40	0.27
C	dX	0.43	−0.04	−0.02
	dY	0.67	−0.06	−0.02
	dZ	1.11	0.29	−0.04
D	dX	0.45	−0.03	0.00
	dY	−0.78	0.09	0.05
	dZ	−0.76	0.53	0.01

The translations are found from eqn 7.30, where the centroid is used as the common point with known values in both systems:

$$\begin{bmatrix} X_T \\ Y_T \\ Z_T \end{bmatrix} = \begin{bmatrix} X \\ Y \\ Z \end{bmatrix} - \lambda R \begin{bmatrix} x \\ y \\ z \end{bmatrix} = \begin{bmatrix} 2091.29 \\ 5456.47 \\ 25.13 \end{bmatrix} - \begin{bmatrix} 617.56 \\ -1535.71 \\ -3859.87 \end{bmatrix} = \begin{bmatrix} 1473.73 \\ 6992.18 \\ 3885.00 \end{bmatrix}$$

8 Observations and statistical concepts

8.1 Introduction

Surveying and photogrammetry are concerned with observing, reducing (transforming), analysing and interpreting data. The 'observations' (or 'measurements') are usually more in number than are actually required to solve the problem—i.e. there are what are termed '*redundant observations*'.

For example:

(i) If a distance has been measured twice, there is one redundant observation.
(ii) If all three angles of a triangle have been observed, there is one redundant observation.

In the first case the redundancy is very obvious; in the second case it is evident only because of our knowledge that two angles determine the shape of a triangle. Although each angle was measured only once, there is nevertheless a redundancy. In any set of observations containing redundant observations it is possible to select a subset which may be used to determine the other components. Any two angles of a triangle may be used to derive the other angle. However, different results will be obtained depending on which two angle are selected. A unique result is obtained only by imposing an additional criterion (such as the least-squares condition) on the observations. In other words, original observations are not consistent with each other. They are subject to variations due to changes in observer, equipment, or conditions.

In order to appreciate these variations more fully it is worth taking a closer look at the processes involved in measurement.

Example 1 The measurement of a distance

Consider the simplest case of the measurement of a line using a tape stretched along the ground. The procedure involves a series of identifiable steps:

1. setting the tape in a straight line between the end points;
2. aligning the zero mark at one end;
3. observing the reading at the other end;
4. recording the observation in a field book.

Note that the process of alignment of graduations on the tape with marks on the ground takes place at both ends of the line; and implicit in the recorded value is the subtraction of one reading from the other. This becomes more obvious if a section of the tape at the zero end is missing, and a reading other than zero is recorded for that end as well. The main point to bear in mind is that the final value recorded is the result of a series of physical operations. The number of these operations becomes greater with more complicated situations—such as a line on sloping ground, or across a small valley, or where the distance is measured with EDM equipment, etc.

Example 2 The measurement of an angle

Here again the process may be broken down into a number of discrete steps:

1. centring/levelling the theodolite over the ground mark;
2. focusing and bisecting the first target;
3. setting the micrometer for the horizontal circle reading;
4. reading (including interpolating) the direction;
5. recording the direction;
6. steps 2 to 5 for the second target;
7. subtracting one value from the other to give the angle.

Example 3 The measurement of parallax with a parallax-bar

1. alignment of the photographs;
2. alignment of the stereoscope and parallax-bar;
3. setting the floating mark on the ground;
4. reading the micrometer;
5. recording the observation.

In all these examples the numerical outcome—called the 'observation'—is a result of many successive elementary operations, such as aligning, pointing, bisecting, interpolating, setting, recording, etc. Other less obvious processes which might also be included are aspects such as standardization and calibration, correction for atmospheric conditions, lens distortion, etc. The application of these additional processes requires first of all that more observations are made of other quantities—e.g. the tape must be compared with some standard, or temperature and pressure must be observed, etc.

By now it is obvious that because of changing circumstances observations of quantities are bound to vary, with respect both to each other and to the true value of that quantity. The magnitude of the variations may be reduced by taking more care, using a better instrument, using automatic recording, etc., but some variation is unavoidable. It is natural consequence of the steps taken in making observations. It follows also that the 'true' value of a measured quantity is unattainable. It is only possible to obtain an 'estimate' of the true value by observing that quantity. Mathematically, therefore, an observation is regarded as a variable, and the difference between it and the so-called 'true' value is known as the 'error' in that observation.

Thus, if t is the true value of a quantity, and x is the observed value, then the *error* e is defined as

$$e = x - t \tag{8.1}$$

Since the true value is never known it is not possible to establish e. However, if a good estimate ($\bar{x}$) of the true value is obtained it may be used as a reference value instead. The variation (v)—also called a *residual*—is then defined as

$$v = \bar{x} - x \tag{8.2}$$

The 'discrepancy' ($= -v$) is equivalent to e; but residual is more commonly used since the more reliable estimate $\bar{x}$ is obtained by adding the residual v to the observation x—i.e. it is used in the sense of a correction or 'adjustment' to the observation.

Therefore

$$\bar{x} = x + v \tag{8.3}$$

In the past the study of variations in measurements led to what was referred to as the 'theory of errors'; however, because of the factors concerning 'true' value mentioned above, the subject is now referred to as 'theory of observations'. On the other hand the term 'error' is widely used and, in order to avoid any confusion, it will still be used in this text. The reader should bear in mind, however, the qualifications associated with its use. Errors may be classified under three headings, and each of these will be discussed in turn.

8.2 Types of error

Errors may be classified as gross errors, systematic errors, or random errors.

Gross errors are likely to be the largest which occur in a set of observations, and they are usually the ones that cause least trouble. In general they are easily detected and therefore easily removed. They arise as a result of simple carelessness, or ignorance, or failure of equipment. Examples are:

1. adopting the wrong value for a bench mark;
2. reading a theodolite incorrectly;
3. recording metres instead of feet with an EDM instrument;
4. identifying the wrong photo-control point;
5. neglecting grid constants in calculating t-T corrections;
6. loss of count in digitizing equipment.

Gross errors should be detected, and the corresponding observations rejected, as soon as possible after they have been made. Therefore, once it is recognized which type of gross error might occur, procedures should be devised which expose these errors. For example, with regard to the errors mentioned above, the following procedures could be adopted:

1. close the level line on another bench mark;

2. compare face-left readings with face-right readings;
3. measure a known distance with the EDM instrument;
4. establish photo-control points in pairs, or triplets;
5. compare with a graph showing variation of *t*-*T* across the projection;
6. start by observing a reference point, and check back to it at frequent intervals.

It is often the case that more than one procedure may be used to detect gross error, and the method adopted will depend on circumstances prevailing at the time.

Systematic errors are those which occur according to some regular pattern. Their occurrence may therefore be modelled mathematically, and such a model may be used to compensate for the errors. The pattern could be a function of either the observer, equipment, environment, or position within the spatial domain of the experiment (survey); and the problem initially is that of establishing the source of the error. Again, some examples will help to illustrate particular sources of systematic error:

1. neglecting instrument/reflector constant in EDM measurements;
2. a theodolite which is not centred over the ground mark;
3. a comparator with non-orthogonal axes;
4. the performance of an observer early or late in the day;
5. the difference between two observers in setting the floating mark on the ground in a photogrammetric model;
6. height differences without refraction corrections applied;
7. transformations from spheroid to projection without applying scale-factor.

Example 1 is caused by carelessness on the part of the observer; 3 by faulty equipment; 2 by faulty technique; 4 and 5 by physiological factors; 6 by environmental conditions; and 7 by an incorrect mathematical model. Futhermore, 1 and 5 have a constant effect on the observations, whereas the rest have effects which are a function of parameters such as direction, position, time, or meteorological conditions, etc. All, however, are systematic errors, and may be modelled mathematically by means of suitable equations. A problem which is never solved completely is that of establishing appropriate refinements to the equations so that all systematic errors are accounted for: nevertheless, this remains the ultimate objective.

As an alternative to mathematical modelling, it is often possible (and sometimes preferable) to calibrate and adjust the equipment, or devise observational procedures which compensate for systematic error. However, it is also true to say that, with the developments in electronics and computer technology, more and more of the systematic effects are being catered for by computation—either on-line or off-line.

Random errors are those variations remaining in observations after gross errors and systematic effects have been removed. Once observations are free of these two errors they may be regarded as samples of a *random variable* (also referred to as a *stochastic variable*)—the quantity being measured. A random variable is a variable which can have several possible values, and in surveying and photogrammetry the number of values is infinite. Associated with each value is its *probability*—the probability that that particular value will be recorded. Thus, in the same way that specific mathematical models are used to deal with systematic error, so probability models are used to cope with the problem of random error; or—to avoid the use of the term 'error'—probability theory is used to obtain the best value of a random variable, and to assess its reliability. Some elementary statistical concepts relating to probability theory are introduced in the next section.

8.3 Statistical concepts

One of the most important objectives in surveying and photogrammetry is to derive the best estimate of a random variable from a particular set of observations. The set of all possible values is termed the *population*, and the particular set obtained in practice is the *sample set*. The sample set is used to derive *parameters* which infer properties about the population as a whole; and, because only a sample is used, the parameters are only estimates of these properties. A sample may be either a *random sample* or one which is biased in some respect. The former is normally sought but seldom attained. However, once gross errors and known systematic effects have been removed, the assumption is made that a random sample remains. For example, consider a distance that has been measured 21 times, and the following values recorded:

4.618, 4.617, 4.615, 4.614, 4.616, 4.617, 4.616, 4.619,
4.617, 4.616, 4.619, 4.618, 4.617, 4.618, 4.617, 4.615,
4.621, 4.616, 4.616, 4.618, 4.617

The mean obtained from the 21 values is the *sample mean*, which is an estimate of the *population mean* (or 'true' value) for that quantity. It is an estimate of the best (or 'most probable') value for the distance. An estimate of the reliability of the sample mean is also required, and here it is necessary to study the collective behaviour (or 'dispersion') of the random sample. One way of doing this is to construct a *histogram* representing the frequency of occurrence of specific values, as shown in Figure 8.1*a*.

The histogram may also be constructed to indicate relative frequency, where relative frequency is obtained by dividing frequency of occurrence by the total number of observations. Figure 8.1*b* shows this type of histogram. The similarities between the two histograms are obvious, and it is clear that the frequency distribution is centred on a value close to 4.617. If it were possible to plot the total population then the histogram would appear as a more regular

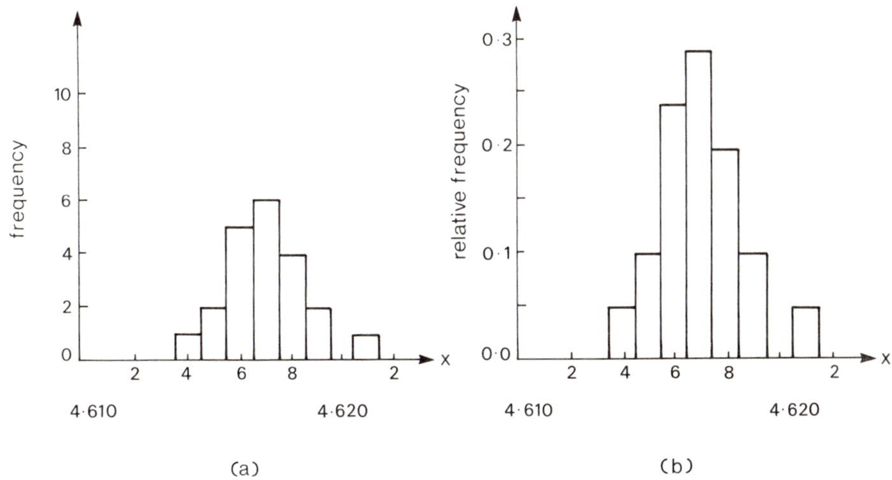

Figure 8.1 The histogram.

and symmetric figure. In such a case the relative frequency distribution is known as a *probability distribution*, and the shape of the figure becomes so stable that it is more convenient to use a mathematical model (the *probability density function*) to represent it. Many different types of probability density function exist, but the particular one used in surveying and photogrammetry is the *normal distribution*, described mathematically by eqn (8.4), and represented

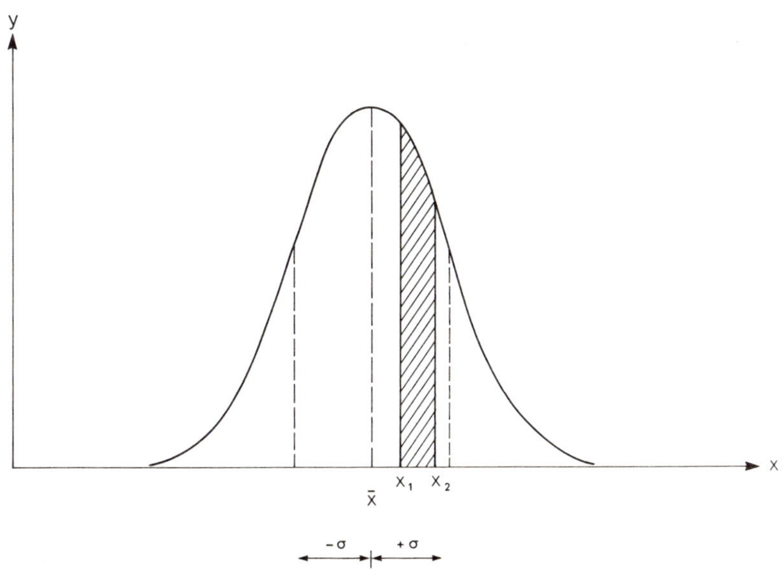

Figure 8.2 Normal distribution curve.

graphically in Figure 8.2:

$$y = \frac{1}{\sigma(2\pi)^{1/2}} \exp\left(-\frac{1}{2}\left(\frac{x - \bar{x}}{\sigma}\right)^2\right) \tag{8.4}$$

where $e(=2.71828)$ is the base of the natural system of logarithms.

The probability that an observation lies between x_1 and x_2 is the area under the curve between these two values, and is indicated as the shaded region. The central value ($\bar{x}$) is the mean value (or most probable value), and is the value adopted for the parameter representing the 'true' value. The parameter representing dispersion is *standard deviation* (σ), and it is calculated using the formula

$$\sigma = \pm\left(\frac{\Sigma v^2}{r}\right)^{1/2} \tag{8.5}$$

where v is given by eqn (8.2), and r is the number of redundancies (or degrees of freedom)—in this case 20.

σ is in fact the value at the point of inflection of the curve, and it indicates the precision or accuracy of a particular set of observations.

Variance (σ^2)—equal to standard deviation squared—is also frequently used. It has particular application in least-squares theory, as will become evident later.

Other expressions which are equivalent to standard deviation are also used, such as 'standard error', 'mean square error', or 'root mean square error'. Standard error is more appropriate in the context of accuracy (involving a large sample), whereas standard deviation relates better to precision (or a small sample). In practice the expressions are used rather loosely, and they are appropriate only when normal or near-normal distributions are being considered.

The shape of the frequency curve indicates whether it represents observations of high or low precision (accuracy). Figure 8.3*a* shows a number of curves for different sets of observations belonging to different populations.

The distributions are seen to vary according to changes in $\bar{x}$ and σ. These differences make it awkward to compare different normal distributions. Therefore, the solution is to regard each distribution as having a value $\bar{x}$ equal to 0 and establishing deviations from $\bar{x}$ in units of σ, i.e. $\sigma = \pm 1$. The distribution is then known as the *standard normal distribution*, and is a probability distribution. It is shown in Figure 8.3*b*, and is represented by the probability density function

$$y = \frac{1}{(2\pi)^{1/2}} \cdot \exp\left(-\frac{t^2}{2}\right) \tag{8.6}$$

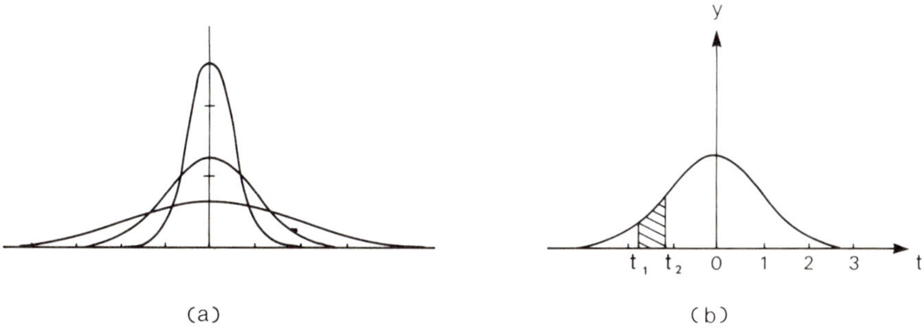

Figure 8.3: *a*, frequency curves; *b*, standard normal distribution.

where t is the stochastic variable x expressed in standard units, i.e.

$$t = \frac{x - \bar{x}}{\sigma} \tag{8.7}$$

The total area under the standard normal curve is unity, and therefore the area under the curve between the ordinates t_1 and t_2 is the probability that x lies between t_1 and t_2. The advantage of such a curve is that it is only necessary to have a single area table for the estimation of probabilities of all normal distributions. The table can be for probabilities falling inside a given value for t, when they are referred to as *confidence levels*, or outside a given value for t, when they are called *significance levels*, denoted by α. Thus a confidence level is equal to $(1 - \alpha)$. It is common to have tabulated values for the cumulative area under the curve and, since the total area is unity, the corresponding levels of confidence or significance are easily obtained. Such a table may be used to find the percentages of observations falling within 1σ, 2σ, 3σ, etc.:

$$\text{The percentage falling within} \begin{cases} 1\sigma & \text{is} \quad 68.26, \\ 2\sigma & \text{is} \quad 95.44, \\ 3\sigma & \text{is} \quad 99.74. \end{cases}$$

Or, expressed in another way, the probabilities of observations exceeding $1\sigma, 2\sigma, 3\sigma$ are 0.3174, 0.0456, 0.0026 respectively.

or,	1	in	3	observations is likely to exceed	1σ,
	1	in	20	observations is likely to exceed	2σ,
	1	in	400	observations is likely to exceed	3σ.

This information is useful from the point of view of establishing criteria for the rejection of observations.

A thorough analysis of observations, or quantities derived from observation, includes carring out certain statistical tests. It may be necessary to assess whether the observations belong to a normally distributed population, or even the same normal distribution, or whether one of the observations contains a gross error and therefore must be rejected.

The identification of gross errors (or 'outliers') in observations is important in surveying and photogrammetry, particularly in the age of automated data capture when there is less reliance on intuition than in the past. A number of procedures are possible, including one based on repeated measurements, when for example an observation may be rejected because it does not satisfy some criterion such as the 1 in 400 (0.9974 probability) criterion—i.e. its residual is greater in absolute terms than 3σ, where σ is determined from the sample set. The danger of using such a criterion is evident when the number of observations is small. An alternative is a criterion such as that suggested by Chauvenet, which simply states that the level of significance (α) should be chosen to be $1/(2n)$, where n is the number of repeated observations. Thus for three observations α would be 0.167, and from a table of probabilities it is seen that an observation with residual greater than 1.38σ would be rejected; for 10 observations, those with residuals greater than 2.81σ would be rejected. Other more rigorous tests for outliers are described in texts on mathematical statistics, such as [21]. The results of a least-squares computation may also be used for identifying outliers, and the ratio of a least-squares residual to its standard error may be used as the basis for such a test. Further details of these procedures may be found in [14] and [27]. Other statistical tests to assess whether observations belong to the same population or not may be found in the same publications.

8.4 Multivariate distributions

In the previous section the properties of only one stochastic variable were considered. Most situations in surveying and photogrammetry involve more than one variable.

For example:

(i) observing a number of angles at a control station;
(ii) measuring a bearing and distance to a point;
(iii) measuring the xy-coordinates of a point with a comparator.

The statistical concepts used so far apply just as much to these multidimensional situations. If two normally distributed variables have been observed, the histogram for the two-dimensional distribution may be constructed in the third dimension and it may be represented in two dimensions using contours. For example, if the xy-coordinates of a point are measured many times the observations may be represented graphically as shown in Figure 8.4. A grid of class boundaries may be drawn and the frequency of occurrence in each square evaluated. Contours of equal probability may then be drawn, and the surface represented by them becomes the probability surface. The intersection of this

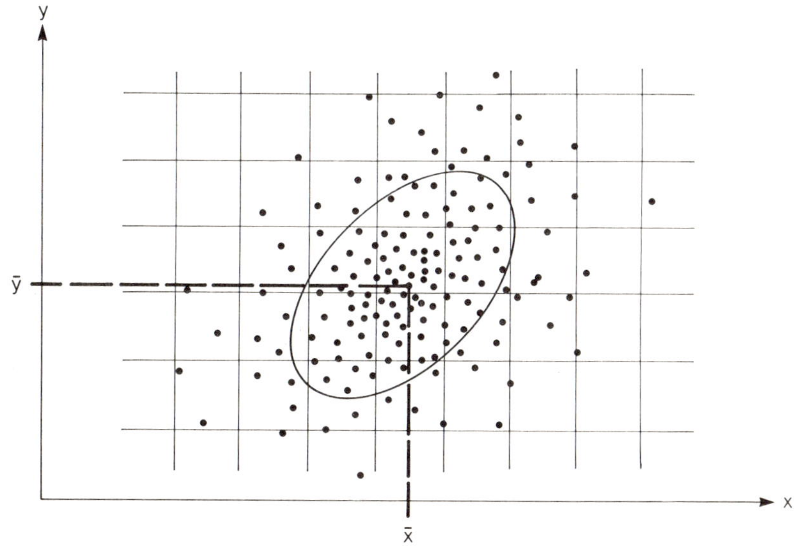

Figure 8.4 Two-dimensional distribution.

surface with any vertical plane will produce the normal distribution curve, and the intersection with a horizontal plane will produce an ellipse.

The common centre of all the ellipses will be $\bar{x}, \bar{y}$, and one of these ellipses will be the *standard ellipse*. The orientation of the ellipses indicates whether or not there is *correlation* (statistical dependence) between the two variables. If the standard ellipse is taken to represent the surface its orientation will indicate any such correlation. The possibilities that might arise are shown in Figure 8.5.

In order to be able to state the extent of the correlation, *covariance* is defined as

$$\sigma_{xy} = \frac{[(x-\bar{x})(y-\bar{y})]}{r} \tag{8.8}$$

where r is the number of redundant observations, and [ ] signifies summation over all values.

If x and y are independent σ_{xy} will always be 0, but the converse is not necessarily true.

Note that σ_{xy} may be positive or negative, therefore $(\sigma_{xy})^{1/2}$ has no meaning.

The *correlation coefficient* (ρ)—which varies between $+1$ and -1—may also be used to express correlation. It is defined as

$$\rho = \frac{\sigma_{xy}}{\sigma_x \sigma_y} = \frac{\sigma_{xy}}{(\sigma_x^2 \sigma_y^2)^{1/2}} = \frac{[v_x v_y]}{([v_x^2][v_y^2])^{1/2}} \tag{8.9}$$

Not that $|\sigma_{xy}| \leqslant \sigma_x \sigma_y$, where | | signifies modulus.

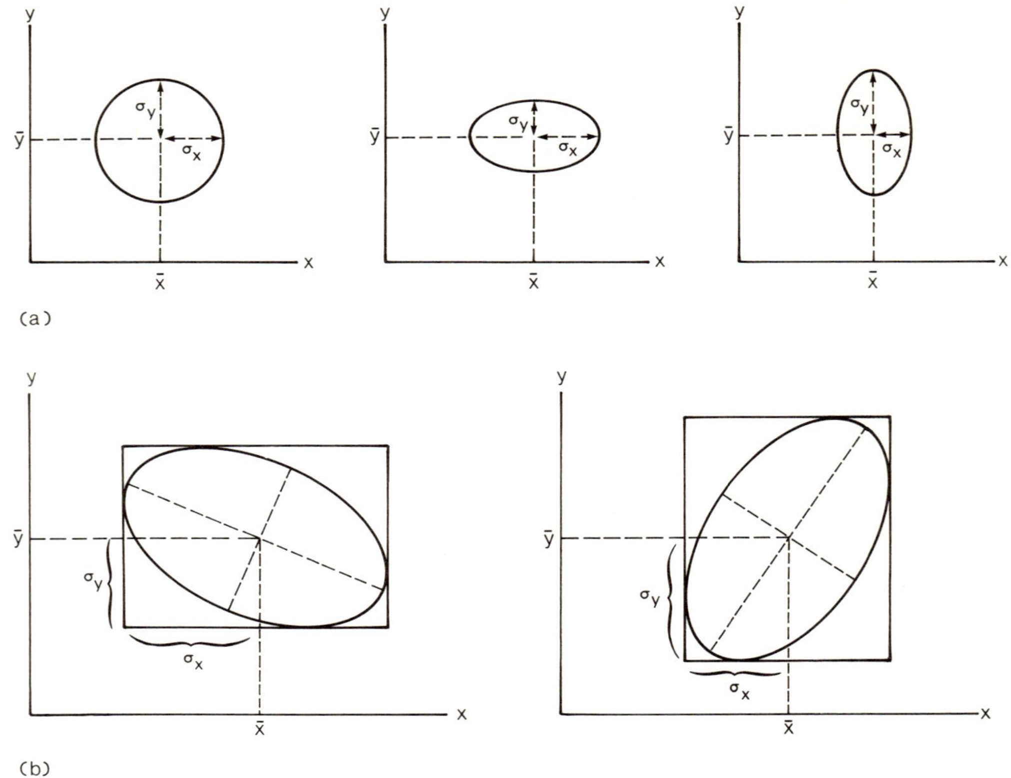

Figure 8.5 Two-dimensional correlation: *a*, *x* and *y* independent (no correlation) (i) $\sigma_x = \sigma_y$, (ii) $\sigma_x > \sigma_y$, (iii) $\sigma_x < \sigma_y$; *b*, *x* and *y* correlated, (i) negative correlation, (ii) positive correlation.

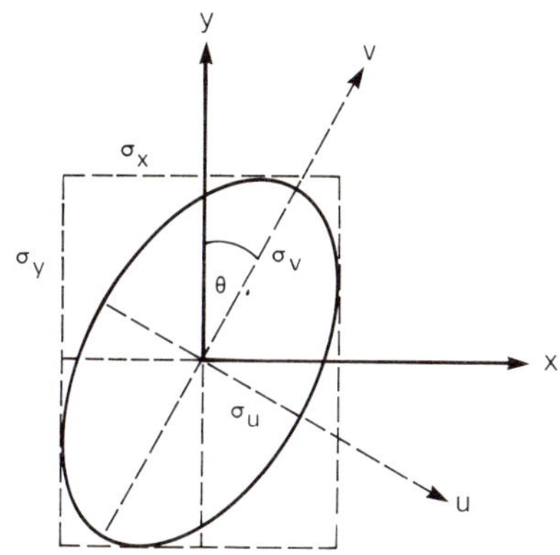

Figure 8.6

It is possible to replace the correlated variables x and y with two which are not correlated (u and v), by a simple rotation (θ) of the coordinate system. Figure 8.6 illustrates the situation, and the angle θ determines the orientation of the ellipse. Further details are given later in section 8.6.

The same principles may be applied to situations involving three-dimensional coordinates—xyz-coordinates in geodesy and photogrammetry—when the standard ellipsoid may be conceived. Further discussion on error ellipsoids is contained in [3] and [27].

8.5 The propagation of variances and covariances

The statistical concepts described so far have been concerned with the actual observations. In practice observations are normally used to derive other quantities; angles and distances are observed in order to derive rectangular coordinates, or y-parallaxes are observed to derive the elements of relative orientation in photogrammetry. These derived quantities may in turn be used to establish values of other quantities. Therefore, it is useful to examine how the error properties of the original observations are transferred to the derived quantities—i.e. assess the stochastic properties of a variable which is dependent on a number of other variables whose stochastic properties are known.

The simplest case is where the variable is a linear function of another variable, as expressed by eqn (8.10), and shown by Fig. 8.7.

$$y = ax \tag{8.10}$$

where x is a stochastic variable, and a is a constant.

All increments $\mathrm{d}x$ will be transformed into increments $\mathrm{d}y$, thus

$$\mathrm{d}y = a \cdot \mathrm{d}x$$

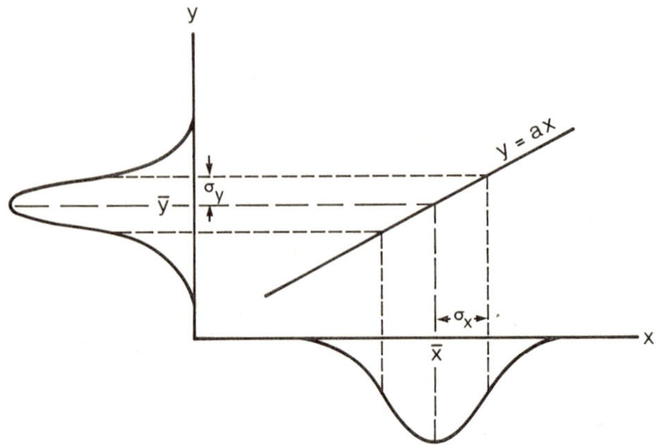

Figure 8.7

and residuals v_x will be transformed into residuals v_y, i.e.

$$v_y = av_x$$

Therefore the distribution for x will be transformed by a simple scale factor, and if the distribution for x is a normal distribution the distribution for y will also be normal i.e.

$$\left.\begin{aligned} \bar{y} &= a\bar{x} \\ \sigma_y &= a\sigma_x \end{aligned}\right\} \tag{8.11}$$

If the function is

$$y = ax + b, \quad \text{where } b \text{ is another constant,}$$

then $\bar{y} = a\bar{x} + b$

and $\sigma_y = a\sigma_x$

and the variance is unchanged.

A function may involve more than one original stochastic variable, such as

$$z = ax + by$$

where x and y are stochastic variables, and a and b are constants. If there are n observations of x and y, then

$$z_i = a(x)_i + b(y)_i \qquad (\text{i} = 1, 2, \ldots, n)$$

and $\bar{z} = a\bar{x} + b\bar{y}$

Also $v_z = a(v_x)_i + b(v_y)_i$

where the v's are with respect to the mean.

Therefore

$$[(v_z)_i^2] = a^2[(v_x)_i^2] + b^2[(v_y)_i^2] + 2ab[(v_x)_i(v_y)_i]$$

which, after division by r (the number of redundancies), becomes

$$\sigma_z^2 a^2 \sigma_x^2 + b^2 \sigma_y^2 2ab\sigma_{xy} \tag{8.12}$$

where σ_{xy} is the covariance between the variables x and y. If x and y are independent σ_{xy} is 0, and the last term disappears.

These results may be generalized for a linear function of any number of stochastic variables, and the equation is often referred to as the 'law of propagation of error'.

For example, if the function is

$$w = ax + by + cz$$

then the variance for w may be written directly as

$$\sigma_w^2 = a^2\sigma_x^2 + b^2\sigma_y^2 + c^2\sigma_x^2 + 2ab\sigma_{xy} + 2ac\sigma_{xz} + 2bc\sigma_{yz}$$

A numerical example to illustrate the application is given below.

Example 8.1

The diagram shows a profile of the terrain and point A has a known height. The height of B is obtained by measuring height differences (Δh) in three stages, as follows:

Stage 1: Δh_1 determined by spirit levelling—$\sigma_1 = 0.02$ m
Stage 2: Δh_2 determined by trigonometric levelling—$\sigma_2 = 0.20$ m
Stage 3: Δh_3 determined by barometric levelling—$\sigma_3 = 0.50$ m

H_A is the known height of point A, and is assumed to be error-free; the observed height differences are also assumed to be uncorrelated. Calculate the accuracy of the derived height of point B.

Solution

The first step is to express the height of B as a function of the variables with known stochastic properties:

$$H_B = H_A + \Delta h_1 + \Delta h_2 + \Delta h_3$$

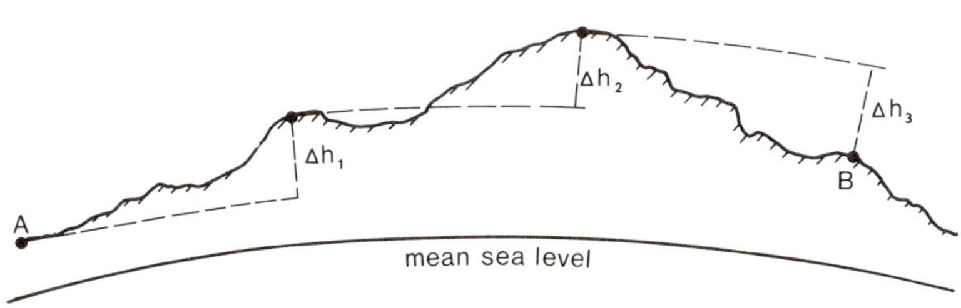

Figure 8.8.

The function is linear, therefore the law of propagation of variances may be applied:

$$\sigma_B^2 = l^2 \cdot \sigma_1^2 + l^2 \cdot \sigma_2^2 + l^2 \cdot \sigma_3^2$$
$$= 4\,\text{cm}^2 + 400\,\text{cm}^2 + 2500\,\text{cm}^2 = 2904\,\text{cm}^2$$

Therefore $\sigma_B = 54\,\text{cm} = 0.54\,\text{m}$

Comment

The example above serves to illustrate that the standard error of the derived variable is *not* simply the sum of the standard errors of the original variables.

A second example establishes a specific law for the propagation of error.

Example 8.2

Consider the function

$$y = x_1 + x_2 + x_3 + \cdots + x_n,$$

where $x_1, x_2, x_3, \ldots, x_n$ are independent variables, but with the same variance i.e.

$$\sigma_1 = \sigma_2 = \sigma_3 = \cdots = \sigma_n,$$

and let the common variance be σ. Then

$$\sigma_y^2 = \sigma^2 + \sigma^2 + \sigma^2 + \cdots + \sigma^2$$

i.e.

$$\sigma_y^2 = n\sigma^2$$
$$\sigma_y = n^{1/2}.\sigma \tag{8.13}$$

The variance of the mean of n independent observations may be found in a similar way, as follows.

Example 8.3

If $x_1, x_2, \ldots, x_n$ are independent observations of the same variable they belong to the same population.

Therefore $\sigma_1 = \sigma_2 = \cdots = \sigma_n = \sigma_x$

The mean is found from

$$\bar{x} = \frac{1}{n}(x_1 + x_2 + \cdots + x_n)$$

Therefore

$$\sigma_{\bar{x}}^2 = \left(\frac{1}{n}\right)^2 (\sigma_1^2 + \sigma_2^2 + \cdots + \sigma_n^2)$$

$$= \frac{n}{n^2}\sigma_x^2 = \frac{\sigma_x^2}{n} = \frac{[vv]}{n(n-1)} \tag{8.14}$$

Thus, the variance of the mean of n observations is equal to the variance of a single

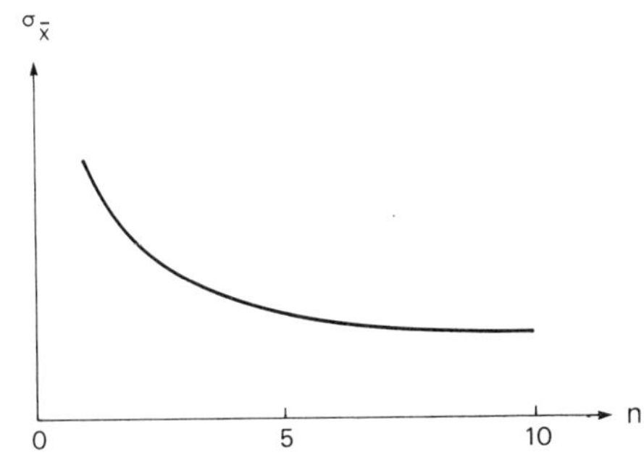

Figure 8.9.

observation divided by the number of observations, or

$$\sigma_{\bar{x}} = \frac{\sigma_x}{(n)^{1/2}} = \left(\frac{[vv]}{n(n-1)}\right)^{1/2} \tag{8.15}$$

Values of $\sigma_{\bar{x}}$ may be plotted against the number of observations, and the result is shown in Fig. 8.9. It is clear from the graph that two observations are better than one, and three observations are much better than one; but with many more observations there are diminishing returns.

If there is more than one linear function of the stochastic variables there will be correlation between the derived variables. For example, consider the following functions:

$$\left.\begin{aligned} y &= ax_1 + bx_2 + e \\ z &= cx_1 + dx_2 + f \end{aligned}\right\} \tag{8.16}$$

where x_1, x_2 are stochastic variables with standard deviations σ_1, σ_2, and covariance σ_{12}; and a, b, c, d, e, f are constants.

In addition to σ_y^2 and σ_z^2 there will be the covariance σ_{yz} between the derived variables.

Consider the error equations

$$v_y = a(v_x)_i + b(v_x)_i$$

$$v_z = c(v_x)_i + d(v_x)_i$$

When these are multiplied together the following equation is derived:

$$v_y v_z = ac(v_x)_1^2 + bd(v_x)_1^2 + (ad + bc)(v_x)_i(v_x)_i$$

Taking the sum over all values, and dividing by r, the following equation is obtained,

$$\sigma_{yz} = ac\sigma_1^2 + bd\sigma_2^2 + (ad + bc)\sigma_{12} \tag{8.17}$$

Summarizing therefore, and considering also eqn (8.12), the results for the two equations in (8.16) would be

$$\left.\begin{aligned}\sigma_y^2 &= a^2\sigma_1^2 + b^2\sigma_2^2 + 2ab\sigma_{12}\\ \sigma_z^2 &= c^2\sigma_1^2 + d^2\sigma_2^2 + 2cd\sigma_{12}\\ \sigma_{yz} &= ac\sigma_1^2 + bd\sigma_2^2 + (ad+bc)\sigma_{12}\end{aligned}\right\} \tag{8.18}$$

which may be generalized for any number of linear functions, and any number of variables in each equation.

Equations (8.18) are the basis for establishing the stochastic properties of derived quantities. If the number of functions is significantly greater than 2 the process becomes tedious, and it is more convenient to use matrix notation.

Therefore, the linear equations may be represented as

$$\boldsymbol{Y} = \boldsymbol{AX} + \boldsymbol{C} \tag{8.19}$$

where

$$\boldsymbol{Y} = \begin{pmatrix} y \\ z \end{pmatrix}, \quad \boldsymbol{A} = \begin{pmatrix} a & b \\ c & d \end{pmatrix}, \quad \boldsymbol{X} = \begin{pmatrix} x_1 \\ x_2 \end{pmatrix}, \quad \boldsymbol{C} = \begin{pmatrix} e \\ f \end{pmatrix}$$

and

$$\boldsymbol{\Sigma}_{XX} = \begin{pmatrix} \sigma_1^2 & \sigma_1^2 \\ \sigma_1^2 & \sigma_2^2 \end{pmatrix} \tag{8.20}$$

is the variance–covariance matrix, which will simply be referred to as the *covariance matrix*.

Therefore, $\boldsymbol{\Sigma}_{YY}$ is found from

$$\boldsymbol{\Sigma}_{YY} = \boldsymbol{\Sigma}_{XX}\boldsymbol{A}^{\mathrm{T}} \tag{8.21}$$

where

$$\boldsymbol{\Sigma}_{YY} = \begin{pmatrix} \sigma_y^2 & \sigma_{yz} \\ \sigma_{yz} & \sigma_z^2 \end{pmatrix}$$

In order to standardize the notation, σ_y^2 may be conveniently written as σ_{yy}, therefore, $\boldsymbol{\Sigma}_{YY}$ may be expressed as

$$\boldsymbol{\Sigma}_{YY} = \begin{pmatrix} \sigma_{yy} & \sigma_{yz} \\ \sigma_{yz} & \sigma_{zz} \end{pmatrix}$$

Note that $\boldsymbol{\Sigma}_{XX}$ and $\boldsymbol{\Sigma}_{YY}$ are symmetric matrices.

If the original variables x_1, x_2 are not correlated $\boldsymbol{\Sigma}_{XX}$ is a diagonal matrix. However, if $\boldsymbol{\Sigma}_{XX}$ is diagonal, it does not follow that $\boldsymbol{\Sigma}_{YY}$ is a diagonal matrix.

Equation (8.21) may be considered as the general law of propagation of covariances, since the elements σ_{yy} and σ_{zz} may be regarded as covariances between fully correlated variables.

8.6 The error ellipse

Brief mention was made in section 8.4 of the error ellipse (or ellipse of probability). Such an ellipse has particular significance when it is associated

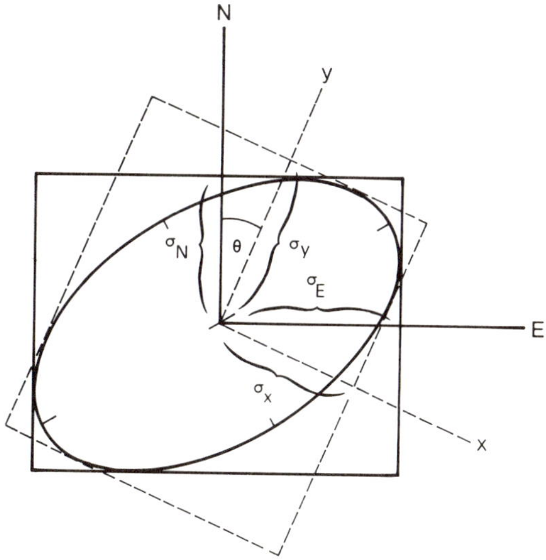

Figure 8.10 The error ellipse.

with planimetric coordinates, such as E, N or x, y. It is often the case that variances and covariances for the coordinates of a point are available, and these may be used to establish the standard error ellipse for the point. The variances define a rectangle about the point, and provide a rather coarse illustration of the dispersion at that point. A more precise illustration is provided by one of many error ellipses contained within the rectangle, and the particular ellipse that is required is specified by the covariance. In order to locate the ellipse it is necessary to determine the orientation of the principal axes, and the values of the maximum and minimum variance which are in the directions of these axes.

Consider Figure 8.10. which illustrates the error ellipse for a point with stochastic properties σ_E^2, σ_N^2, and σ_{EN}.

The corresponding values $\sigma_x^2, \sigma_v^2, \sigma_{xv}$ are found by applying the law of propagation to the orthogonal transformation equations:

$$\left.\begin{aligned} x &= E\cos\theta - N\sin\theta \\ y &= E\sin\theta + N\cos\theta \end{aligned}\right\} \tag{8.22}$$

Thus

$$\left.\begin{aligned} \sigma_x^2 &= \cos^2\theta\sigma_E^2 + \sin^2\theta\sigma_N^2 - 2\sin\theta\cos\sigma\theta_{EN} \\ \sigma_y^2 &= \sin^2\theta\sigma_E^2 + \cos^2\theta\sigma_N^2 + 2\sin\theta\cos\theta\sigma_{EN} \\ \sigma_{xy} &= \sin\theta\cos\theta\sigma_E^2 - \sin\theta\cos\theta\sigma_N^2 + (\cos^2\theta - \sin^2\theta)\sigma_{EN} \end{aligned}\right\} \tag{8.23}$$

If θ_m represents the angle of rotation required to place the axes x, y along the principal axes of the ellipse, as shown in Figure 8.11, then the covariance σ_{xy} is

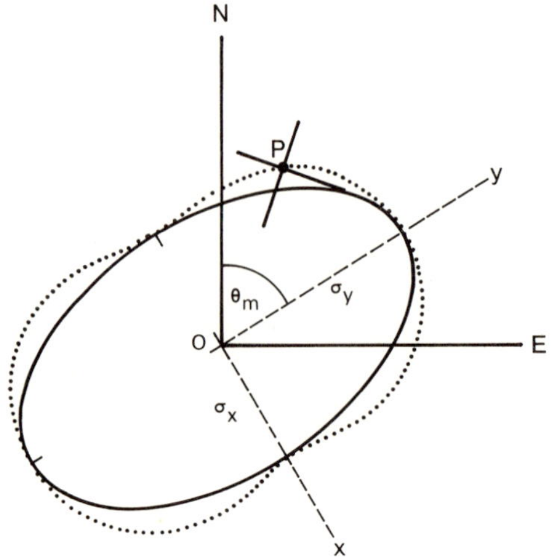

Figure 8.11.

equated to zero and the third equation in (8.23) may be written

$$\sin\theta_m \cos\theta_m \sigma_E^2 - \sin\theta_m \cos\theta_m \sigma_n^2 + (\cos^2\theta_m - \sin^2\theta_m)\sigma_{EN} = 0 \qquad (8.24)$$

Now

$$\cos^2\theta_m - \sin^2\theta_m = \cos 2\theta_m$$

therefore eqn (8.24) may be written

$$\sin\theta_m \cos\theta_m(\sigma_E^2 - \sigma_N^2) + \cos 2\theta_m \sigma_{EN} = 0$$

Therefore

$$2\sin\theta_m \cos\theta_m(\sigma_E^2 - \sigma_N^2) + 2\cos 2\theta_m \sigma_{EN} = 0$$

or

$$\sin 2\theta_m(\sigma_E^2 - \sigma_N^2) + 2\cos 2\theta_m \sigma_{EN} = 0$$

Thus

$$\tan 2\theta_m = \frac{2\sigma_{EN}}{\sigma_N^2 - \sigma_E^2} \qquad (8.25)$$

and so a value for θ_m may be derived.

An alternative derivation of eqn (8.25) involves finding a value for θ_m such that σ_x^2 is a miximum (or minimum)—i.e. by differentiating the first equation in (8.23) with respect to θ, and equating the result to 0.

The quadrant of 2θ is obtained by considering the signs of the numerator and denominator on the right-hand side of eqn (8.25).

Since

$$\tan 2\theta_m = \frac{\sin 2\theta_m}{\cos 2\theta_m}$$

the sign of $\sin 2\theta_m$ will be the same as the sign of $2\sigma_{EN}$, and the sign of $\cos 2\theta_m$ will be the same as that of $(\sigma_N^2 - \sigma_E^2)$.

Once θ_m has been found, values for σ_x^2 and σ_y^2 may be derived by substituting in eqns (8.23). Values may also be derived without establishing a value for θ_m, by calculating the eigenvalues of the covariance matrix Σ_{YY}. Further details are given in [14] and [27].

Example 8.4

The stochastic properties of a coordinated point are expressed by

$$\sigma_E^2 = 86.5\,\text{mm}^2, \qquad \sigma_N^2 = 39.0\,\text{mm}^2, \qquad \sigma_{EN} = -24.4\,\text{mm}^2$$

Calculate the orientation of the standard error ellipse, and derive values for the semimajor and semiminor axes.

Solution

θ_m is found from eqn (8.25) thus

$$\tan 2\theta_m = \frac{2\sigma_{EN}}{\sigma_N^2 - \sigma_E^2} = \frac{2 \times (-24.4)}{39.0 - 86.5} = \frac{-48.8}{-47.5} = 1.02737$$

Therefore $2\theta_m = 45.77°$ or, $225.77°$, and since both numerator and denominator are negative, therefore $2\theta_m = 225.77°$ and $\theta_m = 112.89°$.

Values for σ_x and σ_y are found from eqns (8.23) thus $\sigma_x = 5.4$ mm and $\sigma_y = 9.8$ mm.

Although the error ellipse is a fairly good approximation for the error distribution about the point it is not an exact representation. The locus for σ in any direction is given by the *pedal curve* of the ellipse, shown as a dotted line in Figure 8.11. It is the locus of a point P, which is at a bearing θ, and whose distance from the centre of the ellipse is the standard error σ_θ. Geometrically, its position along OP is defined by the tangent to the ellipse which is at right angles to OP. The pedal curve is seldom used in practice; however, more details may be obtained by consulting [27] and [40].

The concept of an error ellipse may be extended to the three-dimensional situation, in which case it is possible to conceive of an error ellipsoid. Extended further, a multidimensional situation involving an $(n \times n)$ covariance matrix may be imagined and, although the corresponding 'hyperellipsoid' is difficult to conceive, it does appear to have some practical value. The results of computational problems such as large survey networks, or blocks of aerial triangulation in photogrammetry, may be assessed by comparing values such as the largest eigenvalue or largest hyperellipsoid volume.

Coordinate error ellipses are usually called absolute error ellipses, though in fact they represent a description of the stochastic properties relative to the fixed points. The patterns of these error ellipses for points in a large network contain useful information, and serve to help identify where the strengths and weaknesses lie, or what systematic errors are present. On the other hand it is often more convenient in practice to consider the relative errors between points.

Thus if

$$\left.\begin{aligned} E &= E_j - E_i \\ N &= N_j - N_i \end{aligned}\right\} \tag{8.26}$$

then

$$\left.\begin{aligned} \sigma_E^2 &= \sigma_{E_i}^2 + \sigma_{E_j}^2 - 2_{E_iE_j} \\ \sigma_N^2 &= \sigma_{N_i}^2 + \sigma_{N_j}^2 - 1\sigma_{N_iN_j} \\ \sigma_{EN} &= \sigma_{E_iN_i} - \sigma_{E_iN_j} - \sigma_{E_jN_i} + \sigma_{E_jN_j} \end{aligned}\right\} \tag{8.27}$$

which may be used in eqns (8.25) and (8.23) to calculate the orientation and size of the relative error ellipse between points i and j.

Relative error ellipses are usually more useful than the so-called absolute ellipses, because the latter are heavily dependent on the positions of the fixed points, whereas relative ellipses are less so.

It was stated earlier in section 8.3 that, for the case of a single variable, the probability of an observation falling within $-\sigma$ and $+\sigma$ was 0.6826. In the bivariate case it can be shown (see [28]) that the probability of the least-squares estimate falling within the standard error ellipse is 0.394. Furthermore, if n represents a multiple of σ, and p is the probability that the least squares estimates fall within an ellipse defined by $n\,\sigma_{max}$ and $n\,\sigma_{min}$, then the following values apply:

Multiple of σ (n)	Probability (p)
1.000	0.394
2.000	0.865
2.447	0.950
3.000	0.989

Thus if an error ellipse is drawn, which has values for the semimajor and semiminor axes which are 2.447 times larger than σ_{max} and σ_{min} there is 95% confidence that the true values fall within the ellipse.

8.7 Weights and cofactors

The process of weighting observations takes into account the superior quality of one observation (or one set of observations) over another. Superior quality may be due to a better observer, or a better instrument, or better conditions of observation. Each of these three factors will influence the result, and if values for quality are assigned to each then an estimate may be made for the quality of the observation. There are problems, however, associated with the empirical determination of these values, since all three factors will never be constant at the same time. Therefore, it is more convenient to base quality on the variance obtained when a number of observations of the quantity are made. Furthermore, if the mean of a set of observations is used, the quality of the mean may be derived from the number of observations in the set.

It was seen earlier that the variance of the mean of n observations is found from

$$\sigma_{\bar{x}}^2 = \frac{\sigma_x^2}{n}$$

Therefore, if the means from different sets of observations are to be equally reliable (i.e. have the same σ), $\sigma_x^2\, n$ must be the same for each set. Thus if σ_x is the same for each set, equal reliability is assured only if n is the same.

If n is not the same, i.e.

$$\sigma_{\bar{x}_1}^2 = \frac{\sigma_x^2}{n_1}, \sigma_{\bar{x}_2}^2 = \frac{\sigma_x^2}{n_2}, \ldots, \text{etc.}$$

then $\sigma_{\bar{x}_1}^2, \sigma_{\bar{x}_2}^2, \ldots$ may be made the same $(= \sigma_{\bar{x}}^2)$, by multiplying the right-hand sides by $n_1/n, n_2/n, \ldots$ etc., which are called the *weights* of the means $\bar{x}_1, \bar{x}_2, \ldots$, etc. If $n = 1$, then in a set with one observation, $\sigma_{\bar{x}}^2 = \sigma_x^2/1 \times 1/1$, and a single observation has weight 1.

Hence *an observation of weight w is equivalent to w observations of weight* 1.

Therefore, if $\bar{x}$ is derived from n observations of weight 1, it may be regarded as a single observation ($\bar{x}$) of weight n.

Also, if σ_0 is the standard deviation of an observation of weight 1 (i.e. σ_0^2 is referred to as the *unit variance*), and $\sigma_{\bar{x}}$ is the standard deviation of an observation of weight n, then

$$\sigma_{\bar{x}}^2 = \frac{\sigma_0^2}{n}$$

or, more generally, with weights $w_1, w_2, \ldots$,

$$\sigma_1^2 = \frac{\sigma_0^2}{w_1}, \qquad \sigma_2^2 = \frac{\sigma_0^2}{w_2}, \ldots$$

or

$$\frac{w_1}{w_2} = \frac{\sigma_2^2}{\sigma_1^2} \tag{8.28}$$

In other words, *weight is inversely proportional to variance*. This is a very important conclusion, and it serves as a basis for a more rigorous estimation of observed quantities.

It follows also that the *weight matrix* is proportional to the *inverse* of the covariance matrix.

If the original variables are not correlated (so that $\mathbf{\Sigma}_{XX}$ is a diagonal matrix), the elements of the weight matrix may be found from the reciprocals of the elements in $\mathbf{\Sigma}_{XX}$.

For example, if

$$\mathbf{\Sigma}_{XX}=\begin{bmatrix} 2 & & \\ & 3 & \\ & & 4 \end{bmatrix}, \quad \text{then} \quad \boldsymbol{W}_{XX}=\begin{bmatrix} 1/2 & & \\ & 1/3 & \\ & & 1/4 \end{bmatrix} \tag{8.29}$$

If the original variables *are* correlated $\boldsymbol{W}$ must be found from the inverse matrix, as follows

$$\boldsymbol{W}=\mathbf{\Sigma}_{XX}^{-1} \tag{8.30}$$

If $\mathbf{\Sigma}$ is not diagonal $\boldsymbol{W}$ will certainly not be diagonal, and is more likely to be a full matrix, without zero elements.

For example, if

$$\mathbf{\Sigma}=\begin{bmatrix} 2 & -1 & 0 \\ -1 & 2 & -1 \\ 0 & -1 & 2 \end{bmatrix}, \quad \text{then} \quad \boldsymbol{W}=\mathbf{\Sigma}^{-1}=\begin{bmatrix} 0.75 & 0.50 & 0.25 \\ 0.50 & 1.00 & 0.50 \\ 0.25 & 0.50 & 0.75 \end{bmatrix}$$

The *standard error of the weighted mean* of n observations may also be found using the relationship in eqn (8.28).

Since

$$\sigma_0^2=w_1\sigma_1^2=w_2\sigma_2^2=\cdots=w_n\sigma_n^2$$

therefore

$$\sigma_0=(w_1)^{1/2}\sigma_1=\cdots=(w_n)^{1/2}\sigma_n$$

and a residual,

$$v_i=\bar{x}-x_i$$

of an observation (with weight w_i), may be reduced to that of an observation with weight 1 by multiplying by $(w)^{1/2}$

Thus $$(w_i)^{1/2}v_i=(w_i)^{1/2}x-(w_i)^{1/2}x_i$$

and σ_0 may be found from

$$\sigma_0=\left(\frac{[wv^2]}{r}\right)^{1/2} \tag{8.31}$$

Also, since an observation of weight w is equivalent to w observations of weight 1, therefore n observations of varying weight will be equivalent to $[w]$ observations of weight 1.

Therefore

$$\sigma_{\bar{x}} = \left(\frac{[wv^2]}{[w]\cdot r}\right)^{1/2} \tag{8.32}$$

It follows also that, for a single observation of weight w_i,

$$\sigma_x = \left(\frac{[wv^2]}{w_i r}\right)^{1/2} \tag{8.33}$$

and if all observations have equal weight,

$$\sigma_{\bar{x}} = \left(\frac{[v^2]}{nr}\right)^{1/2}, \quad \text{or} \quad \sigma_{\bar{x}} = \left(\frac{[v^2]}{n(n-1)}\right)^{1/2}$$

which is the result obtained earlier in eqn (8.15).

It was seen earlier that weights could be considered as relative quantities, whereas variances and covariances are expressed in absolute terms. However, the latter may be used in a purely relative sense by dividing (or multiplying) by a scalar quantity, termed the *reference variance* (σ^2). Thus if all variances and covariances are divided by this reference variance they become what are termed *cofactors*, denoted by the symbol g.

Therefore, for the stochastic variables x and y,

$$g_{xx} = \frac{\sigma_{xx}}{\sigma^2}, \qquad g_{yy} = \frac{\sigma_{yy}}{\sigma^2}, \qquad g_{xy} = \frac{\sigma_{xy}}{\sigma^2} \tag{8.34}$$

At any subsequent stage the absolute variances and covariances may be recovered by multiplying by the reference variance. The convenience of using a reference variance arises if variances and covariances are awkward numbers—either very large or very small—and the computation is to be carried out with limited resources. The example below illustrates the procedure, although it is not complicated enough to demonstrate the advantages.

Example 8.5

Consider the following linear function

$$z = 2x + y + 3$$

where x and y are stochastic variables which are uncorrelated, and which have the following variances:

$$\sigma_x^2 = \sigma_{xx} = 10\,\text{cm}^2; \qquad \sigma_y^2 = \sigma_{yy} = 15\,\text{cm}^2$$

When variances and covariances are used

$$\sigma_{zz} = (4 \times 10) + (1 \times 15) = 55\,\text{cm}^2$$

When cofactors are used, let

$$\sigma^2 = 5\,\text{cm}^2, \quad \text{therefore} \quad g_{xx} = 10/5 = 2, \quad \text{and} \quad g_{yy} = 15/5 = 3$$

Therefore $g_{zz} = (4 \times 2) + (1 \times 3) = 11$ and the variance may be re-established by multiplying by σ^2, thus

$$\sigma_{zz} = \sigma^2 g_{zz} = 5\,\text{cm}^2 \times 11 = 55\,\text{cm}^2$$

Although this example does not demonstrate any particular advantage in using cofactors, it does illustrate the principle involved. The advantages become apparent when more complex problems are solved, as will be seen later.

If matrix notation is used, and the set of linear equations is represented as

$$\boldsymbol{Y} = \boldsymbol{AX} + \boldsymbol{C}$$

then, in the same way that eqn (8.21) was derived, the law of propagation of cofactors may be expressed as

$$\boldsymbol{G}_{YY} = \boldsymbol{A}\boldsymbol{G}_{XX}\boldsymbol{A}^{\mathrm{T}} \tag{8.35}$$

where $\boldsymbol{G}_{XX}$ and $\boldsymbol{G}_{YY}$ are the cofactor matrices for the original and derived variables respectively.

Note

(i) that, as with covariance matrices, the cofactor matrices $\boldsymbol{G}$ are always symmetric;

(ii) that the weight matrix $\boldsymbol{W}$ is the inverse of the cofactor matrix, thus

$$\boldsymbol{W} = \boldsymbol{G}^{-1} \tag{8.36}$$

(iii) that $\boldsymbol{G}$ is the general expression, whereas $\boldsymbol{\Sigma}$ is specific; therefore, by regarding σ^2 as unity $\boldsymbol{\Sigma}$ may be replaced by $\boldsymbol{G}$.

Throughout the remainder of the text $\boldsymbol{G}$ may be used to denote stochastic properties and, to simplify the notation, the subscripts XX may be omitted when the matrix refers to original variables. If there is any doubt they will be restored. Similarly, if there is a requirement to state specific variances and covariances, then $\boldsymbol{\Sigma}$ will be used. Thus, unless otherwise indicated, $\boldsymbol{\Sigma}$ and $\boldsymbol{G}$ will be used to represent *a priori* covariance and cofactor matrices respectively.

The examples given below serve to illustrate the laws of propagation of covariances and cofactors.

Example 8.6

The three angles of a triangle have been observed, by setting up a theodolite at each point and observing the directions to the other two points.

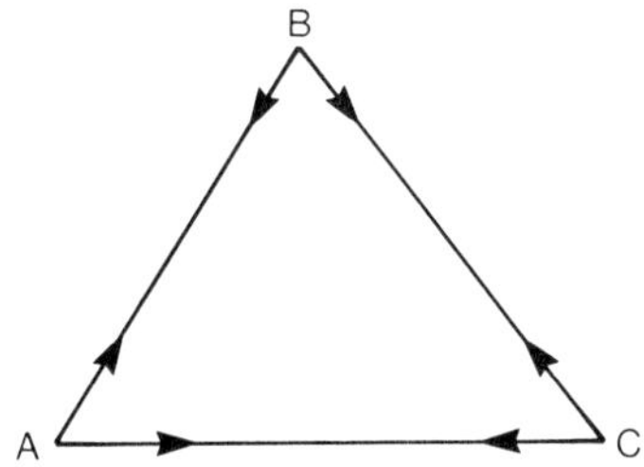

Figure 8.12

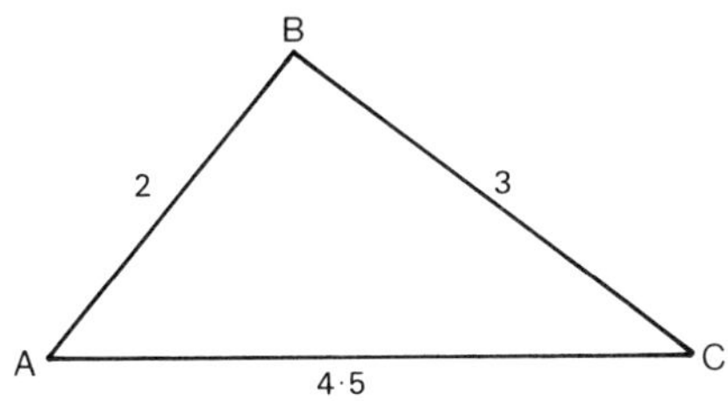

Figure 8.13

(a) If σ_d is the standard deviation for each observed direction, and the observations are uncorrelated, establish a formula for the standard deviation of the misclosure of the triangle. In other words, what would be the expected misclosure, expressed in terms of σ_d?

(b) Let s'' be the standard deviation of an observed direction over a distance of 1 km, and let σ_d for each observation be inversely proportional to the length of the line sighted. If the side lengths of the triangle are as indicated, find the following in terms of s:
   (i) the standard deviation of each angle;
   (ii) the standard deviation of the misclosure of the triangle; and
   (iii) the standard deviation of the adjusted angles, where the correction to each angle is proportional to the variance of that angle.

(c) Repeat (b) using matrix notation.

Solution:

(a) The misclosure (Δ) is found from

$$\Delta = 180 - (A + B + C) \tag{8.37}$$

where A, B, C are the observed angles.

However, each angle is the difference between two observed directions; therefore

$$\Delta = 180 - ((d_{AC} - d_{AB}) + (d_{BA} - d_{BC}) + (d_{CB} - d_{CA}))$$

and, applying the law of propagation of variances,

$$\begin{aligned} \sigma_\Delta^2 &= \sigma_d^2 + \sigma_d^2 + \sigma_d^2 + \sigma_d^2 + \sigma_d^2 + \sigma_d^2 \\ &= 6\sigma_d^2 \end{aligned}$$

Therefore, $$\sigma_\Delta = 2.45\,\sigma_d$$

(b) σ_d is inversely proportional to distance; therefore for a distance D, $\sigma_d = s/D$, where D is in km.

A table of variances for observed directions may be drawn up as follows:

Side	Distance (km)	σ_d (″)	σ_d^2 (″)^2
AB	2	$0.50\,s$	$0.25\,s^2$
BC	3	$0.33\,s$	$0.10\,s^2$
CA	4.5	$0.22\,s$	$0.05\,s^2$

(i) $A = d_{AC} - d_{AB}$; $B = d_{BA} - d_{BC}$; $C = d_{CB} - d_{CA}$;

$$\sigma_A^2 = 0.05s^2 + 0.25s^2 = 0.30s^2; \quad \text{therefore} \quad \sigma_A = 0.55s$$
$$\sigma_B^2 = 0.25s^2 + 0.10s^2 = 0.35s^2; \quad \text{therefore} \quad \sigma_B = 0.59s$$
$$\sigma_C^2 = 0.10s^2 + 0.05s^2 = 0.15s^2; \quad \text{therefore} \quad \sigma_C = 0.39s$$

(ii) The misclosure is given by eqn (8.37)

$$\Delta = 180 - (A + B + C)$$

Therefore $\sigma_\Delta^2 = 0.30s^2 + 0.35s^2 + 0.15s^2 = 0.80s^2$; therefore

$$\sigma_\Delta = 0.9\ s$$

(iii) Let a, b, c be the corrections to each angle—based on the misclosure and allocated according to the variances of each observed angle (i.e. inversely according to their weights).

Hence the adjusted angles are given by

$$\bar{A} = A + a; \quad \bar{B} = B + b; \quad \bar{C} = C + c;$$

where the corrections a, b, c are calculated according to the ratio

$$\frac{\text{variance of angle}}{\text{variance of misclosure}}$$

Thus

$$a = \frac{0.30s^2 \cdot \Delta}{0.80s^2} = \frac{3 \cdot \Delta}{8}; \quad b = \frac{0.35s^2 \cdot \Delta}{0.80s^2} = \frac{3.5 \cdot \Delta}{8};$$

$$c = \frac{0.15s^2 \cdot \Delta}{0.80s^2} = \frac{1.5 \cdot \Delta}{8}$$

Therefore

$$\bar{A} = A + \frac{3 \cdot \Delta}{8}$$

and applying the law of propagation to this equation,

$$\sigma_{\bar{A}}^2 = \sigma_A^2 + \left(\frac{3}{8}\right)^2 \cdot \sigma_\Delta^2 + 2\left(\frac{3}{8}\right) \cdot \sigma_{A\Delta}$$

Because Δ is derived from A and the other angles, it is correlated with A; therefore a value for $\sigma_{A\Delta}$ must be found. The two quantities are expressed is linear functions of variables with known stochastic properties—i.e. the angles A, B, C—and the covariance is established according to the third of eqns (8.18).

Thus

$$A = A$$
$$\Delta = 180 - (A + B + C)$$

therefore, $\sigma_{A\Delta} = -\sigma_A^2$ (Note: A is not correlated with B or C)

therefore

$$\sigma_{\bar{A}}^2 = \sigma_A^2 + \left(\frac{3}{8}\right)^2 \cdot \sigma_\Delta^2 - \frac{3}{4} \cdot \sigma_A^2$$

$$= \frac{1}{4} \cdot \sigma_A^2 + \left(\frac{3}{8}\right)^2 \cdot \sigma_\Delta^2$$

$$= 0.075 s^2 + \left(\frac{3}{8}\right)^2 \cdot (0.80 s^2) = 0.1875 s^2$$

Therefore $\sigma_{\bar{A}} = 0.43 s$.

Note that $\sigma_{\bar{A}}$ is less than σ_A. This is to be expected since the adjusted angle should be more reliable than the observed angle.

Values for $\sigma_{\bar{B}}$ and $\sigma_{\bar{C}}$ may be found in a similar manner.

(c) When matrix notation is used the equations are established in the form

$$\boldsymbol{Y} = \boldsymbol{AX} + \boldsymbol{C}$$

and the matrices $\boldsymbol{A}$ and $\boldsymbol{\Sigma}_{XX}$ identified and used according to eqn (8.21) in order to obtain $\boldsymbol{\Sigma}_{YY}$.

Thus for the observations taken at point A, B, C:

(i) Equations for the observed angles are

$$\begin{aligned} A &= d_{AC} - d_{AB} \\ B &= \qquad\qquad d_{BA} - d_{BC} \\ C &= \qquad\qquad\qquad\qquad d_{CB} - d_{CA} \end{aligned}$$

Therefore

$$\left[\boldsymbol{A} = \begin{matrix} 1 & -1 & . & . & . & . \\ . & . & 1 & -1 & . & \\ . & . & . & . & 1 & -1 \end{matrix} \right],$$

$$\boldsymbol{\Sigma}_{XX} = s^2 \begin{bmatrix} 0.05 & . & . & . & . & . \\ . & 0.25 & . & . & . & . \\ . & . & 0.25 & . & . & . \\ . & . & . & 0.10 & . & . \\ . & . & . & . & 0.10 & . \\ . & . & . & . & . & 0.05 \end{bmatrix}$$

and

$$\boldsymbol{\Sigma}_{YY} = \boldsymbol{A}\boldsymbol{\Sigma}_{XX}\boldsymbol{A}^{\mathrm{T}} = s^2 \begin{bmatrix} 0.30 & . & . \\ . & 0.35 & . \\ . & . & 0.15 \end{bmatrix}$$

(ii) The equation for the misclosure is

$$\Delta = 180 - (A + B + C)$$

The matrix $\boldsymbol{A}$ is $\quad \boldsymbol{A} = (-1 \quad -1 \quad -1)$

and the matrix $\boldsymbol{\Sigma}_{XX}$ is the matrix $\boldsymbol{\Sigma}_{YY}$ obtained in (i), i.e.

$$\boldsymbol{\Sigma}_{XX} = s^2 \begin{bmatrix} 0.30 & . & . \\ . & 0.35 & . \\ . & . & 0.15 \end{bmatrix}$$

Therefore $\boldsymbol{\Sigma}_{YY} = \boldsymbol{A}\boldsymbol{\Sigma}_{XX}\boldsymbol{A}^{\mathrm{T}} = 0.80s^2$

(iii) The equations for the adjusted angles are

$$\bar{A} = A + a$$
$$\bar{B} = B + b$$
$$\bar{C} = C + c$$

and $\quad a = \dfrac{\sigma_A^2}{\sigma_\Delta^2} \cdot \Delta = \dfrac{3}{8} \cdot \Delta; \quad b = \dfrac{\sigma_B^2}{\sigma_\Delta^2} \cdot \Delta = \dfrac{3.5}{8} \cdot \Delta; \quad c = \dfrac{\sigma_C^2}{\sigma_\Delta^2} \cdot \Delta = \dfrac{1.5}{8} \cdot \Delta$

Therefore

$$\bar{A} = A + \frac{3.0}{8} \cdot \Delta$$

$$\bar{B} = B + \frac{3.5}{8} \cdot \Delta$$

$$\bar{C} = C + \frac{1.5}{8} \cdot \Delta$$

$\boldsymbol{\Sigma}_{xx}$ must still be found:

$$\left.\begin{aligned} A &= A \\ B &= B \\ C &= C \\ \Delta &= 180 - (A + B + C) \end{aligned}\right\} \text{therefore,} \quad \begin{aligned} \sigma_{A\Delta} &= -\sigma_A^2 \\ \sigma_{B\Delta} &= -\sigma_B^2 \\ \sigma_{C\Delta} &= -\sigma_C^2 \end{aligned}$$

Therefore

$$\Sigma_{XX} = s^2 \begin{bmatrix} 0.30 & . & . & -0.30 \\ . & 0.35 & . & -0.35 \\ . & . & 0.15 & -0.15 \\ -0.30 & -0.35 & -0.15 & 0.80 \end{bmatrix}$$

and

$$\Sigma_{YY} = A\Sigma_{XX}A^{\mathrm{T}} = s^2 \begin{bmatrix} 0.1875 & -0.1312 & -0.0562 \\ & 0.1969 & -0.0656 \\ \text{symmetric} & & 0.1219 \end{bmatrix}$$

Thus, for example, if s is equal to 3″

$$\Delta_{YY} = \begin{bmatrix} 1.69 & -1.18 & -0.50 \\ & 1.77 & -0.59 \\ \text{symm.} & & 1.10_{(\mathrm{sec}^2)} \end{bmatrix}$$

and $$\sigma_{\bar{A}} = 1.3'', \quad \sigma_{\bar{B}} = 1.3'', \quad \sigma_{\bar{C}} = 1.0''$$

Example 8.7

A photogrammetric model is to be levelled analytically using three height control points, numbered 1, 2 and 3 in Figure 8.14.

P and Q are other points which have been observed photogrammetrically, and whose terrain heights are required. Establish the accuracy of the derived terrain heights of P and Q, in terms of the accuracy of height observation in the photogrammetric model.

The assumptions made in the solution to this problem are:

(i) the observed photogrammetric heights are independent and of equal weight;
(ii) the standard deviation of setting the floating mark on the model surface is 1/5000 of the flying height (Z); and
(iii) the given heights for the control points are error free.

Solution

Final heights for P and Q are given by the equation

$$H = h + \Delta \tag{8.38}$$

where h is the observed value, and Δ is the correction to the observed value, derived from analytical levelling.

Applying the law of propagation of cofactors,

$$g_{HH} = g_{hh} + g_{\Delta\Delta} + 2g_{h\Delta} \tag{8.39}$$

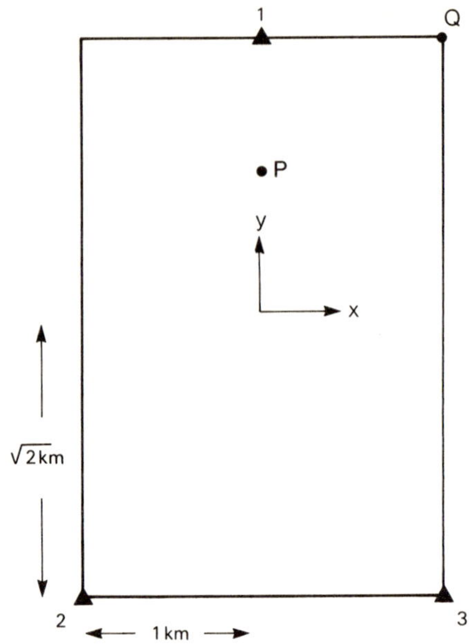

Figure 8.14

Let $g_{hh} = 1$;

$$\left[\text{i.e.}\quad g_{hh} = \frac{\sigma_{h^2}}{\sigma^2} = \frac{(Z/5000)^2}{\sigma^2} = 1,\right.$$

therefore the reference variance is

$$\left.\sigma^2 = (Z/5000)^2\right]$$

$g_{\Delta\Delta}$ must be found from

$$\Delta = ax + by + c \tag{8.40}$$

where x, y are known coordinates (regarded as constants), and a, b, c are the parameters for analytical levelling—two tilts and a shift in height; they are derived from observations to control points, therefore they are stochastic variables.

Applying the law of propagation of cofactors to eqn (8.40) gives

$$g_{\Delta\Delta} = x^2 g_{aa} + y^2 g_{bb} + g_{cc} + 2xy g_{ab} + 2x g_{ac} + 2y g_{bc} \tag{8.41}$$

$g_{h\Delta}$ may be found from

$$\left.\begin{aligned} h &= h \\ \Delta &= ax + by + c \end{aligned}\right\} \tag{8.42}$$

The parameters a, b, c, are derived from observations to control and, since the observations at P and Q are independent of those at control, therefore the observations at P and Q are not correlated with a, b and c. Therefore $g_{h\Delta} = 0$

In eqn (8.41) the cofactors $g_{aa}, g_{bb}, g_{cc}, g_{ab}, g_{ac}, g_{bc}$ must be found.

Parameter a is found from

$$a = \frac{\Delta_3 - \Delta_2}{x_3 - x_2} = \frac{\Delta_3 - \Delta_2}{2 \times 10^3}, \tag{8.43}$$

where the coordinates are in metres.

Now $\Delta_2 = H_2 - h_2$, where H_2 is the given value, and h_2 is the observed value. Therefore, $g_{\Delta_2\Delta_2} = g_{h_2h_2} = 1$; similarly for $g_{\Delta_3\Delta_3}$.

Therefore

$$g_{aa} = \frac{1+1}{(2 \times 10^3)^2} = \frac{1}{2 \times 10^6} \tag{8.44}$$

Parameter b is found from

$$b = \frac{1}{2(2^{1/2} \times 10^3)} \cdot \left(\frac{\Delta_2 + \Delta_3}{2} - \Delta_1 \right) \tag{8.45}$$

Therefore,

$$g_{bb} = \frac{1}{4(2 \times 10^6)} \cdot \left(\frac{1}{4} + \frac{1}{4} + 1 \right) = \frac{1.5}{8 \times 10^6} \tag{8.46}$$

Parameter c is found from

$$c = \frac{\Delta_1 + \frac{1}{2}(\Delta_2 + \Delta_3)}{2} = \tfrac{1}{4}(2\Delta_1 + \Delta_2 + \Delta_3) \tag{8.47}$$

therefore

$$g_{cc} = \tfrac{1}{16}(4 + 1 + 1) = \tfrac{3}{8} \tag{8.48}$$

When the equations (8.43), (8.45), and (8.47) are considered in pairs the mixed cofactors may be found, and the results are:

$$\left.\begin{aligned}
g_{ab} &= \frac{1}{(2 \times 10^3)} \cdot \frac{1}{4(2^{1/2} \times 10^3)} - \frac{1}{(2 \times 10^3)} \cdot \frac{1}{4(2^{1/2} \times 10^3)} = 0 \\
g_{ac} &= \frac{1}{(2 \times 10^3)} \cdot \frac{1}{16} - \frac{1}{(2 \times 10^3)} \cdot \frac{1}{16} = 0 \\
g_{bc} &= -\frac{1}{2(2^{1/2} \times 10^3)} \cdot \frac{2}{4} + \frac{1}{4(2^{1/2} \times 10^3)} \cdot \frac{1}{4} + \frac{1}{4(2^{1/2} \times 10^3)} \cdot \frac{1}{4} \\
&= -\frac{1}{4(2^{1/2} \times 10^3)} + \frac{1}{8(2^{1/2} \times 10^3)} \\
&= -\frac{1}{8(2^{1/2} \times 10^3)}
\end{aligned}\right\} \tag{8.49}$$

Thus, because of the geometry of the control points, there is no correlation between a and b and between a and c. This example serves to illustrate how derived quantities may turn out to be uncorrelated even though their equations contain the same stochastic variables.

When these values for g_{aa}, g_{bb}, g_{cc}, g_{bc}, found in eqns (8.44), (8.46), (8.48), and (8.49) respectively, are substituted in eqn (8.41), the following expression for $g_{\Delta\Delta}$ is obtained:

$$g_{\Delta\Delta} = x^2\left(\frac{1}{2\times 10^6}\right) + y^2\left(\frac{1.5}{8\times 10^6}\right) + \frac{3}{8} - 2y\left(\frac{1}{8(2^{1/2}\times 10^3)}\right) \tag{8.50}$$

When this expression is substituted in eqn (8.39), and the coordinates for points P and Q introduced, the following values for accuracy of the heights are obtained:

for P:

$$g_{HH} = 1 + 0 + \tfrac{3}{32} + \tfrac{3}{8} - \tfrac{1}{8}$$
$$= 1.34$$

for Q:

$$g_{HH} = 1 + \tfrac{1}{2} + \tfrac{3}{8} + \tfrac{3}{8} - \tfrac{1}{4}$$
$$= 2.0$$

The mixed cofactor $g_{H_P H_Q}$ may be found from

$$H_P = h_P + \Delta_P = h_P + ax_P + by_P + c$$
$$H_Q = h_Q + \Delta_Q = h_Q + ax_Q + by_Q + c$$

Therefore,

$$g_{H_P H_Q} = x_P x_Q, g_{aa} + y_P y_Q \cdot g_{bb} + g_{cc} + (y_P + y_Q)\cdot g_{bc} \qquad \text{(Note: } g_{ab} = g_{ac} = 0\text{)}$$
$$= 0 + \frac{1.5}{8} + \frac{3}{8} - \frac{1.5}{8}$$
$$= 0.375$$

These cofactors may be converted to variances by multiplying by the reference variance; thus

for P: $\sigma_{H^2} = \sigma^2 g_{HH} = (Z/5000)^2\, 1.34$

for Q: $\sigma_{H^2} = \sigma^2 g_{HH} = (Z/5000)^2\, 2.00$

For example, if the flying height is 5000 m, the variances for P and Q would be 1.34 m^2 and 2.00 m^2 respectively, giving standard deviations of 1.15 m and 1.41 m. Thus the effect of extrapolation from the control points is clear.

If matrix notation is used, the matrices may be identified from the following equations, which are derived from eqn (8.40):

$$\left.\begin{aligned} \Delta_P &= 0.a + \frac{(2^{1/2}\times 10^3)}{2}\cdot b + c \\ \Delta_Q &= 10^3\cdot a + (2^{1/2}\times 10^3)\cdot b + c \end{aligned}\right\} \tag{8.51}$$

Therefore

$$G_{\Delta\Delta}=\begin{bmatrix} 0 & \left(\dfrac{2^{1/2}\times 10^3}{2}\right) & 1 \\ 10^3 & (2^{1/2}\times 10^3) & 1 \end{bmatrix} \cdot G_{XX} \cdot \begin{bmatrix} 0 & 10^3 \\ \left(\dfrac{2^{1/2}\times 10^3}{2}\right) & \dfrac{(2^{1/2}\times 10^3)}{1} \\ 1 & 1 \end{bmatrix}$$

where

$$G_{XX}=\begin{bmatrix} g_{aa} & g_{ab} & g_{ac} \\ g_{ab} & g_{bb} & g_{bc} \\ g_{ac} & g_{bc} & g_{cc} \end{bmatrix} = \begin{bmatrix} \dfrac{1}{2\times 10^4} & \cdot & \cdot \\ \cdot & \dfrac{1.5}{8\times 10^6} & -\dfrac{1}{8(2^{1/2}\times 10^3)} \\ \cdot & -\dfrac{1}{8(2^{1/2}\times 10^3)} & \dfrac{3}{8} \end{bmatrix}$$

Therefore

$$G_{\Delta\Delta}=\frac{1}{32}\begin{pmatrix} 11 & 12 \\ 12 & 32 \end{pmatrix}.$$

The final heights are obtained from eqn (8.38)

$$\left.\begin{aligned} H_P &= h_P \quad + \Delta_P \\ H_Q &= \quad h_Q \quad + \Delta_Q \end{aligned}\right\} \tag{8.52}$$

i.e.

$$A=\begin{pmatrix} 1 & . & 1 & . \\ . & 1 & . & 1 \end{pmatrix}, \qquad Y=\begin{pmatrix} H_P \\ H_Q \end{pmatrix}$$

and G_{XX} is formed from $G_{\Delta\Delta}$; thus

$$G_{XX}=\frac{1}{32}\begin{bmatrix} 32 & . & . & . \\ . & 32 & . & . \\ . & . & 11 & 12 \\ . & . & 12 & 32 \end{bmatrix}$$

Therefore,

$$G_{HH}=AG_{XX}A^{\mathrm{T}}=\begin{bmatrix} 1. & 1. \\ .1 & .1 \end{bmatrix}\frac{1}{32}\begin{bmatrix} 32 & . & . & . \\ . & 32 & . & . \\ . & . & 11 & 12 \\ . & . & 12 & 32 \end{bmatrix}\begin{bmatrix} 1 & . \\ . & 1 \\ 1 & . \\ . & 1 \end{bmatrix}$$

$$=\frac{1}{32}\begin{pmatrix} 43 & 12 \\ 12 & 64 \end{pmatrix}=\begin{pmatrix} 1.34375 & 0.375 \\ 0.375 & 2.0 \end{pmatrix}$$

Therefore

$$\Sigma_{HH} = G_{HH} \cdot \sigma^2 = \begin{pmatrix} 1.34375 & 0.375 \\ 0.375 & 2.0 \end{pmatrix} .(Z/5000)^2$$

and for a flying height of 5000m,

$$\Sigma_{HH} = \begin{pmatrix} 1.34 & 0.38 \\ 0.38 & 2.00 \end{pmatrix} \text{m}^2$$

8.8 Non-linear functions of stochastic variables

If x is a stochastic variable with a normal distribution, and $y = f(x)$ is a nonlinear function of that variable, then y will not be normally distributed. Fig. 8.15 illustrates such a relationship. However, if $y = f(x)$ is continuous then part of the curve may be approximated by a straight line; and since the practical range of observations is finite there is some justification in concentrating on the specific part which is of interest. This part may be represented by a tangent at that point, shown dotted in the diagram; thus the real distribution of y in that vicinity is represented approximately by a normal distribution, also shown dotted.

The point on the curve chosen to represent the tangent is an estimate of the 'true' value of x. It could be a particular observation, or some other approximation for the 'true' value, such as the mean. For example, in the intersection problem in surveying, the eastings and northings values for the new point are non-linear functions of the observed bearings. An estimate of the

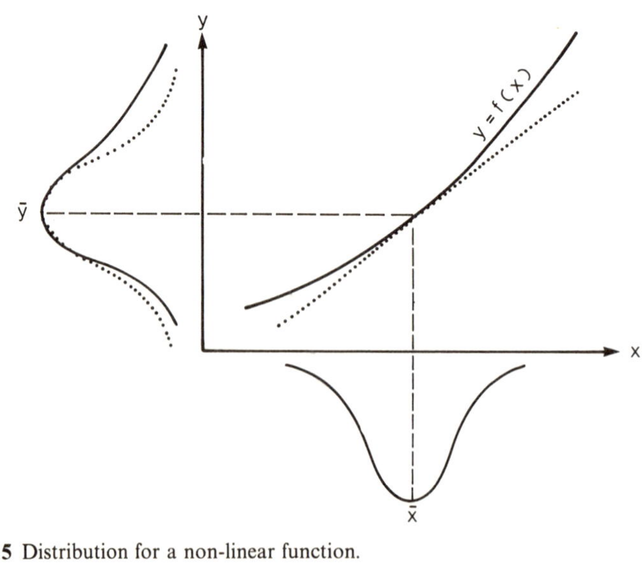

Figure 8.15 Distribution for a non-linear function.

coordinates of the intersected point may be made by scaling the values off an existing map; or alternatively, by using two specific observations to calculate values.

The equation to the tangent is obtained by expanding the function in terms of a Taylor series, and adopting only the linear terms of the series. In other words, the function is differentiated and expanded with respect to each of the stochastic variables in turn, and only the terms with zero-order and first-order derivatives are adopted. For example, the Taylor series expansion of a non-linear function of a single variable, $y = f(x)$, could be expressed in the following way

$$y = y_0 + \left(\frac{\mathrm{d}f}{\mathrm{d}x}\right)_0 \cdot \mathrm{d}x + \frac{1}{2!}\left(\frac{\mathrm{d}^2 f}{\mathrm{d}x^2}\right)_0 \cdot (\mathrm{d}x)^2 + \frac{1}{3!}\left(\frac{\mathrm{d}^3 f}{\mathrm{d}x^3}\right)_0 \cdot (\mathrm{d}x)^3 + \cdots \quad (8.53)$$

where the subscript 0 indicates that the expression in parenthesis is evaluated using an approximate value for the variable—i.e., the expressions denoted by subscript 0 become known constants.

If only the linear terms of the equation are adopted it becomes

$$y = y_0 + \left(\frac{\mathrm{d}f}{\mathrm{d}x}\right)_0 \cdot \mathrm{d}x \quad (8.54)$$

For example, consider the function

$$y = 2 + x + 3x^2$$

The linearized form is

$$\begin{aligned} y &= y_0 + (0 + 1 + 6x)_0 \mathrm{d}x \\ &= y_0 + (1 + 6x)_0 \mathrm{d}x \end{aligned}$$

or $y = a + b \cdot \mathrm{d}x$ where a and b are known constants, and $\mathrm{d}x$ is a stochastic variable.

Applying the law of propagation of variances gives

$$\sigma_y^2 = b^2 \sigma_x^2$$

The same result would have been achieved by considering only that part of the equation containing the stochastic variable, i.e. let $y = y_0 + \mathrm{d}y$ then $\mathrm{d}y = b \cdot \mathrm{d}x$ and applying the law of propagation to this equation, gives

$$\sigma_y^2 = b^2 \sigma_x^2$$

The same principles are used for more general cases where y is a function of several stochastic variables, i.e. $y = (x_1 x_2 x_3)$ where x_1, x_2, x_3 are stochastic variables.

Expansion by Taylor series gives

$$y = f(x_1 x_2 x_3)_0 + \left(\frac{\partial f}{\partial x_1}\right)_0 \cdot \mathrm{d}x_1 + \frac{1}{2!}\left(\frac{\partial^2 f}{\partial x_1^2}\right)_0 \cdot (\mathrm{d}x_1)^2 + \cdots$$

$$+\left(\frac{\partial f}{\partial x_2}\right)_0\cdot dx_2+\frac{1}{2!}\left(\frac{\partial^2 f}{\partial x_2^2}\right)_0\cdot(dx_2)^2+\cdots$$

$$+\left(\frac{\partial f}{\partial x_3}\right)_0\cdot dx_3+\frac{1}{2!}\left(\frac{\partial^2 f}{\partial x_3^2}\right)_0\cdot(dx_3)^2+\cdots \tag{8.55}$$

where $\dfrac{\partial f}{\partial x_1}, \dfrac{\partial f}{\partial x_2}, \ldots$ etc. are partial derivatives.

If the constant term and non-linear terms are ignored, the following linear equation is obtained:

$$dy=\left(\frac{\partial f}{\partial x_1}\right)_0\cdot dx_1+\left(\frac{\partial f}{\partial x_2}\right)_0\cdot dx_2+\left(\frac{\partial f}{\partial x_3}\right)_0\cdot dx_3 \tag{8.56}$$

The expressions $(\partial f/\partial x_1)_0, (\partial f/\partial x_2)_0, (\partial f/\partial x_3)_0$ are the constant coefficients, and dy, dx_1, dx_2, dx_3, are the new stochastic variables.

When the law of propagation is applied to eqn (8.56) the following is obtained

$$\sigma_y^2=\left(\frac{\partial f}{\partial x_1}\right)_0^2\sigma_x^2+\left(\frac{\partial f}{\partial x_2}\right)_0^2\sigma_x^2+\left(\frac{\partial f}{\partial x_3}\right)_0^2\sigma_x^2$$

$$+2\left(\frac{\partial f}{\partial x_1}\right)_0\left(\frac{\partial f}{\partial x_2}\right)_0\sigma_{x_1x_2}+2\left(\frac{\partial f}{\partial x_1}\right)_0\left(\frac{\partial f}{\partial x_3}\right)_0\sigma_{x_1x_3}$$

$$+2\left(\frac{\partial f}{\partial x_2}\right)_0\left(\frac{\partial f}{\partial x_3}\right)_0\sigma_{x_2x_3} \tag{8.57}$$

If there are two functions, $y=f_y(x_1x_2x_3)$ and $z=f_z(x_1x_2x_3)$ then both may be linearized, and the variances σ_y^2 and σ_z^2 found. The covariance σ_{yz} may also be found according to the established rules for two derived variables (see eqn (8.18)).

Note that cofactors g could equally have been used instead of variances and covariances.

In the general case, when a number of variables $y_1, y_2, \ldots y_m$, are expressed as functions of stochastic variables $x_1, x_2, \ldots x_n$, the linearized equations of the form of eqn (8.56) may be expressed in matrix notation as

$$\boldsymbol{\Delta y}=\boldsymbol{J}\cdot\boldsymbol{\Delta x} \tag{8.58}$$

where

$$\boldsymbol{\Delta y}=\begin{pmatrix} dy_1\\ dy_2\\ \vdots\\ dy_m\end{pmatrix}, \quad \boldsymbol{\Delta x}=\begin{pmatrix} dx_1\\ dx_2\\ \vdots\\ dx_n\end{pmatrix}$$

and $\boldsymbol{J}$, known as the *jacobian matrix*, is

$$J = \begin{bmatrix} \dfrac{\partial y_1}{\partial x_1} & \dfrac{\partial y_1}{\partial x_2} & \cdots & \dfrac{\partial y_1}{\partial x_n} \\ \dfrac{\partial y_2}{\partial x_1} & \dfrac{\partial y_2}{\partial x_2} & \cdots & \dfrac{\partial y_2}{\partial x_n} \\ \vdots & & & \vdots \\ \dfrac{\partial y_m}{\partial x_1} & \dfrac{\partial y_m}{\partial x_2} & \cdots & \dfrac{\partial y_m}{\partial x_n} \end{bmatrix}$$

For example, consider the two functions

$$y_1 = 6 + x_1 + 3x_2^2$$
$$y_2 = 2x_1 + x_3^2 + 4x_3^2$$

The jacobian matrix is

$$J = \begin{pmatrix} 1 & 6x_2 & 0 \\ 2 & 3x_2^2 & 8x_3 \end{pmatrix}$$

and the law of propagation of variances and covariances may be expressed as

$$\Sigma_{YY} = J\Sigma_{XX}J^{\mathrm{T}} \tag{8.59}$$

or the law of propagation of cofactors may be expressed as

$$G_{YY} = JG_{XX}J^{\mathrm{T}} \tag{8.60}$$

Some examples to illustrate these concepts are given below.

Example 8.8

A rectangular plot of land is measured and is found to be 40 m long and 20 m wide; the measurements have standard deviations of 2 cm and 1 cm respectively. If the dimensions are uncorrelated, what is the accuracy of the area of the plot?

Solution

The mathematical model for the area (A) is

$$A = xy$$

which is a non-linear function of the stochastic variables x and y. Linearization produces

$$\mathrm{d}A = x\mathrm{d}y + y\mathrm{d}x$$

and applying the law of propagation

$$\sigma_A^2 = x^2\sigma_y^2 + y^2\sigma_x^2 + 2xy\sigma_{xy}$$

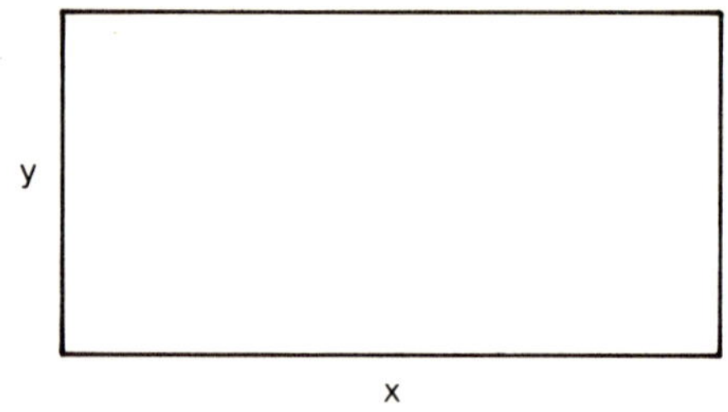

Figure 8.16.

When values are substituted,

$$\sigma_A^2 = 1600\,\text{m}^2\cdot 1\,\text{cm}^2 + 400\,\text{m}^2\cdot 4\,\text{cm}^2 + 0$$
$$= 3200\,\text{m}^2\text{cm}^2$$

Therefore, $$\sigma_A = 56\,\text{m}\cdot\text{cm} = 0.56\,\text{m}^2$$

Example 8.9

A line AB has known coordinates for the point A, and measurements of the bearing α and distance D for the line. The coordinates E_A, N_A and the observations α and D are stochastic variables. Derive expressions for the stochastic properties of the coordinates E_B, N_B, which are calculated from the data.

Solution

The mathematical model for the coordinates of B is

$$E_B = E_A + D\sin\alpha$$
$$N_B = N_A + D\cos\alpha$$

The equations are non-linear, and when they are linearized they become

$$\left.\begin{aligned} \mathrm{d}E_B &= \mathrm{d}E_A + D\cos\alpha\,\mathrm{d}\alpha + \sin\alpha\,\mathrm{d}D \\ \mathrm{d}N_B &= \mathrm{d}N_A - D\sin\alpha\,\mathrm{d}\alpha + \cos\alpha\,\mathrm{d}D \end{aligned}\right\} \qquad (8.61)$$

Applying the law of propagation,

$$\left.\begin{aligned} \sigma_{E_B}^2 &= \sigma_{E_A}^2 + (D\cos\alpha)^2\sigma_\alpha^2 + (\sin\alpha)^2\sigma_D^2 + 2D\cos\alpha\,\sigma_{E_A\alpha} + 2\sin\alpha\,\sigma_{E_AD} \\ &\quad + 2D\cos\alpha\sin\alpha\,\sigma_{\alpha D} \\ \sigma_{N_B}^2 &= \sigma_{N_A}^2 + (-D\sin\alpha)^2\sigma_\alpha^2 + (\cos\alpha)^2\sigma_D^2 - 2D\sin\alpha\,\sigma_{N_A\alpha} \\ &\quad + 2\cos\alpha\,\sigma_{N_AD} - 2D\sin\alpha\cos\alpha\,\sigma_{\alpha D} \\ \sigma_{E_BN_B} &= -D^2\sin\alpha\cos\alpha\,\sigma_\alpha^2 + \sin\alpha\cos\alpha\,\sigma_D^2 + \sigma_{E_AN_A} + (D\cos^2\alpha \\ &\quad - D\sin^2\alpha)\sigma_{\alpha D} - D\sin\alpha\,\sigma_{E_A\alpha} + \cos\alpha\,\sigma_{E_AD} \\ &\quad + D\cos\alpha\,\sigma_{N_A\alpha} + \sin\alpha\,\sigma_{N_AD} \end{aligned}\right\} \qquad (8.62)$$

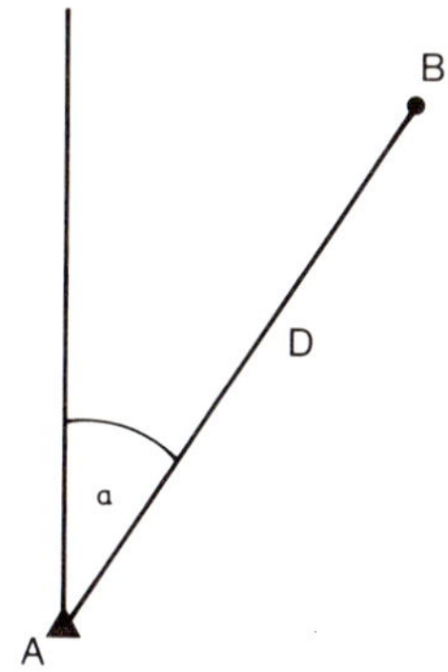

Figure 8.17.

or

$$G_{YY} = JG_{XX} J^{\mathrm{T}}$$

where

$$J = \begin{pmatrix} 1 & . & D\cos\alpha & \sin\alpha \\ . & 1 & -D\sin\alpha & \cos\alpha \end{pmatrix}, \quad G_{XX} = \begin{bmatrix} \sigma^2_{E_A} & \sigma_{E_A N_A} & \sigma_{E_A \alpha} & \sigma_{E_A D} \\ & \sigma^2_{N_A} & \sigma_{N_A \alpha} & \sigma_{N_A D} \\ & & \sigma^2_{\alpha} & \sigma_{\alpha D} \\ \text{symmetric} & & & \sigma^2_D \end{bmatrix},$$

and $G_{YY} = \begin{pmatrix} \sigma^2_{E_B} & \sigma_{E_B N_B} \\ \sigma_{E_B N_B} & \sigma^2_{N_B} \end{pmatrix}$

Comment:

The solution above (eqn (8.62)) appears complicated because the stochastic variables are assumed to be correlated. In practice some variables are regarded as uncorrelated, thus reducing certain terms to 0 and simplifying the result.

Example 8.10

The two endpoints of a line AB have known coordinates with stochastic properties. Derive expressions for the stochastic properties of the bearing α and distance D, calculated from the coordinates.

Solution

(a) The mathematical model for the bearing α is

$$\tan\alpha = \frac{\Delta E}{\Delta N}$$

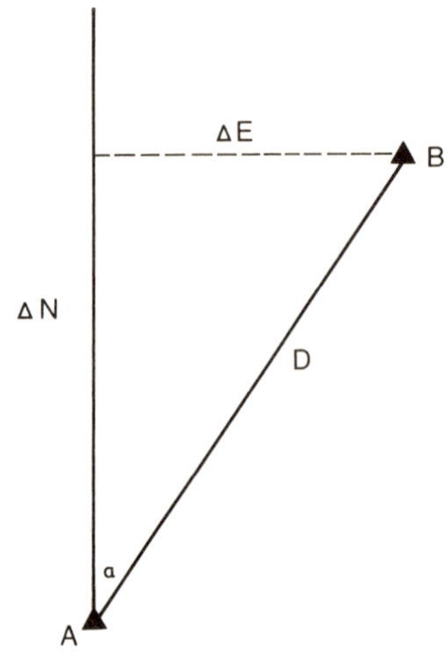

Figure 8.18.

which, on linearization, becomes

$$\frac{1}{\cos^2\alpha}\cdot \mathrm{d}\alpha = \frac{1}{\Delta N}\cdot \mathrm{d}(\Delta E) - \frac{\Delta E}{\Delta N^2}\cdot \mathrm{d}(\Delta N)$$

Now, $\left.\begin{aligned}\Delta E &= E_B - E_A\\ \Delta N &= N_B - N_A\end{aligned}\right\}$ therefore $\left.\begin{aligned}\mathrm{d}(\Delta E) &= \mathrm{d}E_B - \mathrm{d}E_A\\ \mathrm{d}(\Delta N) &= \mathrm{d}N_B - \mathrm{d}N_A\end{aligned}\right\}$

Therefore,

$$\begin{aligned}\mathrm{d}\alpha &= \frac{\cos^2\alpha}{\Delta N}(\mathrm{d}E_B - \mathrm{d}E_A) - \cos^2\alpha\frac{\Delta E}{\Delta N^2}(\mathrm{d}N_B - \mathrm{d}N_A)\\ &= \frac{\Delta N^2}{D^2}\cdot\frac{1}{\Delta N}(\mathrm{d}E_B - \mathrm{d}E_A) - \frac{\Delta N^2}{D^2}\cdot\frac{\Delta E}{\Delta N^2}(\mathrm{d}N_B - \mathrm{d}N_A)\end{aligned}$$

Therefore

$$\mathrm{d}\alpha = \frac{\cos\alpha}{D}(\mathrm{d}E_B - \mathrm{d}E_A) - \frac{\sin\alpha}{D}(\mathrm{d}N_B - \mathrm{d}N_A) \tag{8.63}$$

Note that the units for $\mathrm{d}\alpha$ are radians.

(b) The mathematical model for the distance D is

$$D^2 = \Delta E^2 + \Delta N^2$$

which, on linearization, becomes

$$2D\cdot \mathrm{d}D = 2\Delta E\cdot \mathrm{d}(\Delta E) + 2\Delta N\cdot \mathrm{d}(\Delta N)$$

Therefore,

$$\mathrm{d}D = \frac{\Delta E}{D}(\mathrm{d}E_B - \mathrm{d}E_A) + \frac{\Delta N}{D}(\mathrm{d}N_B - \mathrm{d}N_A)$$

or

$$\mathrm{d}D = \sin\alpha(\mathrm{d}E_B - \mathrm{d}E_A) + \cos\alpha(\mathrm{d}N_B - \mathrm{d}N_A) \tag{8.64}$$

The law of propagation may be applied to the linear equations (8.63) and (8.64), thus

$$\boldsymbol{G}_{YY} = \boldsymbol{J}\boldsymbol{G}_{XX}\boldsymbol{J}^{\mathrm{T}}$$

where

$$\boldsymbol{J} = \begin{pmatrix} \dfrac{\cos\alpha}{D} & \dfrac{-\cos\alpha}{D} & \dfrac{-\sin\alpha}{D} & \dfrac{\sin\alpha}{D} \\ \sin\alpha & -\sin\alpha & \cos\alpha & -\cos\alpha \end{pmatrix}$$

$$\boldsymbol{G}_{XX} = \begin{bmatrix} \sigma^2_{E_B} & \sigma_{E_BE_A} & \sigma_{E_BN_B} & \sigma_{E_BN_A} \\ & \sigma^2_{E_A} & \sigma_{E_AN_B} & \sigma_{E_AN_A} \\ & & \sigma^2_{N_B} & \sigma_{N_BN_A} \\ \text{symmetric} & & & \sigma^2_{N_A} \end{bmatrix}$$

$$\boldsymbol{G}_{YY} = \begin{pmatrix} \sigma^2_\alpha & \sigma_{\alpha D} \\ \sigma_{\alpha D} & \sigma^2_D \end{pmatrix}$$

Example 8.11

Tacheometric surveying involves the observation of stadia hair intercepts on a vertical staff, and the corresponding horizontal and vertical angles. If the standard deviations for the measurement of an intercept and a vertical angle (v) are 2 mm and 1′ respectively, and the observations are uncorrelated

(a) derive expressions for the accuracy of the horizontal distance (D) and the height difference (H); and

(b) derive values for the accuracy of D and H for distances of 10 m and 50 m, and for a range of vertical angles from 0° to 15°.

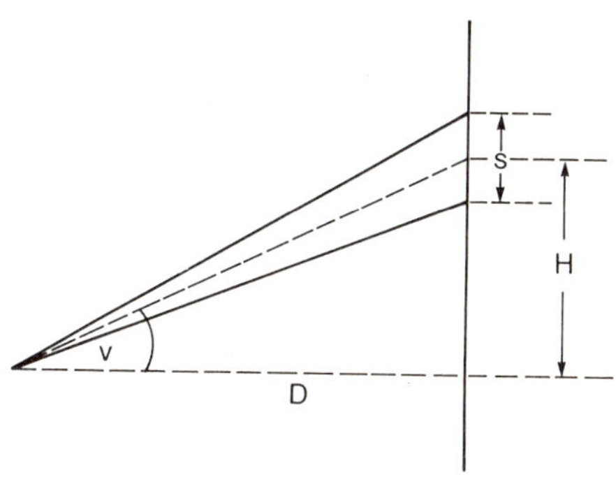

Figure 8.19

Solution

(a) If u and l are the upper and lower stadia intercept readings then s the stadia interval is given by

$$s = u - l \tag{8.65}$$

Thus

$$\sigma_s^2 = \sigma_u^2 + \sigma_l^2. \tag{8.66}$$

and the mathematical models for horizontal distance and height difference are

$$\left.\begin{aligned} D &= 100\, s \cos^2 v \\ H &= 100\, s \sin v \cos v \end{aligned}\right\} \tag{8.67}$$

When the equations are linearized they become

$$\left.\begin{aligned} \mathrm{d}D &= 100 \cos^2 v \cdot \mathrm{d}s - 200 s \sin v \cos v \cdot \mathrm{d}v \\ \mathrm{d}H &= 100 \sin v \cos v \cdot \mathrm{d}s + 100\, s(\cos^2 v - \sin^2 v) \cdot \mathrm{d}v \end{aligned}\right\} \tag{8.68}$$

and, applying the law of propagation,

$$\begin{aligned} \sigma_D^2 &= (100 \cos^2 v)^2 \sigma_s^2 (200\, s\, (\sin v \cos v)^2 \sigma_{v^2}) \\ \sigma_H^2 &= (100 \sin v \cos v)^2 \sigma_s^2 + (100\, s(\cos^2 v - \sin^2 v))^2 \sigma_{v^2}) \end{aligned} \tag{8.69}$$

which may be written

$$\left.\begin{aligned} \sigma_D^2 &= a\sigma_s^2 + b\sigma_v^2 \\ \sigma_H^2 &= c\sigma_s^2 + d\sigma_t^2 \end{aligned}\right\} \tag{8.70}$$

(b) σ_s^2 is calculated from eqn (8.66).

$$\sigma_s^2 = 4 + 4 = 8\ \mathrm{mm}^2$$

and

$$\sigma_v^2 = (1')^2 = (8.46 \times 10^{-8})\mathrm{rad}^2.$$

Values for the terms in eqns (8.70), and final values for σ_D and σ_H, may be tabulated as follows:

s mm	v °	$a\sigma_s^2$ (mm²)	$b\sigma_t^2$ (mm²)	$c\sigma_s^2$ (mm²)	$\mathrm{d}\sigma_t^2$ (mm²)	σ_D^D (mm)	σ_H (mm)
100	0	80000	0	0	8	283	3
	5	78789	0	603	8	281	25
	10	75248	1	2340	7	274	48
	15	69641	2	5000	6	264	71
500	0	80000	0	0	212	283	14
	5	78789	6	603	205	281	28
	10	75248	25	2340	187	274	50
	15	69641	52	5000	158	264	72

Comment

1. There is little change in σ_D for the range of observations considered. The dominant term in both equations is that containing α_s^2—i.e. the main factor influencing the results is the precision in the measurement of the stadia interval.
2. In the case of α_H there is a substantial decrease in accuracy when large vertical angles are used. This again is due to the influence of α_s^2.
3. The assumption above has been that α_s remains constant for the range of observations. In practice it is more likely that it increases significantly for greater distances and for the larger vertical angles. For example, with $s = 500$ mm, and an intercept standard deviation of 3 mm and 4 mm for vertical angles of 5° and 10° respectively, the following results are obtained:

$\sigma_u(\sigma_1)$ (mm)	σ_{u^2} (mm^2)	v °	$a\sigma_s^2$ (mm^2)	$b\sigma_v^2$ (mm^2)	$c\sigma_s^2$ (mm^2)	$d\sigma_v^2$ (mm^2)	σ_D (mm)	σ_H (mm)
3	18	5	177275	6	1357	205	421	40
4	32	10	300992	25	9358	187	549	98

9 Least-squares estimation

9.1 Introduction

The estimation of final values of observed quantities (or the 'adjustment' of observations, as it has been referred to in the past), is a necessary procedure for obtaining the required consistency between observations, and for obtaining estimates of related stochastic variables. It has sense only if there are redundant observations. For example, adopting the mean of 20 parallax-bar readings for a single point is a process of estimation. The mean is the estimate for a single stochastic variable in this case, and although 19 of the observations are redundant they all contribute to the derivation of the final value.

Any functional model representing a geometrical system has a minimum number of independent (functionally) variables that determines the model uniquely. A few examples will illustrate this concept (see Figure 9.1).

(i) 20 parallax-bar readings of a point—one variable determines the model.
(ii) The shape of a plane triangle—two angles (A and B, or A and C, or B and C), or two ratios between three sides determines the model. Note that although the actual variables might change, the number does not.
(iii) The size and shape of a plane triangle—one side plus two angles, or three sides, or two sides plus one angle determines the model.
(iv) A line of levels—If A is a point of known height then three height differences must be observed to establish the heights of B, C, and D. If D is also the starting point then only two height differences are required.

It is clear from these examples that a certain minimum number of observations is required to determine the minimum number of variables in each case. Provided they are the right sort of observations, these two numbers will be equal. For example, in case (ii), two observations of the same angle in the triangle will not determine the shape of the triangle.

If there is more than the minimum number of observations the model is 'overdetermined', and an estimation of final values for the observed quantities is necessary. The estimation process can be approximate (e.g. 'equal shifts' or 'semi-graphic' in surveying), or more rigorous. It should preferably be the latter—i.e., providing the best estimates for the stochastic variables, based on the observations and all available information about the observations. The least-squares principle does this. Stated in the simplest terms—that is for observations which are uncorrelated and of equal precision—the principle is $\sum v^2 = \text{minimum}$, and it has the following effects:

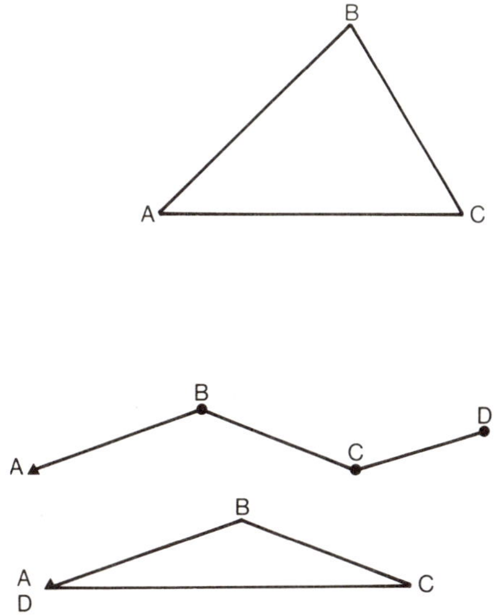

Figure 9.1.

(a) The final estimates ('adjusted' values) are as close as possible to the original observations. (Note that if redundancy is 0 the residuals will all be 0.)
(b) It provides a unique answer to the problem, whatever the method of least-squares estimation.
(c) It has a practical value, in that complex computational problems are solved more easily.

Stated in more general terms, when observations are of varying precision, the principle is

$$E = [wvv] = \text{minimum}$$

where w is the weight and [ ] signifies summation.

In matrix notation, and when observations are also considered to be correlated, the least-squares condition is that which minimises the *quadratic form* of the residuals, viz.

$$\left.\begin{aligned} E &= \boldsymbol{v}^{\mathrm{T}}\boldsymbol{W}\boldsymbol{v} = \text{minimum} \\ \text{or} \quad E &= \boldsymbol{v}^{\mathrm{T}}\boldsymbol{G}^{-1}\boldsymbol{v} = \text{minimum} \end{aligned}\right\} \tag{9.1}$$

where $\boldsymbol{v}$ is the vector of residuals (or corrections to the observations), $\boldsymbol{W}$ is the weight matrix for the observations, and $\boldsymbol{G}$ is the covariance or cofactor matrix for the observations.

It is not a prerequisite of least-squares application that the observations are normally distributed. However, if they are not normally distributed, and account is not taken of such lack of symmetry, the residuals will display some systematic bias, and they will not reflect the true capabilities of the measuring instrument (or the observer). In spite of this, least squares is often applied to observations containing systematic errors—the reason being that it provides a highly efficient computational algorithm, giving a unique answer even in very complicated situations in geodesy, surveying or photogrammetry. This advantage has become more obvious in recent times due to rapid advances in computer technology.

All least-squares problems start with both a mathematical model and a stochastic model.

The *mathematical* (or *functional*) *model* describes the relationship between the observations and the unknowns. The unknowns may be the final estimates of the observed quantities, or other parameters derived from these estimates. Generally each unknown is a function of more than one observed quantity. Also, each observed quantity may contribute to the determination of more than one unknown.

For example, consider a line of levels from A to B to C, and back to A. The final height (H_C) of the point C is a function of the three observed height differences $\Delta h_1, \Delta h_2$ and Δh_3,(Figure 9.2), i.e.

$$H_c = f(\Delta h_1, \Delta h_2, \Delta h_3)$$

and furthermore, Δh_3 contributes to the determination of both H_B and H_c. The mathematical model is often non-linear in other applications; for example,

$$\left.\begin{aligned} E_B &= E_A + D\sin\alpha \\ N_B &= N_A + D\cos\alpha \end{aligned}\right\} \qquad (9.2)$$

and this may cause problems from a computational point of view. Also, the variances and covariances of derived quantities are based on linear functions. Therefore before least squares is used the equations must be made linear, as explained earlier—i.e. approximate values for the unknowns are assigned, and these should be as close as possible to the final values. The new unknowns will be corrections to the assigned values. The process of linearization is merely a preliminary to least-squares estimation—it is not a part of the least-squares procedure as such.

The *stochastic model* (the $\mathbf{G}$ matrix) provides the information concerning the stochastic properties of the original observations. It may be expressed either as variances and covariances, or as cofactors, and very often it is simply the unit

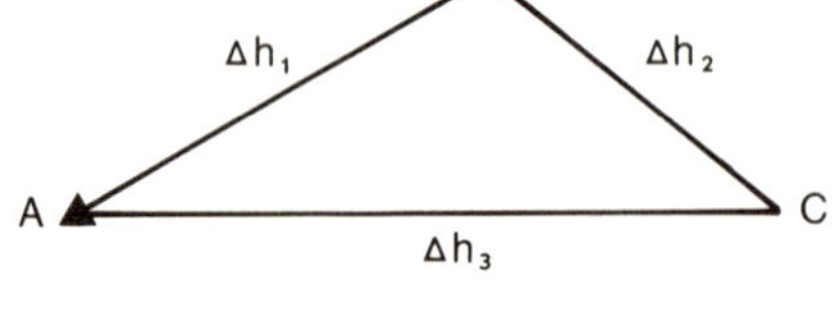

Figure 9.2.

matrix $\boldsymbol{I}$. Before a least-squares computation may be implemented the stochastic model must be specified—even if by default, or lack of information, it is merely $\boldsymbol{I}$. In general it is possible to assess in approximate terms the relative precision of observations, even if it is not possible to establish the covariances between the observations. Thus a diagonal $\boldsymbol{G}$ matrix should not be too difficult to specify.

Several procedures are available for least-squares estimation, but provided the same observational and stochastic data is used the results should be identical. In this text only two of the main procedures used in practice are considered in detail—namely the method using *condition equations*, and the method using *observation equations*.

9.2 Least-squares estimation by condition equations

The mathematical model is based on the conditions that the n adjusted observations must fulfil—i.e. the geometry of the situation is used to set up independent conditions relating the various observations, and if there are r redundant observations there will be r independent condition equations.

The computational procedure is followed more easily by taking a simple example, before developing the general formulae. Consider that the three angles of a triangle have been observed, and that the observations are independent and of equal weight. The number of redundant observations is 1; therefore $r = 1$ and there will be one condition equation to be fulfilled by the adjusted observations, namely

$$\bar{x}_1 + \bar{x}_2 + \bar{x}_3 = 180^\circ \tag{9.3}$$

The equation is linear, therefore the least-squares procedure may be applied directly. Final estimates in eqn (9.3) are replaced by (observations + corrections), i.e.

$$(x_1 + v_1) + (x_2 + v_2) + (x_3 + v_3) = 180^\circ \tag{9.4}$$

or

$$v_1 + v_2 + v_3 = 180^\circ - (x_1 + x_2 + x_3)$$

or

$$v_1 + v_2 + v_3 = t \tag{9.5}$$

where

$$t = 180^\circ - (x_1 + x_2 + x_3) \tag{9.6}$$

and is termed the *misclosure*.

Equation (9.5) is often referred to as the *reduced condition equation*. The v's are the unknown corrections (residuals) to the observations, and t is a known constant.

The least-squares condition (eqn (9.1)) is now imposed, i.e.

$$E = [wvv] = \text{minimum}$$

In this example all the w's are equal to unity, therefore

$$E = v_1^2 + v_2^2 + v_3^2 = \text{minimum} \tag{9.7}$$

If only two of the v's are derived the problem is solved. Therefore v_3 in eqn (9.7) may be substituted by $(t - v_1 - v_2)$, obtained from eqn (9.5). Thus eqn (9.7) becomes

$$E = v_1^2 + v_2^2 + (t - v_1 - v_2)^2 = \text{minimum} \tag{9.8}$$

and, for this condition to be satisfied,

$$\left.\begin{aligned} \frac{\partial E}{\partial v_1} &= 2v_1 - 2(t - v_1 - v_2) = 0 \\ \frac{\partial E}{\partial v_2} &= 2v_2 - 2(t - v_1 - v_2) = 0 \end{aligned}\right\} \tag{9.9}$$

which may be reduced to

$$\left.\begin{aligned} 2v_1 + 1v_2 &= t \\ 1v_1 + 2v_2 &= t \end{aligned}\right\} \tag{9.10}$$

Equations (9.10) are two consistent equations with a unique solution, found to be

$$v_1 = v_2 = \tfrac{1}{3}t \tag{9.11}$$

and, when these values are substituted in eqn (9.5),

$$v_3 = t - \tfrac{1}{3}t - \tfrac{1}{3}t = \tfrac{1}{3}t \tag{9.12}$$

Therefore, the final estimates become

$$\left.\begin{aligned} \bar{x}_1 &= x_1 + v_1 = x_1 + \tfrac{1}{3}t \\ \bar{x}_2 &= x_2 + v_2 = x_2 + \tfrac{1}{3}t \\ \bar{x}_3 &= x_3 + v_3 = x_3 + \tfrac{1}{3}t \end{aligned}\right\} \tag{9.13}$$

which should satisfy the condition equation (9.3).

The problem can also be solved by using a different technique, which has advantages in more complex situations.

An unknown *correlate* or multiplier (referred to as a *Lagrange multiplier*) is attached to each condition equation. Thus eqn (9.5) is written

$$2k([v] - t) = 0 \tag{9.14}$$

where k is the Lagrange multiplier, and the 2 has been introduced purely for convenience, as will be seen later. Equation (9.7) may be rewritten as

$$E = [v^2] - 2k([v] - t) = \text{minimum} \tag{9.15}$$

without loss of validity, since the term involving k is equal to 0.

For E to be a minimum,

$$\frac{\partial E}{\partial v_1}=\frac{\partial E}{\partial v_2}=\frac{\partial E}{\partial v_3}=0$$

Therefore

$$\left.\begin{aligned}\frac{\partial E}{\partial v_1}&=2v_1-2k=0; \qquad \text{therefore} \quad v_1=k\\ \frac{\partial E}{\partial v_2}&=2v_2-2k=0; \qquad \text{therefore} \quad v_2=k\\ \frac{\partial E}{\partial v_3}&=2v_3-2k=0; \qquad \text{therefore} \quad v_3=k\end{aligned}\right\} \tag{9.16}$$

Substituting in eqn (9.5), $k+k+k=t$

therefore

$$3k=t \tag{9.17}$$

which is termed the *correlated* (or *normal*) *equation.*

Therefore

$$k=\tfrac{1}{3}t \tag{9.18}$$

and

$$v_1=v_2=v_3=\tfrac{1}{3}t \tag{9.19}$$

This is an obvious solution to the problem, and one which could have been obtained intuitively; nevertheless the least-squares application has been demonstrated.

The final estimates become

$$\left.\begin{aligned}\bar{x}_1&=x_1+v_1=x_1+\tfrac{1}{3}t\\ \bar{x}_2&=x_2+v_2=x_2+\tfrac{1}{3}t\\ \bar{x}_3&=x_3+v_3=x_3+\tfrac{1}{3}t\end{aligned}\right\} \tag{9.20}$$

and these results are the best estimates of the three angles, based on the available data.

Equations (9.20) are linear functions, therefore the law of propagation may be applied in order to obtain the stochastic properties of the final values. Taking the first of eqns (9.20) as an example, and substituting for t from eqn (9.6),

$$\begin{aligned}\bar{x}_1&=x_1+\tfrac{1}{3}(180-(x_1+x_2+x_3))\\ &=\tfrac{2}{3}x_1+(180-x_2-x_3)\end{aligned} \tag{9.21}$$

Therefore,

$$g_{\bar{x}_1\bar{x}_1} = \tfrac{4}{9}g_{11} + \tfrac{1}{9}(g_{22} + g_{33}) \tag{9.22}$$

where g_{11}, g_{22} and g_{33} are the cofactors of the three angles respectively.

Similarly for

$$g_{\bar{x}_2\bar{x}_2}, g_{\bar{x}_3\bar{x}_3} \text{and } g_{\bar{x}_1\bar{x}_2}, g_{\bar{x}_1\bar{x}_3}, g_{\bar{x}_2\bar{x}_3}$$

With observations of equal weight, $g_{11} = g_{22} = g_{33} = g$.
Therefore

$$g_{\bar{x}_1\bar{x}_1} = \tfrac{2}{3}g, \text{etc.} \tag{9.23}$$

A value for g_{vv} may also be found; from eqns (9.6) and (9.19),

$$v_1 = t/3 = \tfrac{1}{3}(180 - (x_1 + x_2 + x_3)) \tag{9.24}$$

therefore,

$$g_{v_1v_1} = \tfrac{1}{9}(g_{11} + g_{22} + g_{33}) = \tfrac{1}{3}g \tag{9.25}$$

Therefore it becomes clear that

$$g_{\bar{x}_1\bar{x}_1} = g_{11} - g_{v_1v_1} \tag{9.26}$$

and similarly for $g_{\bar{x}_2\bar{x}_2}, \ldots,$ etc.

As an example the following numerical values are used for the problem outlined above.

Example 9.1

Let the observed angles of the triangle be

	°	′	″
x_1	56	21	32
x_2	49	52	09
x_3	73	46	28

and let the observations be uncorrelated, and of equal weight.
Derive least squares estimates for the three angles.

Solution

There is one redundant observation; therefore $r = 1$, and the one condition equation is

$$\bar{x}_1 + \bar{x}_2 + \bar{x}_3 = 180^\circ$$

When observations are introduced (eqn (9.4)) this becomes

$$(56^\circ21'32'' + v_1) + (49^\circ52'09'' + v_2) + (73^\circ46'28'' + v_3) = 180^\circ$$

and so, $v_1 + v_2 + v_3 = t = 180^\circ - (56^\circ21'32'' + 49^\circ52'09'' + 73^\circ46'28'')$

$= -9''$

Therefore, the reduced condition eqn (9.5) is

$$v_1 + v_2 + v_3 = -9''$$

From eqn (9.15), $E = [v^2] - 2k([v] + 9) = \text{minimum}$ and since, from eqn (9.16),

$$v_1 = v_2 = v_3 = \tfrac{1}{3}t$$

therefore,

$$v_1 = v_2 = v_3 = -3$$

Therefore from eqn (9.20),

$$\bar{x}_1 = 56^\circ 21'32'' - 3'' = 56^\circ 21'29''$$

$$\bar{x}_2 = 49^\circ 52'09'' - 3'' = 49^\circ 52'06''$$

$$\bar{x}_3 = 73^\circ 46'28'' - 3'' = 73^\circ 46'25''$$

The example above involves only one condition equation, and the weights of all observations are equal to unity. In practice much more complicated situations arise, and it is necessary to devise an algorithm which deals with many condition equations and involves observations of varying weight. Consider the general situation, involving

n observations $\quad x_1, x_2, x_3, \ldots, x_n$
with weights $\quad w_1, w_2, w_3, \ldots, w_n$

and r linear condition equations.

Let the known coefficients in the condition equations be

$$\begin{matrix} a_1, a_2, \ldots, a_n \\ b_1, b_2, \ldots, b_n \\ \vdots \\ r_1, r_2, \ldots, r_n \end{matrix}$$

and let the constant terms for each equation be $(c_a, c_b, \ldots, c_r)$ respectively.

Therefore the system of condition equations is

$$\left.\begin{aligned} a_1\bar{x}_1 + a_2\bar{x}_2 + \cdots + a_n\bar{x}_n &= c_a \\ b_1\bar{x}_1 + b_2\bar{x}_2 + \cdots + b_n\bar{x}_n &= c_b \\ &\vdots \\ r_1\bar{x}_1 + r_2\bar{x}_2 + \cdots + r_n\bar{x}_n &= c_r \end{aligned}\right\} \tag{9.27}$$

which, in the reduced form (i.e. substituting $(x_1 + v_1)$ for $\bar{x}_1$, etc.), becomes

$$\left.\begin{aligned} a_1v_1 + a_2v_2 + \cdots + a_nv_n &= t_a \\ b_1v_1 + b_2v_2 + \cdots + b_nv_n &= t_b \\ &\vdots \\ r_1v_1 + r_2v_2 + \cdots + r_nv_n &= t_r \end{aligned}\right\} \tag{9.28}$$

Let

$$\left.\begin{aligned} A &= a_1 v_1 + a_2 v_2 + \cdots + a_n v_n - t_a = 0 \\ B &= b_1 v_1 + b_2 v_2 + \cdots + b_n v_n - t_b = 0 \\ &\vdots \\ R &= r_1 v_1 + r_2 v_2 + \cdots + r_n v_n - t_r = 0 \end{aligned}\right\} \tag{9.29}$$

and, when the least-squares condition and Lagrange multipliers are introduced, E becomes

$$E = [wvv] - 2k_a A - 2k_b B - \cdots - 2k_r R \tag{9.30}$$

For E to be a minimum,

$$\frac{\partial E}{\partial v_1} = \frac{\partial E}{\partial v_2} = \cdots = \frac{\partial E}{\partial v_n} = 0$$

Therefore

$$\left.\begin{aligned} \frac{\partial E}{\partial v_1} &= 2w_1 v_1 - 2(a_1 k_a + b_1 k_b + \cdots + r_1 k_r) = 0 \\ \frac{\partial E}{\partial v_2} &= 2w_2 v_2 - 2(a_2 k_a + b_2 k_b + \cdots + r_2 k_r) = 0 \\ &\vdots \\ \frac{\partial E}{\partial v_n} &= 2w_n v_n - 2(a_n k_a + b_n k_b + \cdots + r_n k_r) = 0 \end{aligned}\right\} \tag{9.31}$$

so that

$$\left.\begin{aligned} v_1 &= \frac{1}{w_1}(a_1 k_a + b_1 k_b + \cdots + r_1 k_r) \\ v_2 &= \frac{1}{w_2}(a_2 k_a + b_2 k_b + \cdots + r_2 k_r) \\ &\vdots \\ v_n &= \frac{1}{w_n}(a_n k_a + b_n k_b + \cdots + r_n k_r) \end{aligned}\right\} \tag{9.32}$$

When these expressions for the v's are substituted into eqns (9.28) the first equation becomes

$$\frac{a_1}{w_1}(a_1 k_a + b_1 k_b + \cdots + r_1 k_r) + \frac{a_2}{w_2}(a_2 k_a + b_2 k_b + \cdots + r_2 k_r)$$

$$+ \cdots + \frac{a_n}{w_n}(a_n k_a + b_n k_b + \cdots + r_n k_r) = t_a \tag{9.33}$$

which may be written

$$\left.\begin{array}{l}\left[\dfrac{a^2}{w}\right]k_a+\left[\dfrac{ab}{w}\right]k_b+\cdots+\left[\dfrac{ar}{w}\right]k_r=t_a \\ \text{and the other equations may be found in a similar way, thus} \\ \left[\dfrac{ab}{w}\right]k_a+\left[\dfrac{b^2}{w}\right]k_b+\cdots+\left[\dfrac{br}{w}\right]k_r=t_b \\ \vdots \\ \left[\dfrac{ar}{w}\right]k_a+\left[\dfrac{br}{w}\right]k_b+\cdots+\left[\dfrac{r^2}{w}\right]k_r=t_r\end{array}\right\} \quad (9.34)$$

Equations (9.34) are the *correlate equations* (or *normal equations*), and there are as many equations as unknown Lagrange multipliers. However, the equations are consistent and a unique solution may be obtained.

Note the symmetry of the coefficients in the equations.

The solution of eqns (9.34) will provide values for the ks which, when substituted in eqns (9.32), will give values for the vs. The adjusted observations and their accuracies may be found in the same way as outlined earlier. However, in this case the amount of algebra is far greater, and it is more convenient to use matrix notation. In the next section the least-squares algorithm is derived using matrix notation, and this is followed with some simple examples to help illustrate the procedure.

9.3 Least-squares estimation by condition equations, and the use of matrix notation

Although the development of the equations below is in matrix notation, reference may be made to the corresponding equations for the simple example in the previous section, via the equation numbers given in italics.

Let the linear mathematical model be

$$\boldsymbol{U}\bar{\boldsymbol{x}}=\boldsymbol{c} \qquad (\mathit{9.3}) \qquad (9.35)$$

where $\boldsymbol{U}$ is the matrix ($r \times n$) of known coefficients; $\bar{\boldsymbol{x}}$ is the vector ($n \times 1$) of unknowns (the adjusted observations); and $\boldsymbol{c}$ is the vector ($r \times 1$) of constants.

The final estimates are the original observations plus corrections, thus

$$\bar{\boldsymbol{x}}=\boldsymbol{x}+\boldsymbol{v} \qquad (9.36)$$

where $\boldsymbol{x}$ is the vector ($n \times 1$) of observations; and $\boldsymbol{v}$ is the vector ($n \times 1$) of corrections (residuals).

Therefore eqn (9.35) may be written

$$\boldsymbol{U}(\boldsymbol{x}+\boldsymbol{v})=\boldsymbol{c} \qquad (\mathit{9.4}) \qquad (9.37)$$

and thus $$Uv = c - Ux$$

which may be written in the form of reduced condition equations as

$$Uv = t \qquad (9.5) \qquad (9.38)$$

where $$t = c - Ux \qquad (9.6) \qquad (9.39)$$

is the vector $(r \times 1)$ of misclosures.

Introducing the least-squares condition,

$$E = v^{T}G^{-1}v = \text{minimum} \qquad (9.7)$$

where G is the cofactor matrix $(n \times n)$ for the observations.

When Lagrange multipliers are introduced, the equation becomes

$$E = v^{T}G^{-1}v - 2k^{T}(Uv - t) \qquad (9.15) \qquad (9.40)$$

where k is the vector of unknown correlates $(r \times 1)$.

When eqn (9.40) is differentiated partially with respect to the variables v, and the partial derivative are equated to 0, the following result is obtained:

$$\frac{\partial E}{\partial v} = G^{-1}v + (G^{-1})^{T}v - 2U^{T}k$$

$$= 2(G^{-1}v - U^{T}k) = 0 \qquad (9.16) \qquad (9.41)$$

Therefore $$G^{-1}v = U^{T}k$$

or

$$v = GU^{T}k \qquad (9.16) \qquad (9.42)$$

which may be substituted in eqn (9.38), to give

$$\left.\begin{aligned} UGU^{T}k &= t \\ \text{or} \quad Nk &= t, \quad \text{where} \quad N = UGU^{T} \end{aligned}\right\} \qquad (9.17) \qquad (9.43)$$

Therefore

$$k = N^{-1}t = (UGU^{T})^{-1}t \qquad (9.18) \qquad (9.44)$$

or, if eqn (9.44) is substituted in eqn (9.42),

$$v = GU^{T}N^{-1}t = GU^{T}(UGU^{T})^{-1}t \qquad (9.19) \qquad (9.45)$$

The final estimates are obtained from eqn (9.36),viz.

$$\bar{x} = x + v \qquad (9.20)$$

This completes the determination of the least-squares estimates of the observed variables. The actual application to specific problems requires the use of only some of the equations—in particular eqns (9.35), (9.37), (9.38), (9.43), (9.45) and (9.36). A check on the numerical solution is ensured by substituting the derived values of v in eqn (9.38), and verifying that the equations are satisfied. However, this does not check the formation of the equations in the first place.

The stochastic properties of the derived quantities may also be found, as follows.

When eqn (9.45) is substituted in eqn (9.36) the following expression for $\bar{x}$ is obtained:

$$\bar{x} = x + GU^{\mathrm{T}}N^{-1}t \qquad (9.21) \qquad (9.46)$$

and a substitution for t may be made from eqn (9.39), thus:

$$\begin{aligned}\bar{x} &= x + GU^{\mathrm{T}}N^{-1}(c - Ux)\\ &= x + GU^{\mathrm{T}}N^{-1}c - GU^{\mathrm{T}}N^{-1}Ux\\ &= GU^{\mathrm{T}}N^{-1}c + (I - GU^{\mathrm{T}}N^{-1}U)x \qquad (9.21) \qquad (9.47)\end{aligned}$$

Equation (9.47) is a linear function of the stochastic variables (x), which are contained in the second term on the right-hand side. Therefore the law of propagation (eqn (8.35)) may be applied:

$$\begin{aligned}G_{\bar{x}\bar{x}} &= (I - GU^{\mathrm{T}}N^{-1}U)G(I - GU^{\mathrm{T}}N^{-1}U)^{\mathrm{T}}\\ &= (G - GU^{\mathrm{T}}N^{-1}UG)(I - GU^{\mathrm{T}}N^{-1}U)^{\mathrm{T}} \qquad (9.22) \qquad (9.48)\end{aligned}$$

Now, for any matrix A, $\quad (I - A)^{\mathrm{T}} = I - A^{\mathrm{T}}$

therefore

$$\begin{aligned}G_{\bar{x}\bar{x}} &= (G - GU^{\mathrm{T}}N^{-1}UG)(I - U^{\mathrm{T}}N^{-1}UG)\\ &= G - GU^{\mathrm{T}}N^{-1}UG - GU^{\mathrm{T}}N^{-1}UG\\ &\quad + GU^{\mathrm{T}}N^{-1}UGU^{\mathrm{T}}N^{-1}UG\\ &= G - GU^{\mathrm{T}}N^{-1}UG - GU^{\mathrm{T}}N^{-1}UG + GU^{\mathrm{T}}N^{-1}UG\end{aligned}$$

Therefore

$$G_{\bar{x}\bar{x}} = G - GU^{\mathrm{T}}N^{-1}UG \qquad (9.49)$$

A value for G_{vv} may also be found from eqn (9.45), thus

$$\begin{aligned}v &= GU^{\mathrm{T}}N^{-1}t\\ &= GU^{\mathrm{T}}N^{-1}(c - Ux)\\ &= GU^{\mathrm{T}}N^{-1}c - GU^{\mathrm{T}}N^{-1}Ux \qquad (9.24) \qquad (9.50)\end{aligned}$$

Therefore

$$\begin{aligned}G_{vv} &= (GU^{\mathrm{T}}N^{-1}U)G(GU^{\mathrm{T}}N^{-1}U)^{\mathrm{T}}\\ &= (GU^{\mathrm{T}}N^{-1}U)G(U^{\mathrm{T}}N^{-1}UG)\\ &= GU^{\mathrm{T}}N^{-1}UG \qquad (9.25) \qquad (9.51)\end{aligned}$$

Therefore, from eqns (9.49) and (9.51), it is clear that

$$G_{\bar{x}\bar{x}} = G - G_{vv} \qquad (9.26) \qquad (9.52)$$

Finally, the covariance matrix for the least-squares estimates may be found from

$$\Sigma_{\bar{x}\bar{x}} = \sigma^2 \cdot G_{\bar{x}\bar{x}}$$

Example 9.2

Let the data be the same as that described in Example 9.1, with the additional information that the standard deviations for the three measured angles are 1″, 2″ and 3″ respectively. The problems is to derive

(a) least-squares estimates for the three angles, and
(b) stochastic properties for these derived quantities.

Solution

(a) The mathematical model is eqn (9.3), viz.

$$\bar{x}_1 + \bar{x}_2 + \bar{x}_3 = 180°$$

Therefore, from eqn (9.35) the following may be identified:

$$U = (1 \quad 1 \quad 1), \qquad \bar{x} = \begin{bmatrix} \bar{x}_1 \\ \bar{x}_2 \\ \bar{x}_3 \end{bmatrix}, \quad c = (180°)$$

Also,

$$x = \begin{bmatrix} x_1 \\ x_2 \\ x_3 \end{bmatrix} = \begin{bmatrix} 56°21'32'' \\ 49°52'09'' \\ 73°46'28'' \end{bmatrix}$$

and, from eqn (9.39),

$$t = c - Ux = 180° - (1 \quad 1 \quad 1) \begin{bmatrix} 56°21'32'' \\ 49°52'09'' \\ 73°46'28'' \end{bmatrix}$$

$$= -9''$$

Let the reference variance, $\sigma^2 = 1 \text{ sec}^2$; therefore

$$G = \Sigma = \begin{bmatrix} 1 & & \\ & 4 & \\ & & 9 \end{bmatrix}$$

and, from eqn (9.43),

$$N = UGU^{\mathrm{T}} = (1 \quad 1 \quad 1)\begin{bmatrix} 1 & & \\ & 4 & \\ & & 9 \end{bmatrix}\begin{bmatrix} 1 \\ 1 \\ 1 \end{bmatrix} = (1 \quad 1 \quad 1)\underbrace{\begin{bmatrix} 1 \\ 4 \\ 9 \end{bmatrix}}_{GU^{\mathrm{T}}} = 14$$

Therefore, $N^{-1} = 1/14$.

From eqn (9.45), $$v = GU^{\mathrm{T}}N^{-1}t = \begin{bmatrix} 1 \\ 4 \\ 9 \end{bmatrix}(1/14)(-9) = \underbrace{\begin{bmatrix} 1/14 \\ 4/14 \\ 9/14 \end{bmatrix}}_{GU^{\mathrm{T}}N^{-1}}(-9) = \begin{bmatrix} -0.6 \\ -2.6 \\ -5.8 \end{bmatrix}$$

Therefore, from eqn (9.36),

$$\bar{x} = \begin{bmatrix} 56°21'32'' \\ 49°52'09'' \\ 73°46'28'' \end{bmatrix} + \begin{bmatrix} -0.6'' \\ -2.6'' \\ -5.8'' \end{bmatrix} = \begin{bmatrix} 56°21'31.4'' \\ 49°52'06.4'' \\ 73°46'22.2'' \end{bmatrix}$$

(b) From eqn (9.49),

$$G_{\bar{x}\bar{x}} = G - GU^{\mathrm{T}}N^{-1}UG$$

Now

$$GU^{\mathrm{T}}N^{-1}UG = \underbrace{\begin{bmatrix} 1/14 \\ 4/14 \\ 9/14 \end{bmatrix}}_{GU^{\mathrm{T}}N^{-1}}(1 \quad 1 \quad 1)\begin{bmatrix} 1 & & \\ & 4 & \\ & & 9 \end{bmatrix}$$

$$= \frac{1}{14}\begin{bmatrix} 1 \\ 4 \\ 9 \end{bmatrix}(1 \quad 4 \quad 9) = \frac{1}{14}\begin{bmatrix} 1 & 4 & 9 \\ 4 & 16 & 36 \\ 9 & 36 & 81 \end{bmatrix}$$

Therefore

$$G_{\bar{x}\bar{x}} = \begin{bmatrix} 1 & & \\ & 4 & \\ & & 9 \end{bmatrix} - \frac{1}{14}\begin{bmatrix} 1 & 4 & 9 \\ 4 & 16 & 36 \\ 9 & 36 & 81 \end{bmatrix}$$

$$= \frac{1}{14}\begin{bmatrix} 13 & -4 & -9 \\ -4 & 40 & -36 \\ -9 & -36 & 45 \end{bmatrix}$$

and, since $\boldsymbol{\Sigma} = \sigma^2 \cdot G$, where $\sigma^2 = 1\ \sec^2$, therefore the covariance matrix for the final angles is

$$\boldsymbol{\Sigma}_{\bar{x}\bar{x}} = 1 \cdot \boldsymbol{G}_{\bar{x}\bar{x}} \sec^2 = \begin{bmatrix} 0.93 & -0.29 & -0.64 \\ & 2.86 & -2.57 \\ \text{symm.} & & 3.21 \end{bmatrix} \sec^2$$

and the standard errors for the final angles are 0.96". 1.69" and 1.79" respectively.

Comment

Note the following:

(i) The final angles differ by a significant amount from those derived in Example 9.1, the reason being the allocation of different weights in Example 9.2.
(ii) The final angles are more accurate then the observed angles, as they should be.
(iii) There is strong correlation between the final angles 2 and 3, due also to the fact that the observations of these two angles are the least reliable, and so an error in one will have a strong influence on the final estimate of the other.

9.4 Least-squares estimation by observation equations

The mathematical model expresses the n final estimates of the observed quantities as functions of the minimum number (m) of parameters required to solve the problem: i.e. there are as many equations as observations, and

$$m = n - r \tag{9.53}$$

where r is the number of redundant observations.

The simple example used before—the three angles of a triangle—will serve to illustrate the procedure.

The parameters corresponding to any two angles of the triangle are required to solve the problem, therefore $m = 2$.

Let the chosen two unknown parameters be p_1 and p_2.

The first step is to express all final estimates as functions of these variables, thus

$$\left.\begin{aligned} \bar{x}_1 &= p_1 \\ \bar{x}_2 &= p_2 \\ \bar{x}_3 &= -p_1 - p_2 + 180 \end{aligned}\right\} \tag{9.54}$$

which are referred to as *observation equations.*

Note that, for this particular example, the unknown parameters on the right-hand side will finally have the same values as some of the estimates on the left. This will not always be the case, and more often than not the values will be different, for example;

(i) in surveying—angle and distance observations expressed as functions of eastings and northings coordinates of the endpoints;
(ii) in photogrammetry—observed height differences expressed as functions of the rotations in absolute orientation.

Equations (9.54) are linear functions, therefore the least squares procedure may be implemented directly. Final estimates on the left hand side are replaced by observations plus corrections.

$$\left.\begin{aligned} x_1 + v_1 &= \quad p_1 \\ x_2 + v_2 &= \qquad\quad p_2 \\ x_3 + v_3 &= -p_1 - p_2 + 180 \end{aligned}\right\} \tag{9.55}$$

Therefore

$$\left.\begin{aligned} v_1 &= \quad p_1 \qquad\quad - x_1 \\ v_2 &= \qquad\quad p_2 - x_2 \\ v_3 &= -p_1 - p_2 - (x_3 - 180) \end{aligned}\right\} \tag{9.56}$$

which are termed the *reduced observation equations.*

The equations may be written

$$\left.\begin{aligned} v_1 &= \quad p_1 \qquad\quad - f_1 \\ v_2 &= \qquad\quad p_2 - f_2 \\ v_3 &= -p_1 - p_2 - f_3 \end{aligned}\right\} \tag{9.57}$$

where f_1, f_2, f_3 are known constants.

The least-squares condition is introduced, viz.

$$E = [wvv] = \text{minimum}$$

which, in this case, becomes

$$E = v_1^2 + v_2^2 + v_3^2 = \text{minimum} \tag{9.58}$$

Therefore

$$\frac{\partial E}{\partial v_1} = \frac{\partial E}{\partial v_2} = \frac{\partial E}{\partial v_3} = 0$$

and, since v_1, v_2 and v_3 are functions of p_1 and p_2,

therefore $$\frac{\partial E}{\partial p_1} = \frac{\partial E}{\partial p_2} = 0$$

Now, substituting expressions for v from eqns (9.57) in eqn (9.58),

$$E = (p_1 - f_1)^2 + (p_2 - f_2)^2 + (-p_1 - p_2 - f_3)^2 \tag{9.59}$$

and therefore,

$$\frac{\partial E}{\partial p_1} = 2(p_1 - f_1) - 2(-p_1 - p_2 - f_3) = 0$$

Therefore

$$2p_1 + p_2 = f_1 - f_3 \tag{9.60}$$

Similarly

$$\frac{\partial E}{\partial p_2} = 2(p_2 - f_2) - 2(-p_1 - p_2 - f_3) = 0$$

Therefore

$$p_1 + 2p_2 = f_2 - f_3 \tag{9.61}$$

Equations (9.60) and (9.61) are the *normal equations*, and there are as many equations as unknowns. Note the symmetry of the coefficients in the equations. The simultaneous solution of the two equations produces values for the unknowns p_1 and p_2.

The stochastic properties of the derived quantities may also be found. In the case of p_1 the solution produces the following equation:

$$p_1 = \tfrac{1}{3}(2f_1 - f_2 - f_3)$$

$$= \tfrac{1}{3}(2x_1 - x_2 - (x_3 - 180)) \tag{9.62}$$

Therefore $$g_{p_1p_1} = \tfrac{4}{9}g_{11} + \tfrac{1}{9}(g_{22} + g_{33})$$

and since

$$g_{11} = g_{22} = g_{33} = g$$

therefore

$$g_{p_1p_1} = \tfrac{2}{3}g \tag{9.63}$$

Similarly, values for $g_{p_2p_2}$ and $g_{p_1p_2}$ may be found.

The cofactors of the final estimates $\bar{x}_1, \bar{x}_2, \bar{x}_3$ may then be found by applying the laws of propagation to the appropriate equations in eqns (9.54).

For example, $\bar{x}_1 = p_1$ therefore

$$g_{\bar{x}_1\bar{x}_1} = g_{p_1p_1} = \tfrac{2}{3}g \tag{9.64}$$

A value for g_{vv} may be found from eqns (9.57):

$$v_1 = p_1 - f_1 = p_1 - x_1$$

therefore

$$\begin{aligned} g_{v_1v_1} &= g_{p_1p_1} + g_{11} - 2g_{p_11} \\ &= g_{p_1p_1} + g_{11} - \tfrac{4}{3}g_{11} \\ &= g_{p_1p_1} - \tfrac{1}{3}g_{11} \\ &= \tfrac{2}{3}g - \tfrac{1}{3}g \\ &= \tfrac{1}{3}g \end{aligned} \tag{9.65}$$

Thus, from eqns (9.63), (9.64) and (9.65), it is clear that

$$g_{\bar{x}_1\bar{x}_1} = g_{p_1p_1} = g_{11} - g_{v_1v_1} \tag{9.66}$$

Similarly for $g_{\bar{x}_2\bar{x}_2}, \ldots,$ etc.

Example 9.3

The data given in Example 9.1 is used again for the simple problem outlined above. The observed angles of the triangle are:

	°	′	″
x_1	56	21	32
x_2	49	52	09
x_3	73	46	28

and the observations are assumed to be uncorrelated, and of equal weight. Derive least-squares estimates for the three angles.

Solution

If p_1 and p_2 are the unknown parameters, the observation equations are those given by eqn (9.54), and when observations are introduced these become

$$\begin{aligned} 56°21'32'' + v_1 &= p_1 \\ 49°52'09'' + v_2 &= p_2 \\ 73°46'28'' + v_3 &= -p_1 - p_2 + 180° \end{aligned}$$

which may be expressed in the reduced form (eqn (9.57)) as

$$\begin{aligned} v_1 &= p_1 \qquad\qquad - 56°21'32'' \\ v_2 &= \qquad\qquad p_2 - 49°52'09'' \\ v_3 &= -p_1 - p_2 - (-106°13'32'') \end{aligned}$$

The solution is found by forming and solving eqns (9.60) and (9.61).
The equations are:

$$\begin{aligned} 2p_1 + p_2 &= 56°21'32'' - (-106°13'32'') = 162°35'04'' \\ p_1 + 2p_2 &= 49°52'09'' - (-106°13'32'') = 156°05'41'' \end{aligned}$$

and the solution is:

$$\begin{aligned} p_1 &= 56°21'29'' \\ p_2 &= 49°52'06'' \end{aligned}$$

Therefore, substituting in eqns (9.54),

$$\begin{aligned} \bar{x}_1 &= 56°21'29'' \\ \bar{x}_2 &= 49°52'06'' \\ \bar{x}_3 &= 73°46'25'' \end{aligned}$$

The problems that arise in practice are much more complicated, and it is necessary to devise an algorithm which can deal with many more parameters and with observations of varying weight.

Consider the general situation, involving n observations, $x_1, x_2, x_3, \ldots, x_n$ with weights $w_1, w_2, w_3, \ldots, w_n$ and r redundant observations.

There will be n observation equations which are functions of m parameters, where $m = n - r$. Let the known coefficients in the linear observation equations be

$$\begin{matrix} a_1 & b_1 & \ldots & m_1 \\ a_2 & b_2 & \ldots & m_2 \\ \vdots & & & \\ a_n & b_n & \ldots & m_n \end{matrix}$$

and let the constant terms for each equation be $c_1, c_2, \ldots, c_n$, respectively.

Therefore the system of observation equations is

$$\left.\begin{aligned} \bar{x}_1 &= a_1p_1 + b_1p_2 + \cdots + m_1p_m + c_1 \\ \bar{x}_2 &= a_2p_1 + b_2p_2 + \cdots + m_2p_m + c_2 \\ &\vdots \\ \bar{x}_n &= a_np_1 + b_2p_2 + \cdots + m_2p_m + c_n \end{aligned}\right\} \tag{9.67}$$

which in the reduced form—i.e. substituting $(x_1 + v_1)$ for $\bar{x}_1$, etc., on the left-hand side—becomes

$$\left.\begin{aligned} v_1 &= a_1p_1 + b_1p_2 + \cdots + m_1p_m - (x_1 - c_1) \\ v_2 &= a_2p_1 + b_2p_2 + \cdots + m_2p_m - (x_2 - c_2) \\ &\vdots \\ v_n &= a_np_1 + b_np_2 + \cdots + m_np_m - (x_n - c_n) \end{aligned}\right\} \tag{9.68}$$

Writing the constant terms as $f_1, f_2, \ldots, f_n$, the equations become

$$\left.\begin{aligned} v_1 &= a_1p_1 + b_1p_2 + \cdots + m_1p_m - f_1 \\ v_2 &= a_2p_1 + b_2p_2 + \cdots + m_2p_m - f_2 \\ &\vdots \\ v_n &= a_np_1 + b_np_2 + \cdots + m_np_m - f_n \end{aligned}\right\} \tag{9.69}$$

and, when the least-squares condition is introduced,

$$\begin{aligned} E = [wvv] &= \text{minimum} \\ &= w_1(a_1p_1 + b_1p_2 + \cdots + m_1p_m - f_1)^2 \\ &\quad + w_2(a_2p_1 + b_2p_2 + \cdots + m_2p_m - f_2)^2 \\ &\quad \vdots \\ &\quad + w_n(a_np_1 + b_np_2 + \cdots + m_np_m - f_n)^2 \end{aligned} \tag{9.70}$$

Therefore

$$\frac{\partial E}{\partial p_1} = \frac{\partial E}{\partial p_2} = \cdots = \frac{\partial E}{\partial p_m} = 0$$

$$\begin{aligned} \frac{\partial E}{\partial p_1} &= 2w_1(a_1p_1 + b_1p_2 + \cdots + m_1p_m - f_1)a_1 \\ &\quad + 2w_2(a_2p_1 + b_2p_2 + \cdots + m_2p_m - f_2)a_2 \\ &\quad \vdots \\ &\quad + 2w_n(a_np_1 + b_np_2 + \cdots + m_np_m - f_n)a_n \\ &= 0 \end{aligned} \tag{9.71}$$

which, if common terms are collected, becomes

$$[wa^2]p_1 + [wab]p_2 + \cdots + [wam]p_m = [waf] \tag{9.72a}$$

and similarly for the other equations,

$$\left.\begin{aligned} [wab]p_1 + [wb^2]p_2 + \cdots + [wbm]p_m &= [wbf] \\ &\vdots \\ [wam]p_1 + [wbm]p_2 + \cdots + [wm^2]p_m &= [wmf] \end{aligned}\right\} \tag{9.72b}$$

Equations (9.72) are the normal equations, which are m in number and contain m unknowns. Again, note the symmetry of the coefficients, and also the role played by the w's compared with the method using condition equations.

The unknown parameters are found by solving the normal equations, and their stochastic properties may also be derived as outlined in the simple example given earlier. However, the algebra involved would be cumbersome, and it is more convenient to use matrix notation.

9.5 Least-squares estimation by observation equations and the use of matrix notation

Although the development of the equations below is in matrix notation, reference may be made to the corresponding equations for the simple example in the previous section, via the equation numbers given in italics.

Let the linear mathematical model be

$$\bar{\boldsymbol{x}} = \boldsymbol{A}\boldsymbol{p} + \boldsymbol{c} \qquad \textit{(9.54)} \qquad (9.73)$$

where $\bar{\boldsymbol{x}}$ is the vector $(n \times 1)$ of adjusted observations; $\boldsymbol{A}$ is the matrix $(n \times m)$ of coefficients of the unknown parameters; $\boldsymbol{p}$ is the vector $(m \times 1)$ of unknown parameters; and $\boldsymbol{c}$ is the vector $(n \times 1)$ of constant terms.

When observations plus corrections are substituted for $\bar{\boldsymbol{x}}$ the equation becomes

$$\boldsymbol{x} + \boldsymbol{v} = \boldsymbol{A}\boldsymbol{p} + \boldsymbol{c} \qquad \textit{(9.55)} \qquad (9.74)$$

Thus the reduced observation equations become

$$\boldsymbol{v} = \boldsymbol{A}\boldsymbol{p} - (\boldsymbol{x} - \boldsymbol{c}) \qquad \textit{(9.56)} \qquad (9.75)$$

which may be written

$$\boldsymbol{v} = \boldsymbol{A}\boldsymbol{p} - \boldsymbol{f} \qquad \textit{(9.57)} \qquad (9.76)$$

where $\boldsymbol{f} = \boldsymbol{x} - \boldsymbol{c}$ is the new vector $(n \times 1)$ of constant terms. (9.77)

The least-squares condition is introduced, viz.

$$\boldsymbol{E} = \boldsymbol{v}^{\mathrm{T}}\boldsymbol{G}^{-1}\boldsymbol{v} = \text{minimum}$$

and, when $\boldsymbol{v}$ is replaced by the expression in eqn (9.76),

$$\begin{aligned}
\boldsymbol{E} &= (\boldsymbol{A}\boldsymbol{p} - \boldsymbol{f})^{\mathrm{T}}\boldsymbol{G}^{-1}(\boldsymbol{A}\boldsymbol{p} - \boldsymbol{f}) = \text{minimum} \\
&= (\boldsymbol{p}^{\mathrm{T}}\boldsymbol{A}^{\mathrm{T}} - \boldsymbol{f}^{\mathrm{T}})\boldsymbol{G}^{-1}(\boldsymbol{A}\boldsymbol{p} - \boldsymbol{f}) \\
&= (\boldsymbol{p}^{\mathrm{T}}\boldsymbol{A}^{\mathrm{T}}\boldsymbol{G}^{-1} - \boldsymbol{f}^{\mathrm{T}}\boldsymbol{G}^{-1})(\boldsymbol{A}\boldsymbol{p} - \boldsymbol{f}) \\
&= \boldsymbol{p}^{\mathrm{T}}\boldsymbol{A}^{\mathrm{T}}\boldsymbol{G}^{-1}\boldsymbol{A}\boldsymbol{p} - \boldsymbol{f}^{\mathrm{T}}\boldsymbol{G}^{-1}\boldsymbol{A}\boldsymbol{p} - \boldsymbol{p}^{\mathrm{T}}\boldsymbol{A}^{\mathrm{T}}\boldsymbol{G}^{-1}\boldsymbol{f} + \boldsymbol{f}^{\mathrm{T}}\boldsymbol{G}^{-1}\boldsymbol{f} \qquad \textit{(9.59)} \qquad (9.78) \\
&= \text{minimum}
\end{aligned}$$

Therefore, for $\boldsymbol{E}$ to be a minimum, $\partial \boldsymbol{E}/\partial \boldsymbol{p} = 0$.

Thus,

$$\frac{\partial \boldsymbol{E}}{\partial \boldsymbol{p}} = \boldsymbol{A}^{\mathrm{T}}\boldsymbol{G}^{-1}\boldsymbol{A}\cdot\boldsymbol{p} + (\boldsymbol{A}^{\mathrm{T}}\boldsymbol{G}^{-1}\boldsymbol{A})^{\mathrm{T}}\cdot\boldsymbol{p} - (\boldsymbol{f}^{\mathrm{T}}\boldsymbol{G}^{-1}\boldsymbol{A})^{\mathrm{T}} - \boldsymbol{A}^{\mathrm{T}}\boldsymbol{G}^{-1}\boldsymbol{f}$$
$$= \boldsymbol{A}^{\mathrm{T}}\boldsymbol{G}^{-1}\boldsymbol{A}\cdot\boldsymbol{p} + \boldsymbol{A}^{\mathrm{T}}\boldsymbol{G}^{-1}\boldsymbol{A}\cdot\boldsymbol{p} - \boldsymbol{A}^{\mathrm{T}}\boldsymbol{G}^{-1}\boldsymbol{f} - \boldsymbol{A}^{\mathrm{T}}\boldsymbol{G}^{-1}\boldsymbol{f}$$
$$= 2(\boldsymbol{A}^{\mathrm{T}}\boldsymbol{G}^{-1}\boldsymbol{A}\cdot\boldsymbol{p}) - 2(\boldsymbol{A}^{\mathrm{T}}\boldsymbol{G}^{-1}\boldsymbol{f})$$
$$= 0$$

Therefore

$$\boldsymbol{A}^{\mathrm{T}}\boldsymbol{G}^{-1}\boldsymbol{A}\cdot\boldsymbol{p} = \boldsymbol{A}^{\mathrm{T}}\boldsymbol{G}^{-1}\boldsymbol{f} \qquad (9.79)$$

which may be written

$$\boldsymbol{N}\boldsymbol{p} = \boldsymbol{F} \qquad \textit{(9.60), (9.61)} \qquad (9.80)$$

where

$$\left.\begin{aligned} \boldsymbol{N} &= \boldsymbol{A}^{\mathrm{T}}\boldsymbol{G}^{-1}\boldsymbol{A} \\ \boldsymbol{F} &= \boldsymbol{A}^{\mathrm{T}}\boldsymbol{G}^{-1}\boldsymbol{f} \end{aligned}\right\} \qquad (9.81)$$

Equation (9.80) is the set of normal equations, and the unknowns may be found from

$$\boldsymbol{p} = \boldsymbol{N}^{-1}\boldsymbol{F} \qquad \textit{(9.62)} \qquad (9.82)$$

The vector $\boldsymbol{v}$ may be found by substitution in eqn (9.76), and the least-squares estimates of the observed variables may then be obtained from eqn (9.36), viz.

$$\bar{\boldsymbol{x}} = \boldsymbol{x} + \boldsymbol{v}$$

which completes the process of least-squares estimation.

A check (derived from eqns (9.76) and (9.79)) on the numerical solution of the parameters may be carried out by verifying that

$$\boldsymbol{A}^{\mathrm{T}}\boldsymbol{G}^{-1}\boldsymbol{v} = 0 \qquad \text{(see eqns (9.76) and (9.79))}$$

This ensures that the computation is correct; however, it does not verify that the formation and linearization of the initial functional model is correct.

The cofactors for the derived unknowns may be found from eqn (9.82), as follows:

$$\begin{aligned} \boldsymbol{p} = \boldsymbol{N}^{-1}\boldsymbol{F} &= (\boldsymbol{A}^{\mathrm{T}}\boldsymbol{G}^{-1}\boldsymbol{A})^{-1}\boldsymbol{A}^{\mathrm{T}}\boldsymbol{G}^{-1}\boldsymbol{f} && \text{from eqn (9.81)} \\ &= (\boldsymbol{A}^{\mathrm{T}}\boldsymbol{G}^{-1}\boldsymbol{A})^{-1}\boldsymbol{A}^{\mathrm{T}}\boldsymbol{G}^{-1}(\boldsymbol{x} - \boldsymbol{c}) && \text{from eqn (9.77),} \end{aligned} \qquad (9.83)$$

The coefficient matrix for $\boldsymbol{x}$ in eqn (9.83) is

$$(\boldsymbol{A}^{\mathrm{T}}\boldsymbol{G}^{-1}\boldsymbol{A})^{-1}\boldsymbol{A}^{\mathrm{T}}\boldsymbol{G}^{-1}$$

Let this matrix be $\boldsymbol{B}$; then

$$\begin{aligned}\boldsymbol{G}_{pp} = \boldsymbol{BGB}^{\mathrm{T}} &= ((\boldsymbol{A}^{\mathrm{T}}\boldsymbol{G}^{-1}\boldsymbol{A})^{-1}\boldsymbol{A}^{\mathrm{T}}\boldsymbol{G}^{-1})\boldsymbol{G}((\boldsymbol{A}^{\mathrm{T}}\boldsymbol{G}^{-1}\boldsymbol{A})^{-1}\boldsymbol{A}^{\mathrm{T}}\boldsymbol{G}^{-1})^{\mathrm{T}}\\ &= (\boldsymbol{A}^{\mathrm{T}}\boldsymbol{G}^{-1}\boldsymbol{A})^{-1}\boldsymbol{A}^{\mathrm{T}}\boldsymbol{G}^{-1}\boldsymbol{G}\boldsymbol{G}^{-1}\boldsymbol{A}(\boldsymbol{A}^{\mathrm{T}}\boldsymbol{G}^{-1}\boldsymbol{A})^{-1}\\ &= (\boldsymbol{A}^{\mathrm{T}}\boldsymbol{G}^{-1}\boldsymbol{A})^{-1}\boldsymbol{A}^{\mathrm{T}}\boldsymbol{G}^{-1}\boldsymbol{A}(\boldsymbol{A}^{\mathrm{T}}\boldsymbol{G}^{-1}\boldsymbol{A})^{-1}\\ &= (\boldsymbol{A}^{\mathrm{T}}\boldsymbol{G}^{-1}\boldsymbol{A})^{-1}\\ &= \boldsymbol{N}^{-1} \qquad (9.84)\end{aligned}$$

i.e. the cofactors for the parameter estimates are given by the inverse of the coefficient matrix of the normal equations.

The corrections to the observations may be found from eqn (9.76); thus

$$\begin{aligned}\boldsymbol{v} &= \boldsymbol{Ap} - \boldsymbol{f}\\ &= \boldsymbol{AN}^{-1}\boldsymbol{F} - \boldsymbol{f} && \text{from eqn (9.82)}\\ &= \boldsymbol{AN}^{-1}\boldsymbol{A}^{\mathrm{T}}\boldsymbol{G}^{-1}\boldsymbol{f} - \boldsymbol{f} && \text{from eqn (9.81)}\\ &= (\boldsymbol{AN}^{-1}\boldsymbol{A}^{\mathrm{T}}\boldsymbol{G}^{-1} - \boldsymbol{I})\boldsymbol{f}\\ &= (\boldsymbol{AN}^{-1}\boldsymbol{A}^{\mathrm{T}}\boldsymbol{G}^{-1} - \boldsymbol{I})(x - c) && \text{from eqn (9.77)} \qquad (9.85)\end{aligned}$$

Therefore the coefficient matrix for $\boldsymbol{x}$ is

$$(\boldsymbol{AN}^{-1}\boldsymbol{A}^{\mathrm{T}}\boldsymbol{G}^{-1} - \boldsymbol{I})$$

and, if this matrix is $\boldsymbol{B}$,

$$\begin{aligned}\boldsymbol{G}_{vv} &= \boldsymbol{BGB}^{\mathrm{T}}\\ &= (\boldsymbol{AN}^{-1}\boldsymbol{A}^{\mathrm{T}}\boldsymbol{G}^{-1} - \boldsymbol{I})\boldsymbol{G}(\boldsymbol{AN}^{-1}\boldsymbol{A}^{\mathrm{T}}\boldsymbol{G}^{-1} - \boldsymbol{I})^{\mathrm{T}}\\ &= (\boldsymbol{AN}^{-1}\boldsymbol{A}^{\mathrm{T}}\boldsymbol{G}^{-1} - \boldsymbol{I})\boldsymbol{G}(\boldsymbol{G}^{-1}\boldsymbol{AN}^{-1}\boldsymbol{A}^{\mathrm{T}} - \boldsymbol{I})\\ &= \boldsymbol{AN}^{-1}\boldsymbol{A}^{\mathrm{T}}\boldsymbol{G}^{-1}\boldsymbol{GG}^{-1}\boldsymbol{AN}^{-1}\boldsymbol{A}^{\mathrm{T}} - \boldsymbol{GG}^{-1}\boldsymbol{AN}^{-1}\boldsymbol{A}^{\mathrm{T}} - \boldsymbol{AN}^{-1}\boldsymbol{A}^{\mathrm{T}}\boldsymbol{G}^{-1}\boldsymbol{G} + \boldsymbol{G}\\ &= \boldsymbol{AN}^{-1}\boldsymbol{NN}^{-1}\boldsymbol{A}^{\mathrm{T}} - \boldsymbol{AN}^{-1}\boldsymbol{A}^{\mathrm{T}} - \boldsymbol{AN}^{-1}\boldsymbol{A}^{\mathrm{T}} + \boldsymbol{G}\\ &= \boldsymbol{G} - \boldsymbol{AN}^{-1}\boldsymbol{A}^{\mathrm{T}} \qquad (9.86)\end{aligned}$$

Now the least-squares estimates of the observed variables are found from eqn (9.73), viz.

$$\bar{\boldsymbol{x}} = \boldsymbol{Ap} + \boldsymbol{c}$$

Therefore, from (eqn (8.35)),

$$\begin{aligned}\boldsymbol{G}_{\bar{x}\bar{x}} &= \boldsymbol{AG}_{pp}\boldsymbol{A}^{\mathrm{T}}\\ &= \boldsymbol{AN}^{-1}\boldsymbol{A}^{\mathrm{T}} \qquad (9.87)\end{aligned}$$

Therefore, it is clear from eqns (9.86) and (9.87) that

$$\boldsymbol{G}_{\bar{x}\bar{x}} = \boldsymbol{G} - \boldsymbol{G}_{vv} \qquad (9.66) \qquad (9.88)$$

which is the same conclusion as that obtained using condition equations.

Example 9.4

The data is the same as that used in Example 9.2, i.e. the observed angles given in the previous example, plus the additional information that the standard deviations for the three measured angles are 1″, 2″ and 3″ respectively. The problem is to derive

(a) least squares estimates for the three angles, and
(b) stochastic properties for these derived quantities.

Solution

(a) The mathematical model is eqn (9.54), viz.

$$\begin{aligned}\bar{x}_1 &= p_1\\ \bar{x}_2 &= p_2\\ \bar{x}_3 &= -p_1 - p_2 + 180^\circ\end{aligned}$$

which, in the reduced form (eqn (9.57)), becomes

$$\begin{aligned}v_1 &= p_1 - 56^\circ 21' 32''\\ v_2 &= p_2 - 49^\circ 52' 09''\\ v_3 &= -p_1 - p_2 - (-106^\circ 13' 32'')\end{aligned}$$

Therefore, the following matrices may be identified:

$$\boldsymbol{x} = \begin{bmatrix}56.35889^\circ\\ 49.86917^\circ\\ 73.77444^\circ\end{bmatrix}, \quad \boldsymbol{v} = \begin{bmatrix}v_1\\ v_2\\ v_3\end{bmatrix}, \quad \boldsymbol{A} = \begin{bmatrix}1 & \\ & 1\\ -1 & -1\end{bmatrix},$$

$$\boldsymbol{p} = \begin{pmatrix}p_1\\ p_2\end{pmatrix}, \quad \boldsymbol{c} = \begin{bmatrix}0\\ 0\\ 180^\circ\end{bmatrix}, \quad \boldsymbol{f} = \boldsymbol{x} - \boldsymbol{c} = \begin{bmatrix}56.35889^\circ\\ 49.86917^\circ\\ -106.22556^\circ\end{bmatrix}$$

Let $\sigma^2 = 1\ \text{sec}^2$; therefore

$$\boldsymbol{G} = \boldsymbol{\Sigma} = \begin{bmatrix}1 & & \\ & 4 & \\ & & 9\end{bmatrix}, \quad \text{and} \quad \boldsymbol{G}^{-1} = \begin{bmatrix}1 & & \\ & 1/4 & \\ & & 1/9\end{bmatrix}$$

Thus, from eqn (9.81),

$$\boldsymbol{N} = \boldsymbol{A}^{\mathrm{T}}\boldsymbol{G}^{-1}\boldsymbol{A} = \begin{pmatrix}1 & & -1\\ & 1 & -1\end{pmatrix}\begin{bmatrix}1 & & \\ & 1/4 & \\ & & 1/9\end{bmatrix}\begin{bmatrix}1 & \\ & 1\\ -1 & -1\end{bmatrix}$$

$$= \underbrace{\begin{pmatrix}1 & & -1/9\\ & 1/4 & -1/9\end{pmatrix}}_{\boldsymbol{A}^{\mathrm{T}}\boldsymbol{G}^{-1}}\begin{bmatrix}1 & \\ & 1\\ -1 & -1\end{bmatrix} = \begin{bmatrix}\frac{10}{9} & \frac{1}{9}\\ \frac{1}{9} & \frac{13}{36}\end{bmatrix} = \frac{1}{36}\begin{pmatrix}40 & 4\\ 4 & 13\end{pmatrix}$$

Therefore,

$$N^{-1}=\frac{36}{504}\begin{bmatrix}13 & -4\\ -4 & 40\end{bmatrix}=\frac{1}{14}\begin{bmatrix}13 & -4\\ -4 & 40\end{bmatrix}$$

Also,

$$F=A^{\mathrm{T}}G^{-1}f=\begin{bmatrix}1 & & -1/9\\ & 1/4 & -1/9\end{bmatrix}\begin{bmatrix}56.35889\\ 49.86917\\ -106.22556\end{bmatrix}$$

$$=\begin{bmatrix}68.16173\\ 24.27013\end{bmatrix}$$

From eqn (9.82),

$$p=N^{-1}F=\frac{1}{14}\begin{pmatrix}13 & -4\\ -4 & 40\end{pmatrix}\begin{pmatrix}68.16173\\ 24.27013\end{pmatrix}$$

$$=\frac{1}{14}\begin{pmatrix}789.02194\\ 698.15833\end{pmatrix}=\begin{pmatrix}56.35871\\ 49.86845\end{pmatrix}$$

and, by substituting in eqn (9.73),

$$\bar{x}=Ap+c=\begin{bmatrix}1 & \\ & 1\\ -1 & -1\end{bmatrix}\begin{pmatrix}56.35871\\ 49.86845\end{pmatrix}+\begin{bmatrix}0\\ 0\\ 180\end{bmatrix}$$

$$=\begin{bmatrix}56.35871\\ 49.86845\\ -106.22716\end{bmatrix}+\begin{bmatrix}0\\ 0\\ 180\end{bmatrix}$$

Therefore,

$$\bar{x}=\begin{bmatrix}56.35871^\circ\\ 49.86845^\circ\\ 73.77284^\circ\end{bmatrix}=\begin{bmatrix}56^\circ 21'31.4''\\ 49^\circ 52'06.4''\\ 73^\circ 46'22.2''\end{bmatrix}$$

(b) The cofactors for the unknowns are given by eqn (9.84), thus

$$G_{pp}=N^{-1}=\frac{1}{14}\begin{pmatrix}13 & -4\\ -4 & 40\end{pmatrix}$$

and the cofactors for the estimates of the observed variables are given by eqn (9.87).

$$G_{\bar{x}\bar{x}}=AN^{-1}A^{\mathrm{T}}=\frac{1}{14}\begin{bmatrix}1 & \\ & 1\\ -1 & -1\end{bmatrix}\begin{pmatrix}13 & -4\\ -4 & 40\end{pmatrix}\begin{pmatrix}1 & & -1\\ & 1 & -1\end{pmatrix}$$

$$=\frac{1}{14}\begin{bmatrix}1 & \\ & 1\\ -1 & -1\end{bmatrix}\begin{pmatrix}13 & -4 & -9\\ -4 & 40 & -36\end{pmatrix}=\frac{1}{14}\begin{bmatrix}13 & -4 & -9\\ & 40 & -36\\ \text{symm.} & & 45\end{bmatrix}$$

Since $\Sigma = \sigma^2 \cdot G$, and $\sigma^2 = 1\ \text{sec}^2$, the covariance matrix for these estimates is

$$\Sigma_{\bar{x}\bar{x}} = I \cdot G_{\bar{x}\bar{x}}\ \text{sec}^2 = \begin{bmatrix} 0.93 & -0.29 & -0.64 \\ & 2.86 & -2.57 \\ \text{symm.} & & 3.21 \end{bmatrix} \text{sec}^2$$

and the standard errors for the estimates are 0.96″, 1.69″ and 1.79″ respectively. The comments at the end of Example 9.2 are also applicable to these results.

The example above involves calculations with numbers which have many significant digits, and this is a common characteristic of mathematical models which are linear at the outset. This does not present a problem for most computers, as generally the capacity will be sufficient for most needs. However, any lingering doubts may be removed by introducing a simple modification to the procedure. Approximate values are assigned to the parameters, and corrections to these values become the new unknowns. The approximate values are additional constants which are included in the calculation of the vector f. It is good practice in these cases to work with values which are at least one order of precision better than those of the observations.

Thus, if p_0 is the vector of approximate values for the parameters; and Δ is the vector of corrections, the new unknowns, then p, the vector of final estimates, is obtained from

$$p = p_0 + \Delta \tag{9.89}$$

and eqn (9.73) becomes

$$\bar{x} = A(p_0 + \Delta) + c \tag{9.90}$$

and, when observations are introduced to the left-hand side,

$$x + v = A(p_0 + \Delta) + c \tag{9.91}$$

which may be written

$$v = A\Delta - (x - c - Ap_0) \tag{9.92}$$

So, the vector of constant terms becomes

$$f = x - c - Ap_0 \tag{9.93}$$

instead of that given by eqn (9.77).

Thus the reduced observation equations become

$$v = A\Delta - f \tag{9.94}$$

which is of the same form as eqn (9.76), but with p replaced by Δ. The solution is therefore

$$\Delta = N^{-1}F \tag{9.95}$$

where N and F are defined by eqns (9.81).

The formulae for the calculation of cofactors remain the same; eqn (9.89) is a linear function, and when the law of propagation is applied the following is obtained:

$$\boldsymbol{G}_{pp} = \boldsymbol{G}_{\Delta\Delta} = \boldsymbol{N}^{-1} \tag{9.96}$$

Similarly, $$\boldsymbol{G}_{\bar{x}\bar{x}} = \boldsymbol{A}\boldsymbol{N}^{-1}\boldsymbol{A}^{\mathrm{T}}.$$

In order to illustrate the use of this technique, Example 9.4 is solved again.

Example 9.5

The problem and the data are the same as that used in Example 9.4.

Solution

(a) The mathematical model is eqn (9.54) (in matrix notation eqn (9.73)):

$$\left.\begin{aligned} \bar{x}_1 &= p_1 \\ \bar{x}_2 &= p_2 \\ \bar{x}_3 &= -p_1 - p_2 + 180^\circ \end{aligned}\right\}, \quad \text{or in matrix notation, } \dot{\boldsymbol{x}} = \boldsymbol{A}\boldsymbol{p} + \boldsymbol{c}$$

If, however, the vector $\boldsymbol{p}$ is replaced by $(\boldsymbol{p}_0 + \boldsymbol{\Delta})$, the equations become

$$\left.\begin{aligned} \bar{x}_1 &= (p_{1_0} + \Delta_1) \\ \bar{x}_2 &= (p_{2_0} + \Delta_2) \\ \bar{x}_3 &= -(p_{1_0} + \Delta_1) - (p_{2_0} + \Delta_2) + 180^\circ \end{aligned}\right\}, \quad \begin{aligned} &\text{or in matrix notation,} \\ &\bar{\boldsymbol{x}} = \mathrm{A}(\boldsymbol{p}_0 + \boldsymbol{\Delta}) + \boldsymbol{c} \end{aligned}$$

Δ_1 and Δ_2 are the new unknowns, which are corrections to approximate values for the parameters.

When observations are introduced to the left-hand side the equations become

$$\left.\begin{aligned} x_1 + v_1 &= (p_{1_0} + \Delta_1) \\ x_2 + v_2 &= (p_{2_0} + \Delta_2) \\ x_3 + v_3 &= -(p_{1_0} + \Delta_1) - (p_{2_0} + \Delta_2) + 180^\circ \end{aligned}\right\}, \quad \begin{aligned} &\text{or in matrix} \\ &\text{notation, } \boldsymbol{x} + \boldsymbol{v} = \boldsymbol{A}(\boldsymbol{p}_0 + \boldsymbol{\Delta}) + \boldsymbol{c} \end{aligned}$$

and, when known constants are grouped together, the equations are

$$\left.\begin{aligned} v_1 &= \Delta_1 - (x_1 - p_{1_0}) \\ v_2 &= \Delta_2 - (x_2 - p_{2_0}) \\ v_3 &= -\Delta_1 - \Delta_2 - (x_3 - 180^\circ + p_{1_0} + p_{2_0}) \end{aligned}\right\}, \quad \begin{aligned} &\text{or in matrix notation,} \\ &\boldsymbol{v} = \boldsymbol{A}\boldsymbol{\Delta} - (\boldsymbol{x} - \boldsymbol{c} - \boldsymbol{A}\boldsymbol{p}_0) \end{aligned}$$

Let the approximate values for the parameters be the observed values. Therefore,

$$\boldsymbol{p}_0 = \begin{pmatrix} p_1 \\ p_2 \end{pmatrix} = \begin{pmatrix} 56^\circ 21' 32'' \\ 49^\circ 52' 09'' \end{pmatrix}$$

Other matrices are the same as in Example 9.4, viz.

$$x = \begin{bmatrix} x_1 \\ x_2 \\ x_3 \end{bmatrix} = \begin{bmatrix} 56°21'32'' \\ 49°52'09'' \\ 73°46'28'' \end{bmatrix}, \qquad A = \begin{bmatrix} 1 & \\ & 1 \\ -1 & -1 \end{bmatrix},$$

$$c = \begin{bmatrix} 0 \\ 0 \\ 180° \end{bmatrix}, \quad \text{and} \quad G = \begin{bmatrix} 1 & & \\ & 4 & \\ & & 9 \end{bmatrix}$$

but, from eqn (9.93),

$$f = x - c - Ap_0 = \begin{bmatrix} 56°21'32'' \\ 49°52'09'' \\ 73°46'28'' \end{bmatrix} - \begin{bmatrix} 0 \\ 0 \\ 180° \end{bmatrix} - \begin{bmatrix} 1 & \\ & 1 \\ -1 & -1 \end{bmatrix} \begin{pmatrix} 56°21'32'' \\ 49°52'09'' \end{pmatrix}$$

$$= \begin{bmatrix} 56 & 21 & 32 \\ 49 & 52 & 09 \\ -106 & 13 & 32 \end{bmatrix} - \begin{bmatrix} 56 & 21 & 32 \\ 49 & 52 & 09 \\ -106 & 13 & 41 \end{bmatrix} = \begin{bmatrix} 0 \\ 0 \\ 9'' \end{bmatrix}$$

As before,

$$N = A^{\mathrm{T}} G^{-1} A = \frac{1}{36} \begin{pmatrix} 40 & 4 \\ 4 & 13 \end{pmatrix}$$

thus

$$N^{-1} = \frac{1}{14} \begin{pmatrix} 13 & -4 \\ -4 & 40 \end{pmatrix}$$

But,

$$F = A^{\mathrm{T}} G^{-1} f = \begin{pmatrix} 1 & & -1/9 \\ & 1/4 & -1/9 \end{pmatrix} \begin{bmatrix} 0 \\ 0 \\ 9 \end{bmatrix} = \begin{pmatrix} -1 \\ -1 \end{pmatrix}$$

From eqn (9.95),

$$\Delta = N^{-1} F = \frac{1}{14} \begin{pmatrix} 13 & -4 \\ -4 & 40 \end{pmatrix} \begin{pmatrix} -1 \\ -1 \end{pmatrix} = \frac{1}{14} \begin{pmatrix} -9 \\ -36 \end{pmatrix} = \begin{pmatrix} -0.64'' \\ -2.57'' \end{pmatrix}$$

Therefore,

$$p = p_0 + \Delta = \begin{pmatrix} 56 & 21 & 32 \\ 49 & 52 & 09 \end{pmatrix} + \begin{pmatrix} -0.64 \\ -2.57 \end{pmatrix} = \begin{pmatrix} 56°21'31.36'' \\ 49°52'06.43'' \end{pmatrix}$$

and, substituting in eqn (9.73),

$$\bar{x} = Ap + c = \begin{bmatrix} 1 & \\ & 1 \\ -1 & -1 \end{bmatrix} \begin{pmatrix} 56 & 21 & 31.36 \\ 49 & 52 & 06.43 \end{pmatrix} + \begin{bmatrix} 0 \\ 0 \\ 180 \end{bmatrix} = \begin{bmatrix} 56°21'31.4'' \\ 49°52'06.4'' \\ 73°46'22.2'' \end{bmatrix}$$

(b) The calculation of cofactors is no different from that given in Example 9.4, and so is not repeated here.

9.6 The estimation of precision using least-squares residuals

In section 8.7 (eqn (8.31)) it was shown that the unit variance σ_0^2 could be found from $\sigma_0^2 = [wv^2]/r$, for uncorrelated observations. In the case of correlated observations the equation becomes

$$\sigma_0^2 = \frac{\boldsymbol{v}^{\mathrm{T}}\boldsymbol{G}^{-1}\boldsymbol{v}}{r} \tag{9.97}$$

where $\boldsymbol{v}$ is the vector of residuals obtained after a least-squares computation.

If the value of σ_0^2 differs by an appreciable amount from unity the implication is that the initial estimates of the variances and covariances is incorrect, or that systematic error or gross error is present in the observations. Caution should be exercized in drawing any firm conclusions, however, since if the redundancy r is small the value of σ_0^2 is statistically not very significant. Provided r is large it is possible to improve on the original estimation of the *a priori* covariance matrix by an iterative process, and an example using a mixture of angle and distance observations is described in [4].

The division (or multiplication) of the covariance matrix by a scalar quantity (e.g. a reference variance σ^2) does not alter the result of a least-squares computation. However, an alteration of the relative values does, and it is important to use the best possible relative values in solving a least-squares problem. On the other hand, applying a reference variance to the original covariance matrix will have the same scalar effect on the assessment of the quality of the results. Therefore, if the original covariance matrix of the observations was divided by a reference variance, the covariance matrices of the derived quantities are obtained by multiplying the *a posteriori* cofactor matrices by the reference variance, as already indicated in the worked examples.

9.7 Non-linear problems and least-squares estimation

The mathematical models which have been used so far for demonstrating least-squares procedures are linear functions of the unknowns. Most functions which occur in surveying and photogrammetric applications are non-linear, and therefore a process of linearization must be carried out in the first instance. Linearization is required whatever least-squares procedure ('condition' equations or 'observation' equations) is being used, and Taylor's theorm is normally used to achieve this. Appendix A describes the process in some detail.

In the case of *condition equations* the reduced equations are still of the form given by eqn (9.38), and the procedure for the determination of the v's is the same thereafter. However, equations which have been linearized usually have coefficients which are not simple integer numbers; the coefficients are partial derivatives of the function, and are evaluated by substituting approximate values for the unknown parameters. For example, in the braced quadrilateral of section 4.2 the side condition (eqn (4.5)) is non-linear, and after linearization (eqn (4.7)) the coefficients are the cotangents of the angles. The example of the braced quadrilateral, with the data given in Table 4.1, is used below to illustrate the least-squares solution of a problem involving a non-linear equation.

Example 9.6

The angles of the network shown in Figure 9.3 have been observed, and the observations which are assumed to be uncorrelated and of equal weight are given below.

Angle	°	′	″
1	84	07	59
2	40	10	36
3	16	11	20
4	39	29	53
5	75	03	02
6	49	15	28
7	30	05	21
8	25	36	25

Derive least-squares estimates for the angles of the network.

Solution

The geometrical shape of the network is determined if a minimum of four angles is observed. Therefore, $r = 8 - 4 = 4$ and four independent conditions must be established to solve the problem.

The four conditions are given by eqn (4.2) and (4.5), the latter being the non-linear side condition. When it is linearized eqn (4.7) is obtained, and when observations are introduced to all four equations the reduced condition equations (eqns (4.10)) are

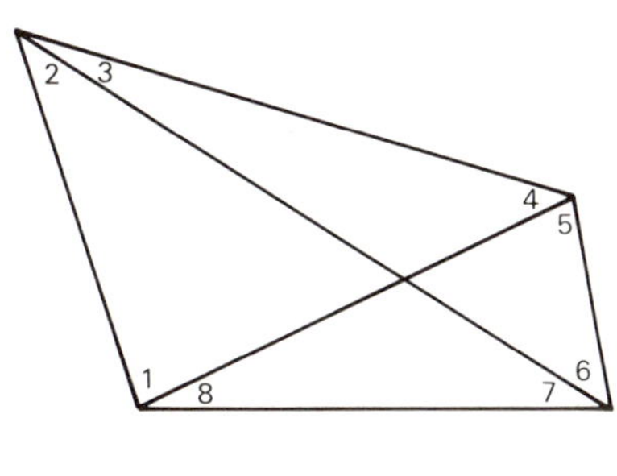

Figure 9.3.

obtained, viz.

$$\left.\begin{aligned} v_1 + v_2 + v_3 + v_4 &= \pi - (1 + 2 + 3 + 4) \\ v_5 + v_6 + v_7 + v_8 &= \pi - (5 + 6 + 7 + 8) \\ v_3 + v_4 + v_5 + v_6 &= \pi - (3 + 4 + 5 + 6) \\ \cot 1 \cdot v_1 - \cot 2 \cdot v_2 + \ldots \text{etc.} &= 1 - \frac{\sin 1 \sin 3 \sin 5 \sin 7}{\sin 2 \sin 4 \sin 6 \sin 8} \end{aligned}\right\} \qquad (9.98)$$

When observed values are introduced the equations become

$$\left.\begin{aligned} v_1 + v_2 + v_3 + v_4 &= 12 \\ v_5 + v_6 + v_7 + v_8 &= -16 \\ v_3 + v_4 + v_5 + v_6 &= 17 \\ u_{41}v_1 + u_{42}v_2 + u_{43}v_3 + \cdots &= 46.14 \end{aligned}\right\} \qquad (9.99)$$

where the units are seconds of arc and the u's are the cotangents, with appropriate sign obtained from eqn (9.98).

Therefore the following matrices may be identified:

$$\boldsymbol{U} = \begin{bmatrix} 1 & 1 & 1 & 1 & & & & \\ & & & & 1 & 1 & 1 & 1 \\ & & 1 & 1 & 1 & 1 & & \\ 0.1028 & -1.1843 & 3.4445 & -1.2132 & 0.2670 & -0.8614 & 1.7258 & -2.0865 \end{bmatrix},$$

$$\boldsymbol{t} = \begin{bmatrix} 12'' \\ -16'' \\ 17'' \\ 46.14'' \end{bmatrix}, \qquad \boldsymbol{G} = \boldsymbol{I}$$

From eqn (9.43),

$$\boldsymbol{N} = \boldsymbol{UGU}^{\mathrm{T}} = \begin{bmatrix} 4 & 0 & 2 & 1.1498 \\ 0 & 4 & 2 & 0.9551 \\ 2 & 2 & 4 & 1.6369 \\ 1.1498 & 0.9551 & 1.6369 & 22.8947 \end{bmatrix}$$

therefore,

$$\boldsymbol{N}^{-1} = \begin{bmatrix} 0.3754 & 0.1279 & -0.2535 & 0.0046 \\ 0.1279 & 0.3934 & -0.2727 & 0.0295 \\ -0.2535 & -0.2727 & 0.5280 & -0.0364 \\ 0.0046 & 0.0295 & -0.0364 & 0.0473 \end{bmatrix}$$

and from eqn (9.45),

$$
\boldsymbol{v} = \boldsymbol{G}\boldsymbol{U}^{\mathrm{T}}\boldsymbol{N}^{-1}\boldsymbol{t} = \boldsymbol{I}\begin{bmatrix} 1 & & & 0.1028 \\ 1 & & & -1.1843 \\ 1 & & 1 & 3.4445 \\ 1 & & 1 & -1.2132 \\ & 1 & 1 & 0.2670 \\ & 1 & 1 & -0.8614 \\ & 1 & & 1.7258 \\ & 1 & & -2.0865 \end{bmatrix}
$$

$$
\times \begin{bmatrix} 0.3754 & 0.1279 & -0.2535 & 0.0046 \\ 0.1279 & 0.3934 & -0.2727 & 0.0295 \\ -0.2535 & -0.2727 & 0.5280 & -0.0364 \\ 0.0046 & 0.0295 & -0.0364 & 0.0473 \end{bmatrix} \begin{bmatrix} 12 \\ -16 \\ 17 \\ 46.14 \end{bmatrix}
$$

$$
= \begin{bmatrix} -1.52 \\ -2.99 \\ 10.94 \\ 5.60 \\ 0.89 \\ -0.40 \\ -6.06 \\ -10.43 \end{bmatrix}
$$

Therefore, from eqn (9.36),

$$
\bar{\boldsymbol{x}} = \boldsymbol{x} + \boldsymbol{v} = \begin{bmatrix} 84 & 07 & 59 \\ 40 & 10 & 36 \\ 16 & 11 & 20 \\ 39 & 29 & 53 \\ 75 & 03 & 02 \\ 49 & 15 & 28 \\ 30 & 05 & 21 \\ 25 & 36 & 25 \end{bmatrix} + \begin{bmatrix} -1.52 \\ -2.99 \\ 10.94 \\ 5.60 \\ 0.89 \\ -0.40 \\ -6.06 \\ -10.43 \end{bmatrix} = \begin{bmatrix} 84 & 07 & 57.48 \\ 40 & 10 & 33.01 \\ 16 & 11 & 30.94 \\ 39 & 29 & 58.60 \\ 75 & 03 & 02.89 \\ 49 & 15 & 27.60 \\ 30 & 05 & 14.94 \\ 25 & 36 & 14.57 \end{bmatrix}
$$

Comment

A comparison with the results obtained in Example 4.1 may be made; the final values for each method and the differences between them are listed below.

Angle	Approximate solution			Least-squares solution			Difference
	°	′	″	°	′	″	″
1	84	08	00.5	84	07	57.48	−3.0
2	40	10	31.5	40	10	33.01	1.5
3	16	11	29.5	16	11	30.94	1.5

Angle	Approximate solution			Least-squares solution			Difference
	°	′	″	°	′	″	″
4	39	29	58.5	39	29	58.60	0.0
5	75	03	05.5	75	03	02.89	−2.5
6	49	15	26.5	49	15	27.60	1.0
7	30	05	14.5	30	05	14.94	0.5
8	25	36	13.5	25	36	14.57	1.0

The results differ by only a few seconds, and in many applications the approximate solution will suffice. However, a disadvantage of the approximate method lies in the fact that a different sequence of solution of the condition equations will produce a different result.

With *observation equations*, and after linearization, the partial derivatives with respect to the unknown parameters are the coefficients; and approximate values for the parameters are assigned in the first instance. These values should be as close as possible to the final estimates, and are usually derived by using a subset of the observations. The new unknowns ($\boldsymbol{\Delta}$) are therefore corrections to these approximate values. Thus, as in the previous section (eqn (9.89)), if $\boldsymbol{p}_0$ is the vector of approximate values for the parameters, and $\boldsymbol{\Delta}$ is the vector of unknown corrections to these approximate values, then

$$\boldsymbol{p} = \boldsymbol{p}_0 + \boldsymbol{\Delta}$$

Furthermore, the least-squares estimates ($\bar{\boldsymbol{x}}$) of the observed quantities are replaced by final discrepancies ($\bar{\boldsymbol{d}}$); the said estimates minus the computed values ($\boldsymbol{x}_0$). The latter are derived from calculations involving the approximate values for the parameters. Therefore,

$$\bar{\boldsymbol{d}} = \bar{\boldsymbol{x}} - \boldsymbol{x}_0 \tag{9.100}$$

and since

$$\bar{\boldsymbol{x}} = \boldsymbol{x} + \boldsymbol{v}$$

therefore

$$\bar{\boldsymbol{d}} = \boldsymbol{x} + \boldsymbol{v} - \boldsymbol{x}_0$$

$$= (\boldsymbol{x} - \boldsymbol{x}_0) + \boldsymbol{v}$$

Therefore

$$\bar{\boldsymbol{d}} = \boldsymbol{d} + \boldsymbol{v} \tag{9.101}$$

where

$$\boldsymbol{d} = \boldsymbol{x} - \boldsymbol{x}_0 \tag{9.102}$$

i.e.

$$\boldsymbol{d} = \text{observed} - \text{computed values}$$

The reduced observation equations are therefore

$$\boldsymbol{v} = \boldsymbol{A}\boldsymbol{\Delta} - \boldsymbol{f}, \qquad \text{where } \boldsymbol{f} = \boldsymbol{d} \tag{9.103}$$

and the solution, as before in eqn (9.95), is given by

$$\boldsymbol{\Delta} = \boldsymbol{N}^{-1}\boldsymbol{F} \tag{9.104}$$

An example illustrating the solution of a non-linear problem is given below, and the same data as that used in Example 4.3 is used.

Example 9.7

Grid bearings have been observed from points A, B, C and D, which have known coordinates. The coordinates are assumed to be error-free. The observations are considered to be uncorrelated, but have standard deviations inversely proportional to the length of the line observed. Calculate least squares estimates for the coordinates of the unknown point P, using the following data:

Coordinates:

	E (m)	N (m)
A	6091.13	3631.61
B	5973.88	1561.00
C	6947.21	178.43
D	7597.31	2063.49

Observations:

	Bearing °	′	″
AP	126	08	26
BP	53	22	37
CP	11	13	10
DP	344	41	38

Solution

The mathematical model for an observed bearing as a function of coordinates is given by eqn (2.55), and the linearized form of it is given by eqn (2.57), viz.

$$\mathrm{d}\alpha_{AP} = p_{AP'}(\mathrm{d}E_P - \mathrm{d}E_A) - q_{AP'}(\mathrm{d}N_P - \mathrm{d}N_A)$$

Provisional (approximate) coordinates of P are established and used to calculate a bearing ($\alpha_{AP'}$) and distance ($D_{AP'}$) between A and the provisional point P', and these are the basis for deriving $\mathrm{d}\alpha_{AP}$, p_{AP} and q_{AP}; where $\mathrm{d}\alpha_{AP}$ = observed − computed bearing; $p_{AP} = \rho \cdot \cos\alpha_{AP'}/D_{AP'}$; $q_{AP} = \rho \cdot \sin\alpha_{AP'}/D_{AP'}$ and $\rho = 206265$.

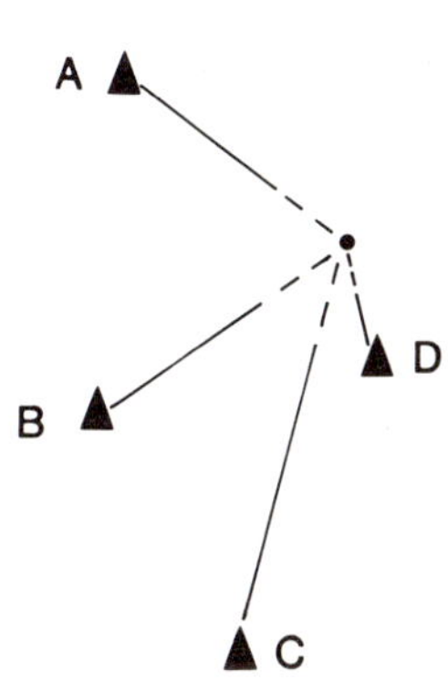

Figure 9.4.

Let the provisional coordinates be established by triangulation intersection using the observations from A and B, and eqns (2.14) and (2.16). Thus the provisional coordinates are

E	N
7437.137 m	2648.625 m

which are used to calculate precise bearings and distances to all the known points; and hence the quantities $d\alpha$, p and q for each of the four lines are derived.

The values are tabulated below.

Line	Computed distance (m)	Computed bearing °	′	″	Observed bearing °	′	″	Obs. minus comp. bearing ($d\alpha''$)	$\frac{\rho \cdot \cos\alpha}{s}$ (p)	$\frac{\rho \cdot \sin\alpha}{s}$ (q)
AP	1666.732	126	08	26	126	08	26	0	−72.99	99.94
BP	1823.198	53	22	37	53	22	37	0	67.49	90.80
CP	2518.311	11	13	05	11	13	10	5	80.34	15.93
DP	606.662	344	41	28	344	41	38	10	327.94	−89.77

When the unknown residual is added the linearized observation equation for the observation from A becomes

$$d\alpha_{AP} + v_{AP} = p_{AP}(dE_P - dE_A) - q_{AP}(dN_P - dN_A) \tag{9.105}$$

and, since the points with known coordinates are error-free, the terms involving corrections to these coordinates are 0. Therefore, the reduced observation equation is

$$v_{AP} = p_{AP}\,dE_P - q_{AP}\,dN_P - d\alpha_{AP} \tag{9.106}$$

Similar equations may be established for the observations from B, C and D, thus producing a set of equations of the form of eqn (9.103), viz.

$$\mathbf{v} = A\boldsymbol{\Delta} - \mathbf{f}$$

where

$$\mathbf{v} = \begin{bmatrix} v_{AP} \\ v_{BP} \\ v_{CP} \\ v_{DP} \end{bmatrix}, \quad A = \begin{bmatrix} -72.99 & -99.94 \\ 67.49 & -90.80 \\ 80.34 & -15.93 \\ 327.94 & 89.77 \end{bmatrix}, \quad \boldsymbol{\Delta} = \begin{pmatrix} dE_P \\ dN_P \end{pmatrix},$$

$$\mathbf{p}_0 = \begin{pmatrix} 7437.137 \\ 2648.625 \end{pmatrix}, \quad \mathbf{f} = \mathbf{d} = \begin{bmatrix} 0 \\ 0 \\ 5'' \\ 10'' \end{bmatrix}$$

Although the standard deviations for the observations are not quoted, information concerning the ratios between them is available. Therefore a cofactor matrix may be

formed, albeit with an unknown reference variance.
Let

$$G = \begin{bmatrix} \frac{1}{1.67^2} & & & \\ & \frac{1}{1.82^2} & & \\ & & \frac{1}{2.52^2} & \\ & & & \frac{1}{0.61^2} \end{bmatrix}, \quad \text{therefore} \quad G^{-1} = \begin{bmatrix} 2.79 & & & \\ & 3.31 & & \\ & & 6.35 & \\ & & & 0.37 \end{bmatrix}$$

From eqn (9.81),

$$N = A^{\mathrm{T}} G^{-1} A = \begin{pmatrix} 110718 & 2834 \\ 2834 & 59749 \end{pmatrix},$$

therefore

$$N^{-1} = \begin{pmatrix} 9.0428 \times 10^{-6} & -4.2892 \times 10^{-7} \\ -4.2892 \times 10^{-7} & 1.6757 \times 10^{-5} \end{pmatrix}$$

and,

$$F = A^{\mathrm{T}} G^{-1} f = \begin{pmatrix} 3764.2 \\ -173.65 \end{pmatrix}$$

Therefore, from eqn (9.104), $\Delta = N^{-1} F = \begin{pmatrix} 0.034 \\ -0.004 \end{pmatrix}$

and the final estimates of the coordinates of P are found from eqn (9.89), viz.

$$p = p_0 + \Delta = \begin{pmatrix} 7437.137 \\ 2648.625 \end{pmatrix} + \begin{pmatrix} 0.034 \\ -0.004 \end{pmatrix} = \begin{pmatrix} 7437.171 \\ 2648.621 \end{pmatrix} \text{m}$$

The solution of non-linear problems relies on having initial approximate values for the parameters, in order to derive values for the elements of the coefficient matrices $\boldsymbol{U}$ and $\boldsymbol{A}$, and the constant term vectors $\boldsymbol{t}$ and $\boldsymbol{f}$. If the approximate values are not close enough to the final least-squares estimates the linearization process will not be an adequate representation of the functional model, and a further iteration will be necessary. In such a case the results from the first least-squares solution are used as approximate values for the next iteration. Iterations should continue until the changes in the parameters and residuals from one iteration to the next are insignificant. In practice most problems are such that it is not difficult to establish approximate values close enough to provide a solution in one or two iterations. Thus 'convergence' is rapid, and this is usually achieved by using a subset of the observations to derive the initial values for the parameters. The exception however, is when the initial values—derived, say, from a set of observations containing a gross error—are so far from the least-squares estimates that the solution converges very slowly, or does not converge at all. On other occasions (e.g. independent model aerial triangulation) the functional model is simplified by making certain assumptions, to the extent that the equations are no longer non-linear. This allows a least-squares solution to be implemented directly, without the need to assign approximate values.

9.8 Combined least squares estimation

The two procedures described so far may be regarded as special cases of a more general approach to solving least-squares estimation problems. For example, it may be desirable to include certain parameters in a mathematical model which otherwise would comprise only observations. Thus an algorithm which can accommodate both types of equation is useful.

If n is the number of observations, and m is the minimum number required to determine the model, then the redundancy r is given by

$$r = n - m$$

and r independent condition equations relating the observations may be established.

If, in addition, unknown parameters are to be included, then a further equation must be established for each parameter. Therefore if p parameters are to be included the number of equations (q) will be

$$q = r + p$$

Since

$$0 \leqslant p \leqslant m$$

therefore

$$r \leqslant q \leqslant (r + m)$$

or

$$r \leqslant q \leqslant n$$

Combining eqns (9.37) and (9.103), the linear (or linearized) equations may be written

or

$$\left.\begin{aligned} \boldsymbol{U}(\boldsymbol{x} + \boldsymbol{v}) &= \boldsymbol{A\Delta} - \boldsymbol{c} \\ \boldsymbol{Uv} &= \boldsymbol{A\Delta} - (\boldsymbol{c} + \boldsymbol{Ux}) \end{aligned}\right\} \qquad (9.107)$$

where
$\boldsymbol{U}$ is a $(\boldsymbol{q} \times n)$ coefficient matrix, and $\boldsymbol{q} < n$;
$\boldsymbol{x}$ is the $(n \times 1)$ vector of observed values;
$\boldsymbol{A}$ is a $(q \times p)$ coefficient matrix, and $q > p$;
$\boldsymbol{v}$ is the $(n \times 1)$ vector of residuals;
$\boldsymbol{\Delta}$ is the $(p \times 1)$ vector of unknown parameters; and
$\boldsymbol{c}$ is the $(q \times 1)$ vector of constants.

There are several alternative approaches to the problem of deriving explicit expressions for $\boldsymbol{\Delta}$ and $\boldsymbol{v}$. Two derivations are given below, which follow a familiar pattern, and which use so-called *equivalent* observations, as suggested in [28].

The first method converts the equation into a form involving unknown parameters only. Let $\boldsymbol{x}_e$ be the vector of equivalent observations, defined by

$$\boldsymbol{x}_e = \boldsymbol{U}\boldsymbol{x} \tag{9.108}$$

and thus the equivalent residuals are

$$\boldsymbol{v}_e = \boldsymbol{U}\boldsymbol{v} \tag{9.109}$$

If $\boldsymbol{G}$ is the cofactor matrix for the observations then $\boldsymbol{G}_e$, the cofactor matrix for the equivalent observations, is obtained by applying the law of propagation to eqn (9.108).

Therefore

$$\boldsymbol{G}_e = \boldsymbol{U}\boldsymbol{G}\boldsymbol{U}^{\mathrm{T}} \tag{9.110}$$

Thus eqn (9.107) becomes

$$\boldsymbol{x}_e + \boldsymbol{v}_e = \boldsymbol{A}\boldsymbol{\Delta} - \boldsymbol{c}$$

or,

$$\boldsymbol{v}_e = \boldsymbol{A}\boldsymbol{\Delta} - (\boldsymbol{c} + \boldsymbol{x}_e) = \boldsymbol{A}\boldsymbol{\Delta} - \boldsymbol{f} \tag{9.111}$$

and, as in eqns (9.81),

$$\boldsymbol{N} = \boldsymbol{A}^{\mathrm{T}}\boldsymbol{G}_e^{-1}\boldsymbol{A}, \qquad \text{and} \qquad \boldsymbol{F} = \boldsymbol{A}^{\mathrm{T}}\boldsymbol{G}_e^{-1}\boldsymbol{F} \tag{9.112}$$

Therefore, as in eqn (9.82), the least squares estimates of the parameters are derived from

$$\boldsymbol{\Delta} = \boldsymbol{N}^{-1}\boldsymbol{F} \tag{9.113}$$

and $\boldsymbol{v}$ may be found by back substitution in eqns (9.111) and (9.109).

The second procedure involves the use of Lagrange multipliers, as in the case of the method using condition equations, and outlined in section 9.3. Thus the equivalent of eqn (9.40), when related to eqn (9.107) above, becomes

$$E = \boldsymbol{v}^{\mathrm{T}}\boldsymbol{G}^{-1}\boldsymbol{v} - 2\boldsymbol{k}^{\mathrm{T}}(\boldsymbol{U}\boldsymbol{v} - \boldsymbol{A}\boldsymbol{\Delta} + \boldsymbol{f}) = \text{minimum} \tag{9.114}$$

where $\boldsymbol{k}$ is of order $(q \times 1)$.

The minimum is achieved by equating the partial derivatives of E to zero, and in this case the derivatives are with respect to both $\boldsymbol{v}$ and $\boldsymbol{\Delta}$.

Thus

$$\frac{\partial E}{\partial \boldsymbol{v}} = 2\boldsymbol{v}^{\mathrm{T}}\boldsymbol{G}^{-1} - 2\boldsymbol{k}^{\mathrm{T}}\boldsymbol{U} = 0$$

$$\frac{\partial E}{\partial \boldsymbol{\Delta}} = -2\boldsymbol{k}^{\mathrm{T}}\boldsymbol{A} = 0$$

Because $\boldsymbol{G}^{-1}$ is symmetric the first equation may be written

$$\boldsymbol{G}^{-1}\boldsymbol{v} = \boldsymbol{U}^{\mathrm{T}}\boldsymbol{k} \tag{9.115}$$

The second equation may be written

$$\boldsymbol{A}^{\mathrm{T}}\boldsymbol{k} = 0 \tag{9.116}$$

Therefore, from eqn (9.115),

$$\boldsymbol{v} = \boldsymbol{G}\boldsymbol{U}^{\mathrm{T}}\boldsymbol{k} \tag{9.117}$$

which may be substituted in eqn (9.107), giving

$$\boldsymbol{U}\boldsymbol{G}\boldsymbol{U}^{\mathrm{T}}\boldsymbol{k} = \boldsymbol{A}\boldsymbol{\Delta} - \boldsymbol{f} \tag{9.118}$$

Using eqn (9.110), this may be converted to

$$\boldsymbol{G}_{\mathrm{e}}\boldsymbol{k} = \boldsymbol{A}\boldsymbol{\Delta} - \boldsymbol{f} \tag{9.119}$$

and solving for $\boldsymbol{k}$,

$$\boldsymbol{k} = \boldsymbol{G}_{\mathrm{e}}^{-1}(\boldsymbol{A}\boldsymbol{\Delta} - \boldsymbol{f}) \tag{9.120}$$

which may be substituted in eqn (9.116); thus

$$\boldsymbol{A}^{\mathrm{T}}\boldsymbol{G}_{\mathrm{e}}^{-1}(\boldsymbol{A}\boldsymbol{\Delta} - \boldsymbol{f}) = 0$$

or,

$$\boldsymbol{A}^{\mathrm{T}}\boldsymbol{G}_{\mathrm{e}}^{-1}\boldsymbol{A}\boldsymbol{\Delta} = \boldsymbol{A}^{\mathrm{T}}\boldsymbol{G}_{\mathrm{e}}^{-1}\boldsymbol{f}$$

i.e.

$$\boldsymbol{N}\boldsymbol{\Delta} = \boldsymbol{F}$$

Thus

$$\boldsymbol{\Delta} = \boldsymbol{N}^{-1}\boldsymbol{F} \tag{9.121}$$

where

$$\boldsymbol{N} = \boldsymbol{A}^{\mathrm{T}}\boldsymbol{G}_{\mathrm{e}}^{-1}\boldsymbol{A}; \quad \text{and} \quad \boldsymbol{F} = \boldsymbol{A}^{\mathrm{T}}\boldsymbol{G}_{\mathrm{e}}^{-1}\boldsymbol{f}$$

and $\boldsymbol{v}$ may be found by back substitution in eqns (9.120) and (9.117).

These results are the same as those derived in the first procedure (eqns (9.112) and (9.113)).

The cofactor matrix for the derived parameters $\boldsymbol{\Delta}$ may be found as follows:

From eqn (9.121),

$$\boldsymbol{\Delta} = \boldsymbol{N}^{-1}\boldsymbol{F} = \boldsymbol{N}^{-1}\boldsymbol{A}^{\mathrm{T}}\boldsymbol{G}_{\mathrm{e}}^{-1}\boldsymbol{f}$$

Also,

$$\begin{aligned}\boldsymbol{f} &= \boldsymbol{c} + \boldsymbol{x}_{\mathrm{e}} \\ &= \boldsymbol{c} + \boldsymbol{U}\boldsymbol{x}, \quad \text{from eqns (9.111) and (9.108),}\end{aligned}$$

where $\boldsymbol{c}$ is a vector of constants.

Therefore $$\boldsymbol{\Delta} = \boldsymbol{N}^{-1}\boldsymbol{A}^{\mathrm{T}}\boldsymbol{G}_{\mathrm{e}}^{-1}(\boldsymbol{c} + \boldsymbol{U}\boldsymbol{x}) \tag{9.122}$$

where the vector of stochastic variables is $\boldsymbol{x}$.

Therefore, applying the law of propagation, and bearing in mind that $\boldsymbol{N}^{-1}$ and $\boldsymbol{G}_{\mathrm{e}}^{-1}$ are symmetric,

$$\begin{aligned}\boldsymbol{G}_{\Delta\Delta} &= (\boldsymbol{N}^{-1}\boldsymbol{A}^{\mathrm{T}}\boldsymbol{G}_{\mathrm{e}}^{-1}\boldsymbol{U})\boldsymbol{G}(\boldsymbol{N}^{-1}\boldsymbol{A}^{\mathrm{T}}\boldsymbol{G}_{\mathrm{e}}^{-1}\boldsymbol{U})^{\mathrm{T}} \\ &= \boldsymbol{N}^{-1}\boldsymbol{A}^{\mathrm{T}}\boldsymbol{G}_{\mathrm{e}}^{-1}\boldsymbol{U}\boldsymbol{G}\boldsymbol{U}^{\mathrm{T}}\boldsymbol{G}_{\mathrm{e}}^{-1}\boldsymbol{A}\boldsymbol{N}^{-1} \\ &= \boldsymbol{N}^{-1}\boldsymbol{A}^{\mathrm{T}}\boldsymbol{G}_{\mathrm{e}}^{-1}\boldsymbol{G}_{\mathrm{e}}\boldsymbol{G}_{\mathrm{e}}^{-1}\boldsymbol{A}\boldsymbol{N}^{-1} \\ &= \boldsymbol{N}^{-1}\boldsymbol{A}^{\mathrm{T}}\boldsymbol{G}_{\mathrm{e}}^{-1}\boldsymbol{A}\boldsymbol{N}^{-1} \\ &= \boldsymbol{N}^{-1}\boldsymbol{N}\boldsymbol{N}^{-1} \\ &= \boldsymbol{N}^{-1}\end{aligned} \tag{9.123}$$

The cofactor matrices for the residuals and the adjusted observations, $\boldsymbol{G}_{vv}$ and $\boldsymbol{G}_{\bar{x}\bar{x}}$, may be found in a similar manner, by applying the law of propagation to eqns (9.117) and (9.36) respectively. The procedure is the same in principle as that outlined in sections 9.3 and 9.5, and the results are

$$\boldsymbol{G}_{vv} = \boldsymbol{G}\boldsymbol{U}^{\mathrm{T}}\boldsymbol{G}_{\mathrm{e}}^{-1}(\boldsymbol{I} - \boldsymbol{A}\boldsymbol{N}^{-1}\boldsymbol{A}^{\mathrm{T}}\boldsymbol{G}_{\mathrm{e}}^{-1})\boldsymbol{U}\boldsymbol{G} \tag{9.124}$$

$$\boldsymbol{G}_{\bar{x}\bar{x}} = \boldsymbol{G} - \boldsymbol{G}\boldsymbol{U}^{\mathrm{T}}\boldsymbol{G}_{\mathrm{e}}^{-1}(\boldsymbol{I} - \boldsymbol{A}\boldsymbol{N}^{-1}\boldsymbol{A}^{\mathrm{T}}\boldsymbol{G}_{\mathrm{e}}^{-1})\boldsymbol{U}\boldsymbol{G} \tag{9.125}$$

Thus, $$\boldsymbol{G}_{\bar{x}\bar{x}} = \boldsymbol{G} - \boldsymbol{G}_{vv} \tag{9.126}$$

as obtained earlier in sections 9.3 and 9.5.

The equations above have been derived for the solution of problems involving both condition equations and observation equations, and may be seen to be more general in application than earlier derivations. If no parameters are to be included in the mathematical model then $\boldsymbol{A}$ is a null matrix and the $A\Delta$ term in eqn (9.107) disappears; thus the equation reverts to the form given by eqn (9.38)—i.e. condition equations. If, on the other hand, $\boldsymbol{U}$ is the unit matrix $\boldsymbol{I}$, then the model reduces to the form of eqn (9.76) (or 9.94)—i.e. observation equations. The principle of generality may be extended even further when, in addition to condition equations and observation equations, the algorithm includes conditions between parameters. These latter equations are referred to as *constraint equations*, and details concerning the manner in which they are incorporated into the solution may be found in [27].

10 Application of least-squares estimation

10.1 Introduction

This chapter concentrates on the application of least-squares adjustment to common problems in surveying and photogrammetry. Attention is focused on the technique using observation equations, and greater emphasis is given to establishing the necessary equations and identifying the appropriate matrices, than to the arithmetic involved. Nevertheless, sufficient intermediate results are provided to allow the interested student to check results at various stages of the calculations. It is assumed that the reader is familiar with the technique of linearization (Appendix A), and with computational procedures using the minimum number of observations to solve specific problems, as outlined in Chapters 2 and 5. Thus, in dealing with particular problems, linearized equations are usually quoted, and approximate values stated, without giving details of the stages involved in producing these results.

For convenience, and ready reference purposes, the more important equations and symbols for the least-squares procedure using observation equations are summarized below.

n is the number of observations;
m is the minimum number of observations required to solve the problem;
r is the number of redundant observations ($r = n - m$).

The mathematical model is

$$\bar{\boldsymbol{x}} = \boldsymbol{A}\boldsymbol{p} + \boldsymbol{c} \tag{10.1}$$

if the equations are linear at the outset; or

$$\bar{\boldsymbol{x}} = \boldsymbol{A}(\boldsymbol{p}_0 + \boldsymbol{\Delta}) + \boldsymbol{c} \tag{10.2}$$

if the technique mentioned in section 9.5 is applied to equations which are linear at the outset; or

$$\bar{\boldsymbol{d}} = \boldsymbol{A}\boldsymbol{\Delta} \tag{10.3}$$

if the non-linear equations have been linearized.

The symbols used in eqns (10.1), (10.2) and (10.3) are defined as follows:

$\bar{\boldsymbol{x}}$ is the vector ($n \times 1$) of adjusted observations;
$\boldsymbol{A}$ is the matrix ($n \times m$) of coefficients of the unknowns;
$\boldsymbol{p}$ is the vector ($m \times 1$) of unknown parameters;
$\boldsymbol{c}$ is the vector ($n \times 1$) of constant terms;

$\boldsymbol{p}_0$ is the vector ($m \times 1$) of approximate values for the parameters;
$\boldsymbol{\Delta}$ is the vector ($m \times 1$) of unknown corrections to the approximate values ($\boldsymbol{p}_0$) of the parameters; and
$\bar{\boldsymbol{d}}$ is the vector ($n \times 1$) of unknown discrepancies.

The unknown residuals are introduced via

$$\left.\begin{aligned}\bar{\boldsymbol{x}} &= \boldsymbol{x} + \boldsymbol{v}\\ \text{or} \quad \bar{\boldsymbol{d}} &= \boldsymbol{d} + \boldsymbol{v}\end{aligned}\right\} \tag{10.4}$$

where $\boldsymbol{x}$ is the vector ($n \times 1$) of observations; $\boldsymbol{d}$ is the vector ($n \times 1$) of known discrepancies (observed-computed values); and $\boldsymbol{v}$ is the vector ($n \times 1$) of residuals.

Therefore the model becomes

$$\boldsymbol{v} = \boldsymbol{A}\boldsymbol{p} - \boldsymbol{f} \tag{10.5}$$

or

$$\boldsymbol{v} = \boldsymbol{A}\boldsymbol{\Delta} - \boldsymbol{f} \tag{10.6}$$

where $\boldsymbol{f}$ is the new vector ($n \times 1$) of constant terms. $\boldsymbol{f}$ is derived from

$$\boldsymbol{f} = (\boldsymbol{x} - \boldsymbol{c}), \text{ in the case of eqn (10.5)} \tag{10.7}$$

$$\left.\begin{aligned}\text{and} \quad \boldsymbol{f} &= (\boldsymbol{x} - \boldsymbol{c} - \boldsymbol{A}\boldsymbol{p}_0)\\ \text{or} \quad \boldsymbol{f} &= \boldsymbol{d}\end{aligned}\right\} \text{ in the case of eqn (10.6)} \tag{10.8}$$

If $\boldsymbol{G}$ is the cofactor matrix for the observations, then the least squares solution is found from

$$\boldsymbol{p} = \boldsymbol{N}^{-1}\boldsymbol{F}, \quad \text{in the case of eqn (10.5)} \tag{10.9}$$

or

$$\boldsymbol{\Delta} = \boldsymbol{N}^{-1}\boldsymbol{F}, \quad \text{in the case of eqn (10.6)} \tag{10.10}$$

where

$$\left.\begin{aligned}\boldsymbol{N} &= \boldsymbol{A}^{\mathrm{T}}\boldsymbol{G}^{-1}\boldsymbol{A}\\ \boldsymbol{F} &= \boldsymbol{A}^{\mathrm{T}}\boldsymbol{G}^{-1}\boldsymbol{f}\end{aligned}\right\} \tag{10.11}$$

In the case of eqn (10.10), the final values for the parameters are found from

$$\boldsymbol{p} = \boldsymbol{p}_0 + \boldsymbol{\Delta} \tag{10.12}$$

and $\boldsymbol{v}$ is derived from eqn (10.5) or (10.6).

Thus $\bar{\boldsymbol{x}}$ may be derived from eqn (10.4).

The cofactor matrices for the derived quantities are found from

$$\boldsymbol{G}_{\mathrm{pp}} = \boldsymbol{G}_{\Delta\Delta} = \boldsymbol{N}^{-1} \tag{10.13}$$

$$\boldsymbol{G}_{\bar{x}\bar{x}} = \boldsymbol{A}\boldsymbol{N}^{-1}\boldsymbol{A}^{\mathrm{T}} \tag{10.14}$$

Finally, variances and covariances for these quantities are obtained from

$$\left.\begin{aligned}\boldsymbol{\Sigma}_{\mathrm{pp}} &= \sigma^2 \cdot \boldsymbol{G}_{\mathrm{pp}}\\ \boldsymbol{\Sigma}_{\Delta\Delta} &= \sigma^2 \cdot \boldsymbol{G}_{\Delta\Delta}\\ \boldsymbol{\Sigma}_{\bar{x}\bar{x}} &= \sigma^2 \cdot \boldsymbol{G}_{xx}\end{aligned}\right\} \tag{10.15}$$

where σ^2 is the reference variance.

10.2 Applications in surveying

Example 10.1

Height differences between points in a small network have been observed and the observations, with corresponding standard deviations, are listed below. Let the observations, which are assumed to be uncorrelated, be numbered as shown in the diagram.

(a) Establish two alternative mathematical models for a least squares adjustment of the network, by the method of observation equations.

(b) Use one of the models to carry out the adjustment.

Line	Observed ht.diff. (m)	Standard deviation (m)
AB	10.00	0.03
AC	15.06	0.06
AD	5.00	0.03
BC	5.00	0.06
BD	– 4.97	0.09
CD	– 10.03	0.06

Solution

(a) A minimum of three height differences are required to solve the network. This becomes clear if a point such as *A* is given an arbitrary height, and the number of

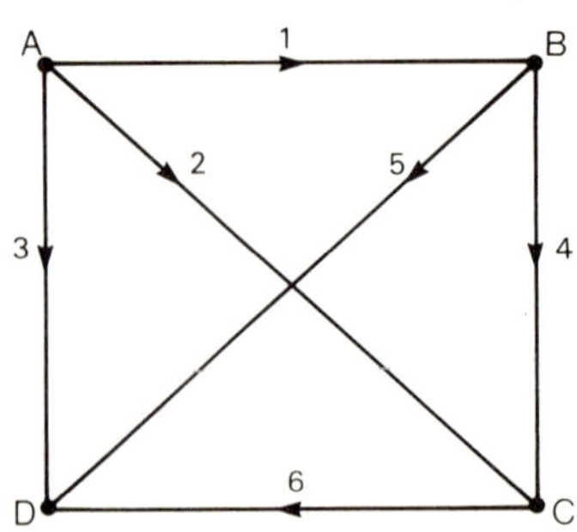

Figure 10.1.

height differences required to establish heights of the remaining points is assessed. Thus $m = 3$, and three parameters are selected as unknowns for the observation equations. The difference between the models below relate to the types of parameters chosen—in the first case it is height differences, and in the second it is actual heights of the points.

(i) Let the three chosen parameters be the height differences AB, AC and AD, and let these be represented by h_{AB}, h_{AC} and h_{AD}. Therefore the equations corresponding to eqn (10.1) are

$$\left.\begin{aligned}\bar{h}_1 &= h_{AB} \\ \bar{h}_2 &= h_{AC} \\ \bar{h}_3 &= h_{AD} \\ \bar{h}_4 &= -h_{AB} + h_{AC} \\ \bar{h}_5 &= -h_{AB} + h_{AD} \\ \bar{h}_6 &= -h_{AC} + h_{AD}\end{aligned}\right\} \tag{10.16}$$

The derivation of the last three equations will become clearer if the condition that a circuit must close is considered in each case. For example, in the case of the fourth equation,

$$h_{AB} + \bar{h}_4 - h_{AC} = 0; \quad \text{therefore} \quad \bar{h}_4 = -h_{AB} + h_{AC}$$

(ii) Let the chosen parameters be the heights of points B, C and D, and let these be H_B, H_C and H_D. Let point A have an arbitrary height of 100 m.

$$\left.\begin{aligned}\bar{h}_1 &= H_B - 100 \\ \bar{h}_2 &= H_C - 100 \\ \bar{h}_3 &= H_D - 100 \\ \bar{h}_4 &= -H_B + H_C \\ \bar{h}_5 &= H_B + H_D \\ \bar{h}_6 &= -H_C + H_D\end{aligned}\right\} \tag{10.17}$$

(b) The first of the two models given in (a) will be used.

The equations are linear, therefore a least-squares adjustment may be implemented directly. However, to simplify the arithmetic involved, approximate values are assigned to the parameters and for convenience the observed values are used. When observations are also introduced on the left-hand side the equations become

$$\begin{aligned}10.00 + v_1 &= (10.00 + \Delta_{AB}) \\ 15.06 + v_2 &= (15.06 + \Delta_{AC}) \\ 5.00 + v_3 &= (5.00 + \Delta_{AD}) \\ 5.00 + v_4 &= -(10.00 + \Delta_{AB}) + (15.06 + \Delta_{AC}) \\ -4.97 + v_5 &= -(10.00 + \Delta_{AB}) + (5.00 + \Delta_{AD}) \\ -10.03 + v_6 &= -(15.06 + \Delta_{AC}) + (5.00 + \Delta_{AD})\end{aligned}$$

where Δ_{AB}, Δ_{AC} and Δ_{AD} are the new unknowns—corrections to approximate values for the parameters.

The equations reduce to

$$\begin{aligned}
v_1 &= \Delta_{AB} && && +0 \\
v_2 &= && \Delta_{AC} && +0 \\
v_3 &= && && \Delta_{AD} +0 \\
v_4 &= -\Delta_{AB} + \Delta_{AC} && && +0.06 \\
v_5 &= -\Delta_{AB} && && +\Delta_{AD} -0.03 \\
v_6 &= && -\Delta_{AC} && +\Delta_{AD} -0.03
\end{aligned}$$

which is in the form given by eqn (10.6)

The following matrices may be identified:

$$\boldsymbol{A} = \begin{bmatrix} 1 & & \\ & 1 & \\ & & 1 \\ -1 & 1 & \\ -1 & & 1 \\ & -1 & 1 \end{bmatrix}, \quad \boldsymbol{\Delta} = \begin{bmatrix} \Delta_{AB} \\ \Delta_{AC} \\ \Delta_{AD} \end{bmatrix}, \quad \boldsymbol{p}_0 = \begin{bmatrix} 10.00 \\ 15.06 \\ 5.00 \end{bmatrix} \text{ m},$$

$$\boldsymbol{f} = \boldsymbol{d} = \begin{bmatrix} 0 \\ 0 \\ 0 \\ -6 \\ 3 \\ 3 \end{bmatrix} \text{cm} \quad \boldsymbol{\Sigma} = {}_{1}\begin{bmatrix} 9 & & & & & \\ & 36 & & & & \\ & & 9 & & & \\ & & & 36 & & \\ & & & & 81 & \\ & & & & & 36 \end{bmatrix} \text{cm}^2$$

Let $\sigma^2 = 9\,\text{cm}^2$,

$$\text{therefore } \boldsymbol{G} = (1/9)\boldsymbol{\Sigma} \quad \text{and} \quad \boldsymbol{G}^{-1} = \begin{bmatrix} 1 & & & & & \\ & 0.25 & & & & \\ & & 1 & & & \\ & & & 0.25 & & \\ & & & & 0.11 & \\ & & & & & 0.25 \end{bmatrix}$$

From eqns (10.11),

$$\boldsymbol{N} = \boldsymbol{A}^{\mathrm{T}}\boldsymbol{G}^{-1}\boldsymbol{A} = \begin{bmatrix} 1.36 & -0.25 & -0.11 \\ -0.25 & 0.75 & -0.25 \\ -0.11 & -0.25 & 1.36 \end{bmatrix}, \quad \text{and} \quad \boldsymbol{F} = \boldsymbol{A}^{\mathrm{T}}\boldsymbol{G}^{-1}\boldsymbol{f} = \begin{bmatrix} 1.17 \\ -2.25 \\ 1.08 \end{bmatrix}$$

$$\boldsymbol{N}^{-1} = \frac{1}{1.19}\begin{bmatrix} 0.96 & 0.37 & 0.14 \\ 0.37 & 1.84 & 0.37 \\ 0.14 & 0.37 & 0.96 \end{bmatrix},$$

therefore from eqn (10.10),

$$\Delta = N^{-1}F = \begin{bmatrix} 0.37 \\ -2.78 \\ 0.31 \end{bmatrix} \text{cm}$$

Therefore, from eqn (10.12), the final values for the parameters are

$$p = p_0 + \Delta = \begin{bmatrix} 10.00 \\ 15.06 \\ 5.00 \end{bmatrix} + \begin{bmatrix} 0.0037 \\ -0.0277 \\ 0.0031 \end{bmatrix} = \begin{bmatrix} 10.0037 \\ 15.0323 \\ 5.0031 \end{bmatrix} \text{m}$$

and from eqn (10.1), the adjusted observations are

$$\bar{x} = Ap + c$$

where it is clear from eqn (10.16) that c is a null vector.
Therefore,

$$\bar{x} = Ap = \begin{bmatrix} 10.004 \\ 15.032 \\ 5.003 \\ 5.029 \\ -5.001 \\ -10.029 \end{bmatrix} \text{m}$$

An example of a non-linear problem—trilateration intersection—is given below, which uses the same data as Example 4.6.

Example 10.2

An EDM instrument has been used to measure distances from known points A, B, C and D, in order to fix the position of point P. The observations are given below, and the two distances measured from A and B have been used to give the following provisional

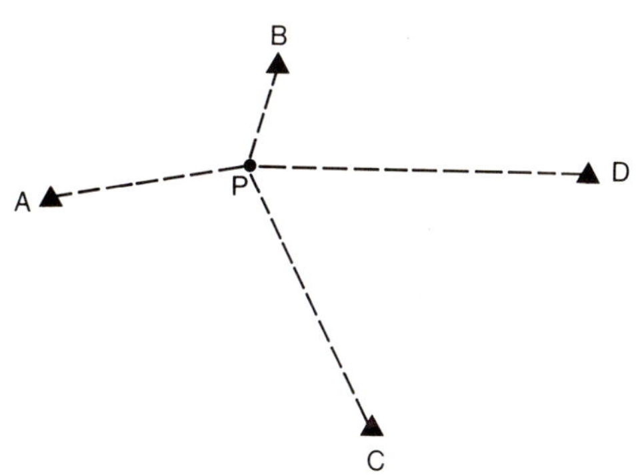

Figure 10.2

coordinates of P:

E	N
7103.332 m	4096.265 m

Assuming that:
(i) the known coordinates are error-free;
(ii) the observations are uncorrelated; and
(iii) the variance of a measured distance is proportional to the distance squared;

calculate the final position of P by least squares.

The given data is:

Point	E (m)	N (m)	Line	Measured Distance (m)
A	5997.73	3912.57	PA	1120.76
B	7279.56	4643.10	PB	574.53
C	7781.63	2593.70	PC	1648.73
D	9002.47	4011.21	PD	1901.22

Solution

The observation equation for a measured distance is given by eqn (2.49), and the linearized form is given by eqn (2.53). When one end of the line is fixed it reduces to the form given by eqn (2.54), viz.

$$\mathrm{d}D = r\cdot \mathrm{d}E_P + s\cdot \mathrm{d}N_P$$

where $\mathrm{d}D$ is the observed minus the computed distance; the latter being based on the provisional coordinates; r, s are known coefficients, derived from the provisional coordinates ($r = \sin\alpha$, $s = \cos\alpha$; where α is the bearing from the known point to P); and $\mathrm{d}E_P, \mathrm{d}N_P$ are the unknown parameters—the corrections to the provisional coordinates.

When the residual is added and the equation re-arranged it becomes

$$v = r\cdot \mathrm{d}E_P + s\cdot \mathrm{d}N_P - \mathrm{d}D$$

The provisional coordinates are used to calculate the required quantities, which are tabulated below.

Line	Computed Bearing ° ′ ″	Computed Distance (m)	Observed Distance (m)	Observed—comp.dist. ($\mathrm{d}D$)	$\sin\alpha$ (r)	$\cos\alpha$ (s)
AP	80 33 58	1120.760	1120.76	0	0.9865	0.1639
BP	197 51 45	574.530	574.53	0	−0.3067	−0.9518
CP	335 42 16	1648.572	1648.73	0.158	−0.4114	0.9114
DP	272 33 52	1901.042	1901.22	0.178	−0.9990	0.0447

When the four equations are expressed in the form of eqn (10.6), the following matrices may be identified:

$$A = \begin{bmatrix} 0.9865 & 0.1639 \\ -0.3067 & -0.9518 \\ -0.4114 & 0.9114 \\ -0.9990 & 0.0447 \end{bmatrix}, \qquad \boldsymbol{\Delta} = \begin{pmatrix} dE_P \\ dN_P \end{pmatrix},$$

$$\boldsymbol{f} = \boldsymbol{d} = \begin{bmatrix} 0 \\ 0 \\ 0.158 \\ 0.178 \end{bmatrix} \text{m}, \qquad \boldsymbol{p}_0 = \begin{pmatrix} 7103.332 \\ 4096.265 \end{pmatrix} \text{m}$$

Absolute variances are not given, but information concerning the relative accuracy is provided, therefore a cofactor matrix may be established.

Let

$$\boldsymbol{G} = \begin{bmatrix} 1.12^2 & & & \\ & 0.57^2 & & \\ & & 1.65^2 & \\ & & & 1.90^2 \end{bmatrix}, \quad \text{therefore} \quad \boldsymbol{G}^{-1} = \begin{bmatrix} 0.797 & & & \\ & 3.08 & & \\ & & 0.367 & \\ & & & 0.277 \end{bmatrix}$$

From eqns (10.11),

$$\boldsymbol{N} = A^{\mathrm{T}}\boldsymbol{G}^{-1}A = \begin{pmatrix} 1.404 & 0.878 \\ 0.878 & 3.117 \end{pmatrix}, \quad \text{and} \quad \boldsymbol{F} = A^{\mathrm{T}}\boldsymbol{G}^{-1}\boldsymbol{f} = \begin{pmatrix} -0.073 \\ 0.055 \end{pmatrix}$$

$$\boldsymbol{N}^{-1} = \frac{1}{3.60376}\begin{pmatrix} 3.117 & -0.878 \\ -0.878 & 1.404 \end{pmatrix},$$

therefore from eqn (10.10),

$$\boldsymbol{\Delta} = \boldsymbol{N}^{-1}\boldsymbol{F} = \begin{pmatrix} -0.0766 \\ 0.0392 \end{pmatrix} \text{m}$$

Therefore, from eqn (10.12), the final values for the parameters (coordinates of P) are

$$\boldsymbol{p} = \boldsymbol{p}_0 + \boldsymbol{\Delta} = \begin{pmatrix} 7103.332 \\ 4096.265 \end{pmatrix} + \begin{pmatrix} -0.077 \\ 0.039 \end{pmatrix} = \begin{pmatrix} 7103.255 \\ 4096.304 \end{pmatrix} \text{m}$$

Comment

The result agrees very well with that obtained semigraphically in Examples 4.4 and 4.6.

Section 2.8 explains how a resection calculation may be carried out using the minimum number of observations, and section 4.4 describes a semigraphic method of compensating for orientation error and deriving final coordinates for the point. The orientation correction represents an additional unknown parameter in a resection problem, and nowhere is it more evident than when observations are made from the unknown point alone. However, it is also present in bearings derived from observations at known points, although as a very much smaller quantity. Section 2.3 describes how an orientation correction may be derived for observations made from a known point. In Example 9.6 it was tacitly assumed that the bearings from A, B, C and D did not include orientation errors. Or, expressed differently, it was assumed that any error in the bearing AP, BP, CP, or DP, was the result of a pointing error to P alone. No great harm was done by this assumption, as the control points were considered fixed anyway, and the solution effectively transferred any such errors to the point P. It should be borne in mind, however, that residuals derived after the solution would also reflect pointing errors to these control points, as well as possible errors in the coordinates of the control points.

When directions are observed from a point to several other points the same orientation error will be present in all the observations. Thus, if the bundle of rays shown in Figure 10.3 is considered, it is clear that all observations have the same zero reference (shown dotted), and therefore all observations have the same orientation correction (θ). The zero reference will often be positioned arbitrarily, and might be several degrees adrift of grid north. The orientation correction is regarded as an additional unknown (parameter) which must be found in the solution of a resection problem. Therefore, instead of eqn (2.55), the mathematical model for an observation to B would be

$$\tan(\beta + \theta) = \frac{\Delta E}{\Delta N} \tag{10.18}$$

where both β and θ are stochastic variables.

On linearization, eqn (10.18) becomes

$$\frac{1}{\cos^2(\beta + \theta)}(\mathrm{d}\beta + \mathrm{d}\theta) = \frac{1}{\Delta N}\cdot \mathrm{d}(\Delta E) - \frac{\Delta E}{(\Delta N)^2}\cdot \mathrm{d}(\Delta N)$$

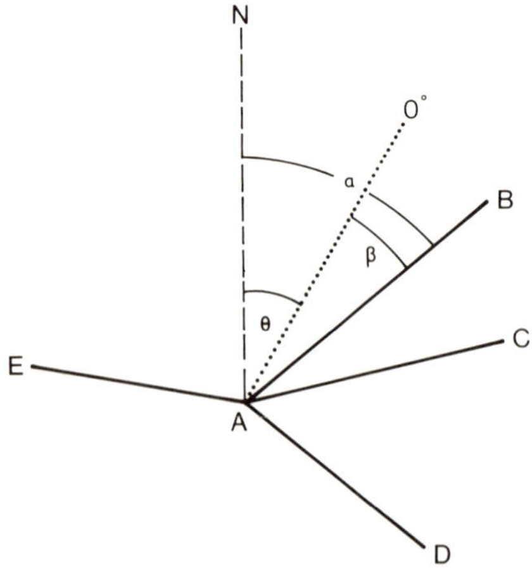

Figure 10.3.

and this reduces to

$$\mathrm{d}\beta + \mathrm{d}\theta = \frac{\cos(\beta+\theta)}{D}\cdot \mathrm{d}(\Delta E) - \frac{\sin(\beta+\theta)}{D}\cdot \mathrm{d}(\Delta N) \qquad (10.19)$$

where $\mathrm{d}\theta$ is the new unknown correction, and D is the length of the line.

An approximate value (θ_0) for the orientation correction is used to evaluate the coefficients, and when this value is applied to the observations approximate bearings are obtained. Thus $\beta + \theta_0 \approx \alpha$, and $\mathrm{d}\beta$ may be replaced by $\mathrm{d}\alpha$, and therefore eqn (10.19) may be written

$$\mathrm{d}\alpha + \mathrm{d}\theta = \frac{\cos\alpha}{D}\cdot \mathrm{d}(\Delta E) - \frac{\sin\alpha}{D}\cdot \mathrm{d}(\Delta N) \qquad (10.20)$$

For a particular observation AB, and with units in seconds of arc,

$$v_{AB} = p_{AB}(\mathrm{d}E_B - \mathrm{d}E_A) - q_{AB}(\mathrm{d}N_B - \mathrm{d}N_A) - \mathrm{d}\theta_A - \mathrm{d}\alpha_{AB} \qquad (10.21)$$

where

$$p_{AB} = \frac{\rho\cdot\cos\alpha_{AB}}{D_{AB}}, \qquad q_{AB} = \frac{\rho\cdot\sin\alpha_{AB}}{D_{AB}}$$

and $\rho = 206265.$

The data used in Examples 2.4 and 4.5 is used again to establish the coordinates of the unknown point by least squares.

Example 10.3

Observations have been made with a theodolite from an unknown point P to four known points A, B, C and D, whose coordinates, regarded as error-free, are given below. Calculate the position of P by least squares if

(i) the observations are uncorrelated,
(ii) the observations have equal weight, and

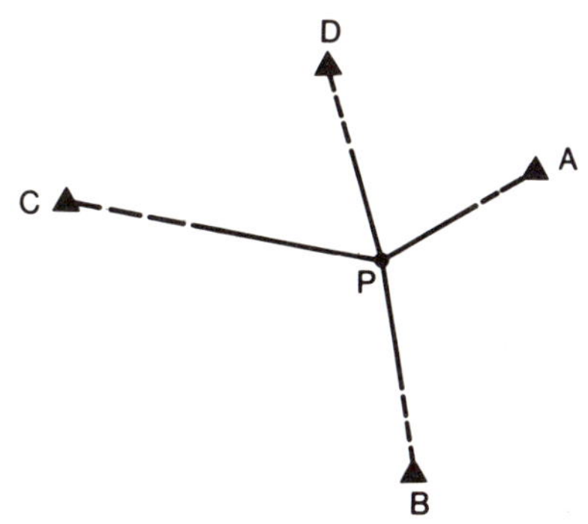

Figure 10.4.

(iii) the approximate coordinates of P, calculated by resection (see Example 2.4) from the observations to A, B, and C, are

E	N
7589.759 m	3666.702 m

The known coordinates and observations are:

	Coordinates: E (m)	N (m)	To	Observations at P: Direction ° ′ ″
A	8246.18	4102.96	A	57 45 00
B	7781.63	2593.70	B	171 13 11
C	5997.73	3912.56	C	280 08 12
D	7279.56	4643.10	D	343 44 11

Solution

Equation (10.21) applied to this resection problem becomes

$$v'' = -p\mathrm{d}E_P + q\mathrm{d}N_P - \mathrm{d}\theta_P - \mathrm{d}\alpha'' \tag{10.22}$$

The approximate coordinates of P are used to calculate joins to A, B, C and D. An approximate orientation correction (θ_0) is derived and found to be $-1°21'29''$. This is applied to the observed directions in order to obtain 'observed' bearings. The results, together with coefficients and constant terms for the observation equations (such as eqn (10.22)), are tabulated below.

Line	Computed distance (m)	Computed bearing ° ′ ″	Observed bearing ° ′ ″	Observed—comp. bearing ($\mathrm{d}\alpha''$)	$\frac{\rho\cos\alpha}{D}$ (p)	$\frac{\rho\sin\alpha}{D}$ (q)
PA	788.168	56 23 31	56 23 31	0	144.8	218.0
PB	1090.022	169 51 42	169 51 42	0	−186.2	33.31
PC	1610.901	278 46 44	278 46 43	−1	19.54	−126.6
PD	1024.488	342 22 30	342 22 42	12	191.9	−60.96

The following matrices may be identified:

$$A = \begin{bmatrix} -144.8 & 218.0 & -1 \\ 186.2 & 33.31 & -1 \\ -19.54 & -126.6 & -1 \\ -191.9 & -60.96 & 1 \end{bmatrix}, \quad \boldsymbol{\Delta} = \begin{bmatrix} \mathrm{d}E_P \\ \mathrm{d}N_P \\ \mathrm{d}\theta_P \end{bmatrix},$$

$$f = \begin{bmatrix} 0 \\ 0 \\ -1 \\ 12 \end{bmatrix}, \quad p_0 = \begin{bmatrix} 7589.759\ \mathrm{m} \\ 3666.702\ \mathrm{m} \\ -1°21'29'' \end{bmatrix}, \quad \boldsymbol{G} = \boldsymbol{I}$$

Therefore,

$$N = A^{\mathrm{T}} G^{-1} A = \begin{bmatrix} 92845 & -11192 & 170.04 \\ & 68377 & -63.75 \\ \text{Symm.} & & 4 \end{bmatrix},$$

and

$$F = A^{\mathrm{T}} G^{-1} f = \begin{bmatrix} -2283.26 \\ -604.92 \\ -11.00 \end{bmatrix}$$

Therefore

$$\Delta = N^{-1} F = \begin{bmatrix} -0.023\,\mathrm{m} \\ -0.014\,\mathrm{m} \\ -2.0'' \end{bmatrix}$$

and the final values for the parameters are

$$p = \begin{bmatrix} E_{po} \\ N_{po} \\ \theta_0 \end{bmatrix} + \begin{bmatrix} \mathrm{d}E_P \\ \mathrm{d}N_P \\ \mathrm{d}\theta \end{bmatrix} = \begin{bmatrix} 7589.736\,\mathrm{m} \\ 3666.688\,\mathrm{m} \\ -1^\circ 21' 31'' \end{bmatrix}$$

Comment

(i) The results are the same as those obtained semigraphically in Example 4.5, though clearly the effort involved in the least-squares solution is considerably less if a computer program is available.

(ii) One of the assumptions adopted was that the observations were uncorrelated, and this is true only to a limited extent. Strictly speaking correlation will be present, because of factors such as similar observing conditions, or instrumental defects, which are more pronounced in certain directions, etc. The extent of the correlation is generally difficult to determine, and with modern instrumentation is likely to be small. More important perhaps is the lack of validity in the second assumption, and a weighting system based on distance is perfectly feasible and easily incorporated. Thus G would be a diagonal matrix rather than a unit matrix.

The example above involved only one additional unknown, because there was only one unknown point to be fixed. Survey networks usually involve several unknown points, and the additional orientation unknown for each new point could result in a significant increase in the total number of unknowns

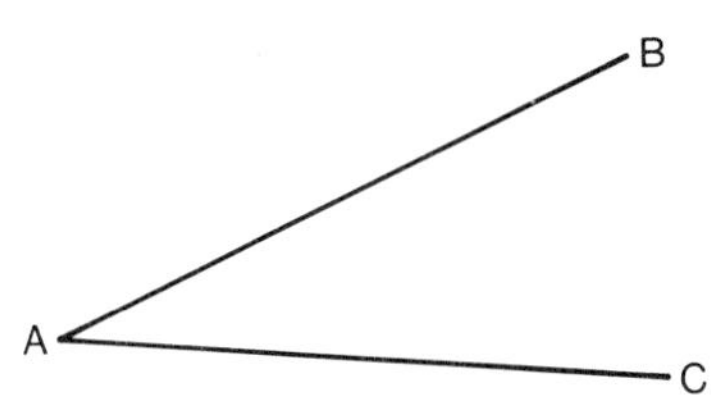

Figure 10.5.

(and hence normal equations) to be solved simultaneously. Thus the additional unknowns could make the difference between a network being solved on a particular computer or not. There are advantages therefore in devising equations which eliminate these unknowns.

The observation equation for an *angle BAC* is found by deriving the difference between the observation equations for the two bearings AB and AC. Thus,

$$\begin{aligned} v_{BAC} = & (p_{AC}(\mathrm{d}E_C - \mathrm{d}E_A) - q_{AC}(\mathrm{d}N_C - \mathrm{d}N_A) - \mathrm{d}\theta_A - \mathrm{d}\alpha_{AC}) \\ & - (p_{AB}(\mathrm{d}E_B - \mathrm{d}E_A) - q_{AB}(\mathrm{d}N_B - \mathrm{d}N_A) - \mathrm{d}\theta_A - \mathrm{d}\alpha_{AB}) \end{aligned} \tag{10.23}$$

which becomes

$$\begin{aligned} v_{BAC} = & (p_{AB} - p_{AC})\mathrm{d}E_A - (q_{AB} - q_{AC})\mathrm{d}N_A - p_{AB}\mathrm{d}E_B \\ & + q_{AB}\mathrm{d}N_B + p_{AC}\mathrm{d}E_C - q_{AC}\mathrm{d}N_C - (\mathrm{d}\alpha_{AC} - \mathrm{d}\alpha_{AB}) \end{aligned} \tag{10.24}$$

where, if the units are radians, $p = \cos\alpha/D$, and $q = \sin\alpha/D$.

Hence if A is the unknown point, and B and C are points with known coordinates which are considered to be error-free, the equation (with units in seconds) reduces to

$$v_{BAC} = (p_{AB} - p_{AC})\,\mathrm{d}E_A - (q_{AB} - q_{AC})\mathrm{d}N_A - (\mathrm{d}\alpha_{AC} - \mathrm{d}\alpha_{AB}) \tag{10.25}$$

where $$p = \rho\cos\alpha/D \quad \text{and} \quad q = \rho\sin\alpha/D.$$

Equation (10.25) may be written in simpler notation as

$$v_{BAC} = t_{BAC}\,\mathrm{d}E_A - u_{BAC}\,\mathrm{d}N_A - \delta\alpha_{BAC} \tag{10.26}$$

where $t_{BAC} = p_{AB} - p_{AC}$, $u_{BAC} = q_{AB} - q_{AC}$, $\delta\alpha_{BAC} = \mathrm{d}\alpha_{AC} - \mathrm{d}\alpha_{AB}$.

It is important to appreciate that, when observation equations for angles are used, the residual for the *derived* angle rather than the *observed* direction, is being minimized by the least-squares procedure. Therefore the variances and covariances appropriate to these angles must be derived and used in the least-squares solution, otherwise the result will be inconsistent with the assumptions made concerning the observations.

The resection problem of Example 10.3 is repeated below, using the method of angles rather than directions.

Example 10.4

Equation (10.26) applied to this resection problem becomes

$$v'' = t\mathrm{d}E_P - u\mathrm{d}N_P - \delta\alpha'' \tag{10.27}$$

The same data as Example 10.3 is used, and the coefficients p and q, which are tabulated below, have the same values. The required coefficients t and u for the angle equations are listed as two additional columns:

Line	Computed distance (m)	Computed bearing ° ′ ″	Observed bearing ° ′ ″	Obs. minus comp. bear. (dα″)
PA	788.168	56 23 31	56 23 31	0
PB	1090.022	169 51 42	169 51 42	0
PC	1610.901	278 46 44	278 46 43	− 1
PD	1024.488	342 22 30	342 22 42	12

$\frac{\rho \cos \alpha}{D}$ (p)	$\frac{\rho \sin \alpha}{D}$ (q)	Angle	t	u	$\delta\alpha''$
144. 8	218. 0	APB	331.0	184. 7	0
− 186. 2	33.31	BPC	− 205.7	159. 9	− 1
19.54	− 126 6	CPD	− 172.4	− 65.64	13
191. 9	− 60.96				

The following matrices may be identified:

$$\boldsymbol{A} = \begin{bmatrix} 331.0 & -184.7 \\ -205.7 & -159.9 \\ -172.4 & 65.64 \end{bmatrix}, \quad \boldsymbol{\Delta} = \begin{pmatrix} \mathrm{d}E_P \\ \mathrm{d}N_P \end{pmatrix}, \quad \boldsymbol{f} = \begin{bmatrix} 0 \\ -1 \\ 13 \end{bmatrix}, \quad \boldsymbol{p}_0 = \begin{pmatrix} 7589.759 \\ 3666.702 \end{pmatrix} \mathrm{m}$$

The $\boldsymbol{G}$ matrix in this case is not a unit matrix, as the angles are *derived* from observations, as follows. If d_A, d_B, d_C are the observed directions, then the 'observed' angles $\theta_{APB}, \theta_{BPC}, \theta_{CPD}$ are derived from

$$\left.\begin{aligned} \theta_{APB} &= -d_A + d_B \\ \theta_{BPC} &= \qquad\;\; -d_B + d_C \\ \theta_{CPD} &= \qquad\qquad\quad\;\; -d_C + d_D \end{aligned}\right\} \tag{10.28}$$

Therefore, the coefficient matrix for the observed variables is

$$\boldsymbol{B} = \begin{bmatrix} -1 & 1 & & \\ & -1 & 1 & \\ & & -1 & 1 \end{bmatrix}$$

The cofactor matrix for the observations is the unit matrix $\boldsymbol{I}$. Therefore the cofactor matrix for the derived angles becomes

$$\boldsymbol{G} = \boldsymbol{B}\boldsymbol{I}\boldsymbol{B}^{\mathrm{T}} = \begin{bmatrix} 2 & -1 & 0 \\ -1 & 2 & -1 \\ 0 & -1 & 2 \end{bmatrix}$$

Therefore

$$\boldsymbol{G}^{-1} = \begin{bmatrix} 0.75 & 0.5 & 0.25 \\ 0.50 & 1.0 & 0.50 \\ 0.25 & 0.5 & 0.75 \end{bmatrix}$$

Therefore,

$$N = A^T G^{-1} A = \begin{pmatrix} 85616.5012 & -8482.0775 \\ -8482.0775 & 67361.2221 \end{pmatrix},$$

and

$$F = A^T G^{-1} f = \begin{pmatrix} -1815.6500 \\ -780.2325 \end{pmatrix}$$

Therefore,

$$\Delta = N^{-1} F = \begin{pmatrix} -.023 \\ -.014 \end{pmatrix} \text{ m}$$

and the final values for the parameters are

$$p = \begin{pmatrix} E_{po} \\ N_{po} \end{pmatrix} + \begin{pmatrix} \mathrm{d}E_P \\ \mathrm{d}N_P \end{pmatrix} = \begin{pmatrix} 7589.759 \\ 3666.702 \end{pmatrix} + \begin{pmatrix} -0.023 \\ -0.014 \end{pmatrix}$$

$$= \begin{pmatrix} 7589.736 \\ 3666.688 \end{pmatrix} \text{m}$$

which are identical to those obtained in Example 10.3.

Whereas angular observations might contain systematic orientation errors, distance observations might contain systematic scale errors (or scale differences) between the measuring instrument and the coordinate system. Section 2.7 explains the trilateration computation based on the minimum number of observations, and section 4.4 describes how the presence of scale error may be identified and compensated for semigraphically. When a least-squares procedure is used to solve a trilateration problem the scale factor unknown may be introduced as an additional parameter (λ). Thus the non-linear equation (eqn (2.49)) may be written

$$(\lambda D)^2 = \Delta E^2 + \Delta N^2 \tag{10.29}$$

which, on linearization, becomes

$$2\lambda D(\lambda \cdot \mathrm{d}D + D \cdot \mathrm{d}\lambda) = 2\Delta E \cdot \mathrm{d}(\Delta E) + 2\Delta N \cdot \mathrm{d}(\Delta N)$$

Therefore

$$\lambda^2 D \cdot \mathrm{d}D = \Delta E \cdot \mathrm{d}(\Delta E) + \Delta N \cdot \mathrm{d}(\Delta N) - \lambda D^2 \cdot \mathrm{d}\lambda$$

or

$$\mathrm{d}D = \frac{\Delta E}{\lambda^2 D} \cdot \mathrm{d}(\Delta E) + \frac{\Delta N}{\lambda^2 D} \cdot \mathrm{d}(\Delta N) - \frac{D}{\lambda} \cdot \mathrm{d}\lambda$$

and, for a line AB, the equation may be written

$$\mathrm{d}D_{AB} = \frac{\sin \alpha_{AB}}{\lambda^2}(\mathrm{d}E_B - \mathrm{d}E_A) + \frac{\cos \alpha_{AB}}{\lambda^2}(\mathrm{d}N_B - \mathrm{d}N_A) - \frac{D_{AB}}{\lambda}\mathrm{d}\lambda$$

or

$$dD_{AB} = r_{AB}(dE_B - dE_A) + s_{AB}(dN_B - dN_A) - t_{AB}\,d\lambda \qquad (10.30)$$

where

$$r_{AB} = \frac{\sin\alpha_{AB}}{\lambda^2}, \quad s_{AB} = \frac{\cos\alpha_{AB}}{\lambda^2}, \quad t_{AB} = \frac{D_{AB}}{\lambda}$$

and an approximate value of λ (normally unity) is entered in order to evaluate the coefficients.

When the residual is added the equation becomes

$$v_{AB} = r_{AB}(dE_B - dE_A) + s_{AB}(dN_B - dN_A) - t_{AB}\,d\lambda - dD_{AB} \qquad (10.31)$$

In order to balance the coefficients in the observation equation it is convenient to replace $d\lambda$ by a larger quantity—for example, $10^5 \cdot d\lambda$. Thus the coefficient would be $-10^{-5}t$.

The parameter for scale-change ($d\lambda$) is not normally included as an unknown. However, if it is, then the set of observation equations would include those for observations between known points. In practice scale-change is usually very small, and if a suitable $\boldsymbol{G}$ matrix is used for the measured distances scale error will be accounted for automatically—without it being determined as a separate unknown. However, in such a case the final v's will also include the scale-factor component, and will appear larger than they should be—thus giving a false impression regarding the accuracy (precision) of measurement. In order to illustrate the determination of λ as an additional unknown, the trilateration problem of Example 10.2 is solved again.

Example 10.5

The given data is:

Point	E (m)	N (m)	Line	Measured Distance (m)
A	5997.73	3912.57	PA	1120.76
B	7279.56	4643.10	PB	574.53
C	7781.63	2593.70	PC	1648.73
D	9002.47	4011.21	PD	1901.22

The approximate coordinates of P, calculated using the observations PA and PB, are

E	N
7103.332m	4096.265m

and the assumptions are:

(i) the known coordinates are error-free;
(ii) the observations are uncorrelated; and
(iii) the variance of a measured distance is proportional to the distance squared.

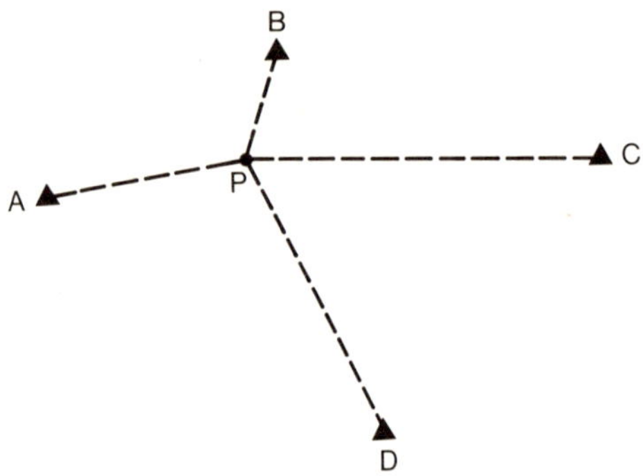

Figure 10.6.

Solution

Equation (10.31) above is used as the linearized observation equation, and the approximate value of λ is taken as unity. The coefficients and constant terms are calculated and tabulated below.

Line	Computed Bearing ° ′ ″	Computed distance (m)	Observed distance (m)	Observed—comp. dist. (dD)	$\frac{\sin\alpha}{\lambda^2}$ (r)	$\frac{\cos\alpha}{\lambda^2}$ (s)	$\frac{D\times 10^{-5}}{\lambda}$ ($10^{-5}t$)
AP	80 33 58	1120.760	1120.76	0	0.9865	0.1639	0.0112
BP	197 51 45	574.530	574.53	0	−0.3067	−0.9518	0.0057
CP	335 42 16	1648.572	1648.73	0.158	−0.4114	0.9114	0.0165
CP	272 33 52	1901.042	1901.22	0.178	−0.9990	0.0447	0.0190

Let $\psi = 10^5\lambda$ and $\mathrm{d}\psi = 10^5.\mathrm{d}\lambda$.

The following matrices may be identified:

$$A = \begin{bmatrix} 0.9865 & 0.1639 & -0.0112 \\ -0.3067 & -0.9518 & -0.0057 \\ -0.4114 & 0.9114 & -0.0165 \\ -0.9990 & 0.0447 & -0.0190 \end{bmatrix}, \quad \Delta = \begin{bmatrix} \mathrm{d}E_P \\ \mathrm{d}N_P \\ \mathrm{d}\psi \end{bmatrix},$$

$$f = \begin{bmatrix} 0 \\ 0 \\ 0.158 \\ 0{\cdot}178 \end{bmatrix} \mathrm{m}, \quad p_0 = \begin{bmatrix} 7103.332\,\mathrm{m} \\ 4096.265\,\mathrm{m} \\ 10^5 \end{bmatrix}$$

Let

$$G^{-1} = \begin{bmatrix} 0.797 & & & \\ & 3.08 & & \\ & & 0.367 & \\ & & & 0.277 \end{bmatrix}, \text{ as before in Example 10.2.}$$

Then

$$N = A^{\mathrm{T}} G^{-1} f = \begin{bmatrix} 1.4039 & 0.8780 & 0.0043 \\ & 3.1171 & 0.0095 \\ \text{Symm.} & & 0.0004 \end{bmatrix},$$

and

$$F = A^{\mathrm{T}} G^{-1} f = \begin{bmatrix} -0.0731 \\ 0.0551 \\ -0.0019 \end{bmatrix}$$

Therefore

$$\boldsymbol{\Delta} = N^{-1} F = \begin{bmatrix} -0.069\,\text{m} \\ 0.053\,\text{m} \\ -5.25 \end{bmatrix}$$

and the final values for the parameters are

$$p = \begin{bmatrix} E_P \\ N_P \\ \psi \end{bmatrix} = p_0 + \boldsymbol{\Delta} = \begin{bmatrix} 7103.263\,\text{m} \\ 4096.318\,\text{m} \\ 99994.75 \end{bmatrix}$$

Therefore

$$\begin{bmatrix} E_P \\ N_P \\ \lambda \end{bmatrix} = \begin{bmatrix} 7103.263\,\text{m} \\ 4096.318\,\text{m} \\ 0.9999475 \end{bmatrix}$$

which agrees well with the results in Examples 4.6 and 10.2.

With the development of electronic distance measurement it has become commonplace to have both angular and distance observations involved in the same problem. Although the observations are of different type they may be combined in the same adjustment provided an appropriate cofactor matrix is used. This amounts essentially to ensuring that an appropriate value for the variance of each observation is used—not only in terms of the precision involved, but also in units which are consistent with those used in the observation equation. The observation equations for bearing and distance are in different units—the one might be seconds of arc and the other metres. Therefore the use of a unit $\boldsymbol{G}$ matrix would invariably lead to an incorrect result. A classic example of this type of oversight is quoted in [9]. The solution to the problem lies in allocating appropriate *a priori* variances, which in this case would be in seconds^2 for bearings, and metres^2 for distances. This effectively means that, when the least-squares procedure is implemented, all observation equations are divided by the *a priori* standard error, and the constant terms become dimensionless.

An example is given below, which involves the two types of observation to fix a point P. To simplify matters it is assumed that orientation errors and scale error are insignificant.

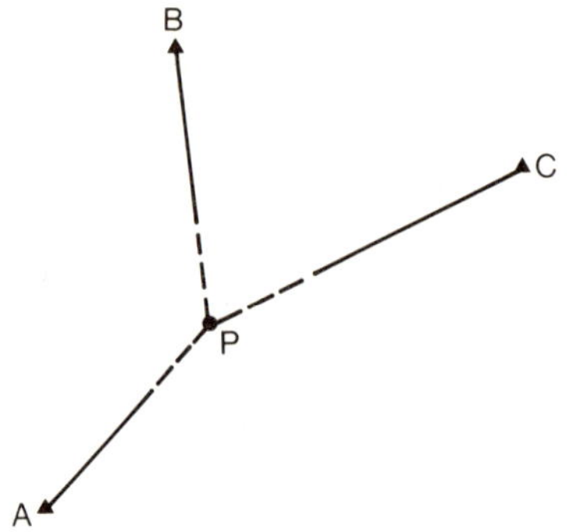

Figure 10.7.

Example 10.6

The given data is:

Point	E	N	Line	Observed bearing	Observed distance
	(m)			° ′ ″	(m)
A	2351.62	3330.16	*AP*	42 57 29	365.932
B	2550.41	4038.72	*BP*	173 27 36	443.667
C	3121.56	3852.91	*CP*	243 54 10	579.661

The approximate coordinates of *P*, calculated from a 'polar' *AP*, are

E	*N*
2600.989 m	3597.968 m

and the assumptions are
(i) the known coordinates are error-free;
(ii) the observations are uncorrelated; and
(iii) the standard deviation of an observed bearing and distance is 10″ and 2 cm respectively.

Solution

The two linearized observation equations for bearing and distance, which are used to solve this problem, are given by eqns (2.60), viz.

$$d\alpha = p\,dE_P - q\,dN_P$$
$$dD = r\,dE_P + s\,dN_P$$

where $d\alpha$ is given in seconds of arc, and

$$p = \frac{\rho\cos\alpha}{D}, \qquad q = \frac{\rho\sin\alpha}{D}, \qquad r = \sin\alpha, \qquad s = \cos\alpha$$

The table of coefficients and constant terms are evaluated and listed below:

Line	Computed bearing ° ′ ″	Computed distance (m)	Observed bearing ° ′ ″	Observed distance (m)	Observed – comp. bear. (dα)″
AP	42 57 29	365.932	42 57 29	365.932	0
BP	173 27 13	443.645	173 27 36	443.667	23
CP	243 54 27	579.646	243 54 10	579.661	– 17

Observed – comp. dist. (d*D*)m	*p*	*q*	*r*	*s*
0	412.5	384.1	0.6815	0.7319
0.022	– 461.9	53.0	0.1140	– 0.9935
0.015	– 156.5	– 319.6	0.8981	– 0.4398

Therefore the following matrices may be identified:

$$A=\begin{bmatrix} 412.5 & -384.1 \\ -461.9 & -53.0 \\ -156.5 & 319.6 \\ 0.6815 & 0.7319 \\ 0.1140 & -0.9935 \\ 0.8981 & -0.4398 \end{bmatrix},\quad \boldsymbol{\Delta}=\begin{pmatrix} \mathrm{d}E_P \\ \mathrm{d}N_P \end{pmatrix},\quad f=\begin{bmatrix} 0 \\ 23 \\ -17 \\ 0 \\ 0.022 \\ 0.015 \end{bmatrix},\quad \boldsymbol{p}_0=\begin{pmatrix} 2600.989 \\ 3597.968 \end{pmatrix}\mathrm{m}$$

$$\boldsymbol{\Sigma}=\begin{bmatrix} 100\,\sec^2 & & & & & \\ & 100\,\sec^2 & & & & \\ & & 100\,\sec^2 & & & \\ & & & 16\times10^{-4}\mathrm{m}^2 & & \\ & & & & 16\times10^{-4}\mathrm{m}^2 & \\ & & & & & 16\times10^{-4}\mathrm{m}^2 \end{bmatrix},$$

and if $\sigma^2=100$, therefore

$$\boldsymbol{G}^{-1}=\begin{bmatrix} 1 & & & & & \\ & 1 & & & & \\ & & 1 & & & \\ & & & 625 & & \\ & & & & 625 & \\ & & & & & 625 \end{bmatrix}$$

Then

$$N = A^{\mathrm{T}}G^{-1}A = \begin{pmatrix} 408803 & -183979 \\ -183979 & 253560 \end{pmatrix}, \quad \text{and} \quad F = A^{\mathrm{T}}G^{-1}f = \begin{pmatrix} -7953.2 \\ -6670.2 \end{pmatrix}$$

Therefore

$$\Delta = N^{-1}F = \begin{pmatrix} -0.046 \\ -0.060 \end{pmatrix}_{\mathrm{m}}$$

and the final values for the parameters are

$$p = \begin{pmatrix} E_p \\ N_p \end{pmatrix} = p_0 + \Delta = \begin{pmatrix} 2600.943 \\ 3597.908 \end{pmatrix}_{\mathrm{m}}$$

Comment

(i) The problem could also be solved with radians as units in the observation equations for the bearings. However, it would then be necessary to have radians^2 as the corresponding units in the G matrix.

(ii) The problem could equally be solved by deriving observation equations for 'observed' angles at the point P, i.e. based on reverse observed bearings. However, in this case the G matrix would include off-diagonal values not equal to 0.

Irrespective of which of the three approaches above is used, the same final result should be obtained.

A problem requiring the solution of a small network, which involves both angular and distance measurement, is given below. The data is the same as that used in the semigraphic solution of Example 4.7.

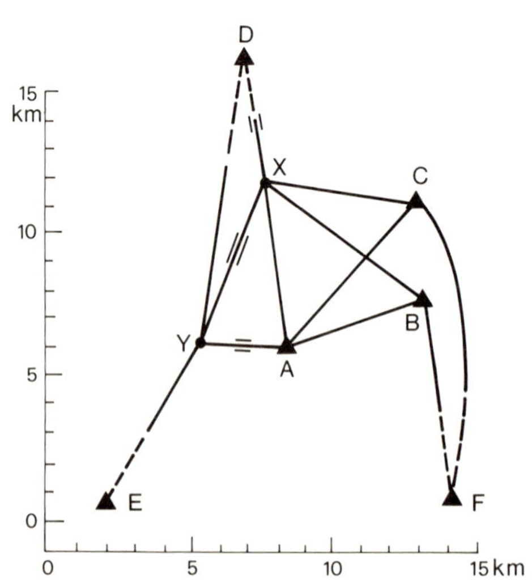

Figure 10.8.

Example 10.7

The given data is:

Known coordinates:

Point	E (m)	N (m)
A	8557.13	5934.72
B	13462.82	7930.45
C	13014.54	11142.59
D	7074.36	16410.66
E	1877.30	516.41
F	14404.00	451.79

Observed distances:

Line	(m)
AY	2901.93
XD	4339.90
XY	6343.80

Observed directions:

At A To	° ′ ″	*At B* To	° ′ ″	*At C* To	° ′ ″
X	353 48 45	A	247 51 36	A	220 33 35
C	40 33 37	X	307 04 46	X	281 04 31
B	67 51 44	F	172 49 35	F	172 35 40
Y	275 17 55				

At X To	° ′ ″	*At Y* To	° ′ ″
C	101 04 24	E	213 41 11
B	127 04 52	D	7 50 41
A	173 48 41	X	20 26 40
Y	200 26 44	A	95 17 50
D	349 15 09		

The assumptions are:

(i) the known coordinates are error-free;

(ii) the observations are uncorrelated; and

(iii) the standard deviation of an observed direction (σ_d) and an observed distance (σ_D) are given by

$$\sigma_d = 10''/(k)^{1/2}, \quad \text{and} \quad \sigma_D = 2(k)^{1/2} \text{ cm respectively;}$$

where k is the length of the line in kilometres.

Solution

The method of bearings will be used, therefore the unknown orientation parameters for each occupied station must be included in the observation equations. However, to simplify matters the scale-error will be regarded as insignificant. Therefore the two types of reduced observation equation which are of interest are derived from eqns (10.21) and (2.53), and are

$$\left.\begin{aligned} v_\alpha &= p_{AB}(\mathrm{d}E_B - \mathrm{d}E_A) - q_{AB}(\mathrm{d}N_B - \mathrm{d}N_A) - \mathrm{d}\theta_A - \mathrm{d}\alpha_{AB} \\ v_D &= r_{AB}(\mathrm{d}E_B - \mathrm{d}E_A) + s_{AB}(\mathrm{d}N_B - \mathrm{d}N_A) \quad - \mathrm{d}D_{AB} \end{aligned}\right\} \tag{10.32}$$

The approximate coordinates of X and Y are calculated using 'polars' AY and YX:

$$E_Y = E_A + D_{AY} \sin \alpha_{AY} = 8557.13 + 2901.93 \sin 275^\circ 17'55'' = 5667.600\,\mathrm{m}$$
$$N_Y = N_A + D_{AY} \cos \alpha_{AY} = 5934.72 + 2901.93 \cos 275^\circ 17'55'' = 6202.703\,\mathrm{m}$$

$$E_x = E_Y + D_{YX} \sin \alpha_{YX} = 5667.600 + 6343.80 \sin 20^\circ 26'40'' = 7883.483\,\mathrm{m}$$
$$N_x = N_Y + D_{YX} \cos \alpha_{YX} = 6202.703 + 6343.80 \cos 20^\circ 26'40'' = 12146.915\,\mathrm{m}$$

Line	Computed bearing ° ′ ″	Computed distance (m)	'Observed' bearing ° ′ ″	Observed distance (m)	Observed-comp. bear. (dα)″
AX	353 48 40	6248.613	353 48 45		5
AC	40 33 37	6854.955	40 33 37		0
AB	67 51 45	5296.105	67 51 44		−1
AY	275 17 55	2901.930	275 17 55		0
BA	247 51 45	5296.105	247 51 36		−9
BX	307 04 46	6993.395	307 04 46		0
BF	172 49 38	7537.650	172 49 35		−3
CA	220 33 37	6854.955	220 33 35		−2
CX	281 04 29	5228.424	281 04 31		2
CF	172 35 42	10780.714	172 35 40		−2
XC	101 04 29	5228.424	101 04 24		−5
XB	127 04 46	6993.395	127 04 52		6
XA	173 48 40	6248.613	173 48 41		1
XY	200 26 40	6343.800	200 26 44		4
XD	349 15 17	4339.839	349 15 09	4339.90	−8
YE	213 41 10	6833.762	213 41 11		1
YD	7 50 47	10304.434	7 50 41		−6
YX	20 26 40	6343.800	20 26 40	6343.80	0
YA	95 17 55	2901.930	95 17 50	2901.93	−5

Observed-comp. dist. (dD)m	p	q	r	s
	32.82	−3.56		
	22.86	19.57		
	14.68	36.08		
	6.56	−70.77		
	−14.68	−36.08		
	17.78	−23.53		
	−27.15	3.42		
	−22.86	−19.57		
	7.58	−38.72		
	−18.87	2.46		
	−7.58	38.72		
	−17.78	23.53		
	−32.82	3.56		
	−30.47	−11.36		
0.061	46.69	−8.86	−0.1864	0.9825
	−25.12	−16.74		
	19.83	2.73		
0	30.47	11.36	0.3493	0.9370
0	−6.56	70.77	0.9957	−0.0923

When these values are associated with the pair of equations (10.32), the 22 observation equations (19 for bearing, and 3 for distance) are seen to have 9 unknown parameters:

$$dE_X, dN_X, dE_Y, dN_Y, d\theta_A, d\theta_B, d\theta_C, d\theta_X, d\theta_Y$$

If the equations for the bearings are followed by those for distances, the following matrices may be identified.

The non-zero coefficients contained in the A matrix are

32.82	3.56			−1			
				−1			
				−1			
		6.56	70.77	−1			
					−1		
17.78	23.53				−1		
					−1		
						−1	
7.58	38.72					−1	
						−1	
7.58	38.72						−1
17.78	23.53						−1
32.82	3.56						−1

$$\begin{bmatrix}
30.47 & -11.36 & -30.47 & 11.36 & & & & -1 & \\
-46.69 & -8.46 & & & & & & -1 & \\
 & & 25.12 & -16.74 & & & & & -1 \\
 & & -19.83 & 2.73 & & & & & -1 \\
30.47 & -11.36 & -30.47 & 11.36 & & & & & -1 \\
 & & 6.56 & 70.77 & & & & & -1 \\
0.1864 & -0.9825 & -0.3493 & -0.9370 & & & & & \\
0.3493 & 0.9370 & -0.9957 & 0.0923 & & & & &
\end{bmatrix}$$

and

$$\boldsymbol{\Delta} = \begin{bmatrix} dE_X \\ dN_X \\ dE_Y \\ dN_Y \\ d\theta_A \\ d\theta_B \\ d\theta_C \\ d\theta_X \\ d\theta_Y \end{bmatrix}, \qquad \boldsymbol{f} = \begin{bmatrix} 5'' \\ 0'' \\ -1'' \\ 0'' \\ -9'' \\ 0'' \\ -3'' \\ -2'' \\ 2'' \\ -2'' \\ -5'' \\ 6'' \\ 1' \\ 4'' \\ -8'' \\ 1'' \\ -6'' \\ 0'' \\ -5'' \\ 0.061\text{ m} \\ 0 \\ 0 \end{bmatrix}, \qquad \boldsymbol{p}_0 = \begin{bmatrix} 7883.483\text{ m} \\ 12146.915\text{ m} \\ 5667.600\text{ m} \\ 6202.703\text{ m} \\ 0 \\ 0 \\ 0 \\ 0 \\ 0 \end{bmatrix}$$

The cofactor matrix for the observations is a diagonal matrix, and the elements along the diagonal are the variances for the observed bearings and distances; the units are $\sec^2$ and m^2 respectively, and the values are

$$16, 15, 19, 34, 19, 14, 13, 15, 19, 9, 19, 14, 16, 16, 23, 15, 10, 16, 34, 17 \times 10^{-4}, 25 \times 10^{-4}, 12 \times 10^{-4}$$

Then

$$\boldsymbol{N} = \boldsymbol{A}^{\mathrm{T}}\boldsymbol{G}^{-1}\boldsymbol{A}; \quad \boldsymbol{F} = \boldsymbol{A}^{\mathrm{T}}\boldsymbol{G}^{-1}\boldsymbol{f}; \quad \text{and} \quad \boldsymbol{\Delta} = \boldsymbol{N}^{-1}\boldsymbol{F} = \begin{bmatrix} 0.156\text{m} \\ -0.029\text{m} \\ 0.017\text{m} \\ 0.002\text{m} \\ 0.3'' \\ 4.3'' \\ 1.1'' \\ 1.1'' \\ 3.7'' \end{bmatrix}$$

Therefore the final values for the parameters are

$$p = p_0 + \Delta = \begin{bmatrix} 7883.639\,\text{m} \\ 12146.886\,\text{m} \\ 5667.617\,\text{m} \\ 6202.705\,\text{m} \\ 0.3'' \\ 4.3'' \\ 1.1'' \\ 1.1'' \\ 3.7'' \end{bmatrix}$$

Comment

These results differ slightly from those obtained semigraphically (Example 4.7). The differences are tabulated below.

Coordinates:

Point	Semigraphic E (m)	Semigraphic N (m)	Least-squares E (m)	Least-squares N (m)	Difference δE (m)	Difference δN (m)
X	7883.74	12146.95	7883.639	12146.886	0.10	0.06
Y	5667.62	6202.73	5667.617	6202.705	0.00	0.02

Orientation corrections:

At	Semigraphic ($''$)	Least squares ($''$)	Difference ($''$)
A	+2	+0.3	2
B	+6	+4.3	2
C	+2	+1.1	1
X	+3	+1.1	2
Y	+5	+3.7	1

The largest difference occurs in the coordinates of X, the first point calculated. An inspection of the error figure reveals that the position chosen is suspect, as too much weight has been given to the ray involving point B. This illustrates the subjective element of semigraphic solutions, and the reason why they are better suited to lower-order work.

The same problem may be solved using angles rather than bearings, as follows.

Example 10.8

The data is the same as that used in Example 10.7.

Solution

The observed directions are used to derive 'observed' angles, which are:

At A Angle	°	′	″	At B Angle	°	′	″	At C Angle	°	′	″
YAX	78	30	50	FBA	75	02	01	FCA	47	57	55
XAC	46	44	52	ABX	59	13	10	ACX	60	30	56
CAB	27	18	07								

At X Angle	°	′	″	At Y Angle	°	′	″
DXC	111	49	15	DYX	12	33	59
CXB	26	00	28	XYA	74	51	10
BXA	46	43	49	AYE	118	23	21
AXY	26	38	03				

The reduced form of the observation equation for an angle is that given by eqn (10.24), which may be written

$$v_{BAC} = t_{BAC}\mathrm{d}E_A - u_{BAC}\mathrm{d}N_A - p_{AB}\mathrm{d}E_B + q_{AB}\mathrm{d}N_B + p_{AC}\mathrm{d}E_C - q_{AC}\mathrm{d}N_C - \delta\alpha_{BAC} \qquad (10.33)$$

where p and q are as defined earlier, and $t_{BAC} = p_{AB} - p_{AC}$; $u_{BAC} = q_{AB} - q_{AC}$; $\delta\alpha_{BAC}$ = observed − computed angle.

The reduced observation equation for a measured distance is the same as that used in the previous example, viz.

$$v_D = r_{AB}(\mathrm{d}E_B - \mathrm{d}E_A) + s_{AB}(\mathrm{d}N_B - \mathrm{d}N_A) - \mathrm{d}D_{AB}$$

If the same approximate coordinates for X and Y are used, the following additional coefficients and constant terms may be derived.

Angle	Computed angle °	′	″	Observed angle °	′	″	Obs. minus comp. angle $(\delta\alpha)''$	t	u
YAX	78	30	45	78	30	50	5		
XAC	46	44	57	46	44	52	−5		
CAB	27	18	08	27	18	07	−1		
FBA	75	02	07	75	02	01	−6		

ABX	59	13	01	59	13	10	9		
FCA	47	57	55	47	57	55	0		
ACX	60	30	52	60	30	56	4		
DXC	111	49	12	111	49	15	3	54.27	−47.58
CXB	26	00	17	26	00	28	11	10.20	15.19
BXA	46	43	54	46	43	49	−5	15.04	19.97
AXY	26	38	00	26	38	03	3	−2.35	14.92
DYX	12	35	53	12	35	59	6	−10.64	−8.63
XYA	74	51	15	74	51	10	−5	37.03	−59.41
AYE	118	23	15	118	23	21	6	18.56	87.51

The values listed above are used together with those calculated in the previous example, and if the angle equations are followed by the distance equations the following matrices may be identified:

$$\mathbf{A} = \begin{bmatrix} 32.82 & 3.56 & -6.56 & -70.77 \\ -32.82 & -3.56 & 0 & 0 \\ 0 & 0 & 0 & 0 \\ 0 & 0 & 0 & 0 \\ 17.78 & 23.53 & 0 & 0 \\ 0 & 0 & 0 & 0 \\ 7.58 & 38.72 & 0 & 0 \\ 54.27 & 47.58 & 0 & 0 \\ 10.20 & -15.19 & 0 & 0 \\ 15.04 & -19.97 & 0 & 0 \\ -2.35 & -14.92 & -30.47 & 11.36 \\ 30.47 & -11.36 & -10.64 & 8.63 \\ -30.47 & 11.36 & 37.03 & 59.41 \\ 0 & 0 & 18.56 & -87.51 \\ 0.1864 & -0.9825 & 0 & 0 \\ 0.3493 & 0.9370 & -0.3493 & -0.9370 \\ 0 & 0 & -0.9957 & 0.0923 \end{bmatrix},$$

$$\mathbf{\Delta} = \begin{bmatrix} dE_X \\ dN_X \\ dE_Y \\ dN_Y \end{bmatrix}, \quad f = \begin{bmatrix} 5'' \\ -5'' \\ -1'' \\ -6'' \\ 9'' \\ 0'' \\ 4'' \\ 3'' \\ 11'' \\ -5'' \\ 3'' \\ 6'' \\ -5'' \\ 6'' \\ 0.061\,\text{m} \\ 0\,\text{m} \\ 0\,\text{m} \end{bmatrix}, \quad \mathbf{p}_0 = \begin{bmatrix} 7883.483 \\ 12146.915 \\ 5667.600 \\ 6202.703 \end{bmatrix} \text{m}$$

The cofactor matrix is *not* a diagonal matrix. It is formed by applying the laws of propagation to the equations which are used to derive the angles, and when combined with the variances for the measured distances the non-zero elements are as given below.

$$
\mathbf{G}=\begin{bmatrix}
50 & -16 \\
-16 & 31 & -15 \\
 & -15 & 34 \\
 & & & 32 & -19 \\
 & & & -19 & 33 \\
 & & & & & 24 & -15 \\
 & & & & & -15 & 34 \\
 & & & & & & & 42 & -19 \\
 & & & & & & & -19 & 33 & -14 \\
 & & & & & & & & -14 & 30 & -16 \\
 & & & & & & & & & -16 & 32 \\
 & & & & & & & & & & & 26 & -16 \\
 & & & & & & & & & & & -16 & 50 & -34 \\
 & & & & & & & & & & & & -34 & 49 \\
 & & & & & & & & & & & & & & 0.0017 \\
 & & & & & & & & & & & & & & & 0.0025 \\
 & & & & & & & & & & & & & & & & 0.0012
\end{bmatrix}
$$

Then

$$
\boldsymbol{N}=\boldsymbol{A}^{\mathrm{T}}\boldsymbol{G}^{-1}\boldsymbol{A};\ \boldsymbol{F}=\boldsymbol{A}^{-1}\boldsymbol{G}^{\mathrm{T}}\boldsymbol{f};\ \text{and}\ \Delta=\boldsymbol{N}^{-1}\boldsymbol{F}=\begin{bmatrix} 0.156 \\ -0.029 \\ 0.017 \\ 0.002 \end{bmatrix}\text{m}
$$

and the final values for the parameters are

$$
\boldsymbol{p}=\boldsymbol{p}_0+\Delta=\begin{bmatrix} 7883.639 \\ 12146.886 \\ 5667.617 \\ 6202.705 \end{bmatrix}\text{m}
$$

Comment

(i) The results are identical to those obtained in Example 10.7.

(ii) Notice that although the observation equations for the observed angles CAB, FBA, and FCA do not contain any terms involving the unknown parameters they are nevertheless included in the solution—the reason being that they are correlated with adjacent angles which do include such parameters.

(iii) In practice, in order to ease the computational burden, a diagonal matrix is often used for the 'observed' angles, and this becomes particularly attractive if the computer storage space is limited. The result, clearly, will be different; however, the difference is usually considered small enough to warrant such a simplification—particularly if the original assumptions are regarded as suspect anyway.

10.3 Applications in photogrammetry

Equation (5.66) represents a general linearized coplanarity equation which may be used for analytical relative orientation. The determinants are constants which are re-evaluated after each computational cycle. In section 7.2 an approximate method was described, which involved the use of six points assumed to be located in the ideal positions. Thus the computer storage requirements were eased considerably, at the expense, however, of increasing the number of required iterations. If the computer resources are more generous, eqn (5.66) may be used as it stands in a least-squares determination of the five parameters. Thus no constraints are imposed on either the number or distribution of the points, which effectively means that incomplete models or mountainous terrain may be handled just as efficiently as more conventional models. The only factors influencing the number of iterations would be the magnitudes of the unknown parameters, and the proximity of the initial estimates to these values. The example below has an uneven distribution of points used for relative orientation, due to the fact that the area covered by the stereopair is of a coastal region. Nevertheless only three iterations were required to achieve a satisfactory solution.

Example 10.9

Rectangular coordinates of points appearing on a pair of overlapping aerial photographs have been observed. The coordinates, which have been reduced to the respective fiducial centres are given in millimetres below. The calibrated focal length (f) of the camera is 152.07 mm.

Point	First photo		Second photo	
	x	y	x	y
404	−0.080	19.116	−93.658	14.574
405	0.511	17.036	−92.987	12.454
804	86.907	19.548	−6.703	13.856
805	88.045	17.716	−5.436	11.973
407	31.046	−34.974	−62.229	−41.107
408	33.030	−37.770	−60.209	−44.011
807	87.934	−34.015	−2.554	−41.132
808	96.142	−34.150	6.200	−41.436
401	1.806	99.196	−91.387	95.285
402	2.811	93.289	−90.699	89.358
801	83.883	83.917	−6.237	79.571
802	85.816	80.344	−4.236	75.940

Assuming that observations have equal weight, use eqn (5.66) to carry out a least-squares determination of the five relative orientation parameters ($dw, d\phi, dk$,

ΔY, ΔZ). Hence use eqns (5.51)–(5.55) to derive independent three-dimensional model coordinates for the 12 points.

Solution

The approximate base length on the photographs is 90 mm, and this is adopted for ΔX, the main component controlling the scale of the model. Thus the final residuals will be given at the scale of the photography. The values for the five parameters after each iteration are given below, and these are followed by the rotation matrix R formed from the final values of dw, dϕ, dk. Rotation parameters are given in radians, and translation parameters in millimetres.

Iteration no.	dw (rad.)	dϕ (rad.)	dk (rad.)	ΔY (mm)	ΔZ (mm)
1	−0.010291	−0.007723	0.012863	4.351	−3.458
2	−0.011675	−0.017049	0.014208	4.121	−3.938
3	−0.011671	−0.017058	0.014219	4.126	−3.937
4	−0.011671	−0.017058	0.014219	4.126	−3.937

$$R = \begin{bmatrix} 0.999756 & -0.014028 & -0.017070 \\ 0.014228 & 0.999831 & 0.011660 \\ 0.016904 & -0.011900 & 0.999786 \end{bmatrix}$$

The rotation matrix is used, together with the values for the elements $\Delta X, \Delta Y, \Delta Z$, to transform (via eqn 5.25) the coordinates of the second photograph. These are then used in eqns (5.51)–(5.55), with the corresponding values for the first photograph, to derive three-dimensional independent model coordinates (mm) in a right-handed coordinate system, with origin in the first projection centre and z-axis as indicated in Figure 5.5. The values, together with the residuals (with respect to the mean, and given in micrometres) for the y-coordinates,—i.e. the residual y-parallaxes—are listed below.

Point	x (mm)	y (mm)	z (mm)	v_y (μm)
404	−0.076	18.144	−144.308	4
405	0.486	16.188	−144.488	2
804	81.344	18.293	−142.336	−3
805	82.516	16.603	−142.520	0
407	29.791	−33.564	−145.923	−3
408	31.713	−36.266	−146.009	−1
807	85.675	−33.135	−148.164	6
808	93.966	−33.380	−148.628	−2
401	1.690	92.850	−142.345	−2
402	2.626	87.155	−142.071	0
801	80.629	80.660	−146.171	−1
802	82.559	77.294	−144.298	0

In Chapter 7 some approximate methods for model connection and absolute orientation were described. The solutions involved determining the parameters sequentially, from a limited number of points distributed according to a set pattern. A more useful algorithm is one which is based on any number of points distributed randomly.

Equation (5.76) represents a three-dimensional similarity transformation, viz.

$$\begin{bmatrix} X \\ Y \\ Z \end{bmatrix} = \lambda \boldsymbol{R} \begin{bmatrix} x \\ y \\ z \end{bmatrix} + \begin{bmatrix} X_T \\ Y_T \\ Z_T \end{bmatrix} \tag{10.34}$$

where $x, y\ z$ are the given coordinates; X, Y, Z are the transformed coordinates; λ is the scale-factor; X_T, Y_T, Z_T are the unknown translations—the XYZ-coordinates of the origin of the xyz system; and $\boldsymbol{R}$ is the rotation matrix, formed from the three unknown rotations w, ϕ, k.

The linearization of eqn (10.34) is described in section 5.11, and the linearized form, when residuals are included, is given by eqn (5.74). The linearized form of eqn (10.34) may also be written as

$$\begin{bmatrix} X \\ Y \\ Z \end{bmatrix} = \begin{bmatrix} \mathrm{d}\lambda & -\mathrm{d}k & \mathrm{d}\phi \\ \mathrm{d}k & \mathrm{d}\lambda & -\mathrm{d}w \\ -\mathrm{d}\phi & \mathrm{d}w & \mathrm{d}\lambda \end{bmatrix} \begin{bmatrix} x \\ y \\ z \end{bmatrix} + \begin{bmatrix} X_T \\ Y_T \\ Z_T \end{bmatrix} \tag{10.35}$$

where the v's are excluded for the time being.

Equation (10.35) may be re-arranged as follows:

$$\begin{bmatrix} X \\ Y \\ Z \end{bmatrix} = \begin{bmatrix} x & 0 & z & -y & 1 & 0 & 0 \\ y & -z & 0 & x & 0 & 1 & 0 \\ z & y & -x & 0 & 0 & 0 & 1 \end{bmatrix} \begin{bmatrix} \mathrm{d}\lambda \\ \mathrm{d}w \\ \mathrm{d}\phi \\ \mathrm{d}k \\ X_T \\ Y_T \\ Z_T \end{bmatrix} \tag{10.36}$$

The three translation parameters (X_T, Y_T, Z_T) may be eliminated temporarily by reducing the coordinates in both systems to a common centroid. Therefore eqn (10.36) becomes

$$\begin{bmatrix} X \\ Y \\ Z \end{bmatrix} = \begin{bmatrix} x & 0 & z & -y \\ y & -z & 0 & x \\ z & y & -x & 0 \end{bmatrix} \begin{bmatrix} \mathrm{d}\lambda \\ \mathrm{d}w \\ \mathrm{d}\phi \\ \mathrm{d}k \end{bmatrix} \tag{10.37}$$

and if several common points are given, a least-squares determination of the four parameters may be carried out. Expressed as observation equations, with n common points, and with the residuals included, eqn (10.37) becomes

$$\begin{bmatrix} v_{x_1} \\ v_{y_1} \\ v_{z_1} \\ \vdots \\ v_{x_n} \\ v_{y_n} \\ v_{z_n} \end{bmatrix} = \begin{bmatrix} x_1 & 0 & z_1 & -y_1 \\ y_1 & -z_1 & 0 & x_1 \\ z_1 & y_1 & -x_1 & 0 \\ \vdots & \vdots & \vdots & \vdots \\ x_n & 0 & z_n & -y_n \\ y_n & -z_n & 0 & x_n \\ z_n & y_n & -x_n & 0 \end{bmatrix} \begin{bmatrix} d\lambda \\ dw \\ d\phi \\ dk \end{bmatrix} - \begin{bmatrix} X_1 \\ Y_1 \\ Z_1 \\ \vdots \\ X_n \\ Y_n \\ Z_n \end{bmatrix} \tag{10.38}$$

i.e.

$$\boldsymbol{v} = \boldsymbol{A} \cdot \boldsymbol{\Delta} - \boldsymbol{f} \tag{10.39}$$

Normal equations formed from these observation equations are

$$\boldsymbol{N\Delta} = \boldsymbol{F}, \qquad \text{where} \quad \boldsymbol{N} = \boldsymbol{A}^{\mathrm{T}}\boldsymbol{A} \quad \text{and} \quad \boldsymbol{F} = \boldsymbol{A}^{\mathrm{T}}\boldsymbol{f}$$

Thus

$$\begin{bmatrix} \Sigma(x^2+y^2+z^2) & 0 & 0 & 0 \\ & \Sigma(y^2+z^2) & -\Sigma xy & -\Sigma xz \\ & & \Sigma(x^2+z^2) & -\Sigma yz \\ \text{symmetric} & & & \Sigma(x^2+y^2) \end{bmatrix} \begin{bmatrix} d\lambda \\ dw \\ d\phi \\ dk \end{bmatrix} = \begin{bmatrix} \Sigma(xX+yY+zZ) \\ (\Sigma yZ - \Sigma zY) \\ (\Sigma zX - \Sigma xZ) \\ (\Sigma xY - \Sigma yX) \end{bmatrix} \tag{10.40}$$

In eqn (10.40) it is clear that $d\lambda$ may be calculated independently, i.e.

$$d\lambda = \frac{\Sigma(xX+yY+zZ)}{\Sigma(x^2+y^2+z^2)} \tag{10.41}$$

and it may be applied to the reduced coordinates.

Therefore the normal equations which must be solved simultaneously are reduced to

$$\begin{bmatrix} \Sigma(y^2+z^2) & -\Sigma xy & -\Sigma xz \\ & \Sigma(x^2+z^2) & -\Sigma yz \\ \text{symm.} & & \Sigma(x^2+y^2) \end{bmatrix} \begin{bmatrix} dw \\ d\phi \\ dk \end{bmatrix} = \begin{bmatrix} (\Sigma yZ - \Sigma zY) \\ (\Sigma zX - \Sigma xZ) \\ (\Sigma xY - \Sigma yX) \end{bmatrix} \tag{10.42}$$

The three equations are solved to provide values for $dw, d\phi, dk$, and so the orthogonal rotation matrix $\boldsymbol{R}$ may be formed. It is applied to the reduced coordinates and the cycle is repeated to determine new increments for $dw, d\phi, dk$. These are added to the previous values to form a new matrix $\boldsymbol{R}$, which again is applied to the reduced coordinates. The iterations continue until the increments are negligible. The three translation parameters X_T, Y_T, Z_T, are the coordinates of the centroid. Alternatively, if the coordinates in the original system (before reduction to the centroid) are to be transformed, the translation parameters may be found from eqn (10.34) as follows:

$$\begin{bmatrix} X_T \\ Y_T \\ Z_T \end{bmatrix} = \begin{bmatrix} X_C \\ Y_C \\ Z_C \end{bmatrix} - \lambda \boldsymbol{R} \begin{bmatrix} x_C \\ y_C \\ z_C \end{bmatrix} \tag{10.43}$$

where subscript C indicates coordinates of the centroid; X_T, Y_T, Z_T are the translation parameters; λ is the final value for the scale factor—the product of values obtained in each iteration; and $\boldsymbol{R}$ is the final rotation matrix—formed from the final values for the rotation parameters w, ϕ, k. Thus all seven parameters are determined, and eqn (10.34) may be used to transform any point given in the original xyz-system.

The method has special advantages because only three equations must be solved simultaneously, and this is a simple matter for any computer. A prerequisite, however, is that the magnitudes of the rotations should be small in order that the solution may converge satisfactorily. Hence the method has particular merit in a model connection, when consecutive models in a strip of aerial photography tend to have a similar scale and orientation. In the case of absolute orientation both scale-change (λ) and azimuth-change (K) are likely to be large. Therefore estimates of these parameters must be made in the first instance, and a transformation (eqn (10.34)) must be carried out using these values before the least-squares solution is implemented. Normally this is not a serious problem, though it could be an inconvenience.

A method, which is often considered more appropriate to the problem of absolute orientation, determines the parameters in two stages as follows.

The *first stage* is based on the two-dimensional similarity equations of section 2.11; thus

$$\left.\begin{aligned} X &= xr - ys + X_T \\ Y &= yr + xs + Y_T \end{aligned}\right\} \tag{10.44}$$

where

$$\left.\begin{aligned} r &= \lambda \cos K \\ s &= \lambda \sin K \end{aligned}\right\} \tag{10.45}$$

and where λ and K may be found from

$$\left.\begin{aligned} \lambda &= (r^2 + s^2)^{1/2} \\ K &= \tan^{-1}(s/r) \end{aligned}\right\} \tag{10.46}$$

Equations (10.44) are linear functions of the unknowns (r, s, X_T, Y_T), and therefore a least-squares solution may be implemented directly, without the need to linearize, estimate parameters, or assign approximate values. Two control points with known XY-coordinates are the minimum required for a solution, but more than two are recommended for greater reliability. Thus, with n control points, the observation equations would be

$$\begin{bmatrix} v_{x_1} \\ v_{y_1} \\ \vdots \\ v_{x_n} \\ v_{y_n} \end{bmatrix} = \begin{bmatrix} x_1 & -y_1 & 1 & 0 \\ y_1 & x_1 & 0 & 1 \\ \vdots & \vdots & \vdots & \vdots \\ x_n & -y_n & 1 & 0 \\ y_n & x_n & 0 & 1 \end{bmatrix} \begin{bmatrix} r \\ s \\ X_T \\ Y_T \end{bmatrix} = \begin{bmatrix} X_1 \\ Y_1 \\ \vdots \\ X_n \\ Y_n \end{bmatrix} \tag{10.47}$$

With r, s, X_T, Y_T solved, values for λ and K may be derived from eqns (10.46), and λ is applied to all coordinates x, y, z. The remaining three parameters (Ω, Φ, Z_T) are obtained in the *second stage*, which is based on the equation

$$Z - z = y\Omega - x\Phi + Z_T \tag{10.48}$$

where x, y, z are the original values multiplied by the scale-factor λ; and Ω, Φ are rotation parameters (assumed to be small).

Equation (10.48) is linear in the unknowns, and three control points with known Z values, and forming a plane, are sufficient to determine the unknowns. With more points, the observation equations are

$$\begin{bmatrix} v_{z1} \\ v_{z2} \\ \vdots \\ v_{z_n} \end{bmatrix} = \begin{bmatrix} y_1 & -x_1 & 1 \\ y_2 & -x_2 & 1 \\ & \vdots & \\ y_n & -x_n & 1 \end{bmatrix} \begin{bmatrix} \Omega \\ \Phi \\ Z_T \end{bmatrix} - \begin{bmatrix} Z_1 - z_1 \\ Z_2 - z_2 \\ \vdots \\ Z_n - z_n \end{bmatrix} \tag{10.49}$$

On completion of the second stage values for all seven parameters have been determined. Therefore an orthogonal matrix $\boldsymbol{R}$ may be formed, and the transformation represented by eqn (10.34) applied to the *original* coordinates x, y, z. These transformed coordinates are the starting point for the next iteration cycle. Successive iterations provide corrections to the parameters, and the iterations are terminated when the corrections are negligible. At the end of each iteration cycle an improved rotation matrix, operating on the original coordinates, should be formed. The other parameters (λ, X_T, Y_T, Z_T) should also be improved absolute parameters.

Note that, if necessary, the translation parameters (X_T, Y_T, Z_T) may be derived independently—by reducing coordinates to a common centroid. Thus the number of unknowns (and hence normal equations), to be solved simultaneously, would be reduced to two in both the cases of eqns (10.47) and (10.48).

Examples illustrating these methods of model connection and absolute orientation are given below.

Example 10.10

The three-dimensional coordinates (in mm) of two adjacent but independent models are given below. Point 2 is the common projection centre, and points C and D are the common terrain points. Join the models together analytically so that all points are given in the coordinate system expressed by XYZ.

		Model 1			Model 2	
Point	X	Y	Z	x	y	z
1	0.00	0.00	170.00			
A	0.25	−97.37	25.69			
B	−0.81	94.86	27.78			

2	90.00	2.90	169.95	0.00	0.00	170.00
C	87.69	−81.38	25.54	−1.32	−88.91	9.82
D	83.66	75.10	29.62	−5.41	82.44	17.88
3				90.00	2.91	172.06
E				84.03	−77.60	10.91
F				80.12	99.68	19.57

Solution

The first stage is to reduce the coordinates of the three common points to their respective centroids. The coordinates of the centroids are found to be

	Model 1			Model 2	
X	*Y*	*Z*	*x*	*y*	*z*
87.117	−1.127	75.037	−2.243	−2.157	65.900

and when the model coordinates are referred to these centroids as the new origins the values are

		Model 1			Model 2	
Point	*X*	*Y*	*Z*	*x*	*y*	*z*
2	2.883	4.027	94.913	2.243	2.157	104.100
C	0.573	−80.253	−49.497	0.923	−86.753	−56.080
D	−3.457	76.227	−45.417	−3.167	84.597	−48.020

The scale factor $d\lambda$ is calculated from eqn (10.41) and found to be 0.912347. When it is applied to the coordinates of the points in the second model the results are

		Model 1			Model 2	
Point	*X*	*Y*	*Z*	*x*	*y*	*z*
2	2.883	4.027	94.913	2.047	1.968	94.975
C	0.573	−80.253	−49.497	0.842	−79.149	−51.164
D	−3.457	76.227	−45.417	−2.889	77.182	−43.811

The coordinates above are used to form and solve the normal equations expressed by eqn (10.42), and the values (in radians) for dw, $d\phi$, dk are found to be

dw	d	dk
−0.021403	0.008863	0.002332

These values are used to form a rotation matrix $\boldsymbol{R}$, given by

$$\boldsymbol{R} = \begin{bmatrix} 0.999958 & -0.002426 & 0.008837 \\ 0.002237 & 0.999768 & 0.021410 \\ -0.008887 & -0.021390 & 0.999732 \end{bmatrix}$$

The translation parameters X_T, Y_T, Z_T are found from eqn (10.43), and when these are used in eqn (10.34) to transform all points in model 2 the results, with associated discrepancies (dx, dy, dz) for the common points, are as given below.

	Model 1			Model 2					
Point	X	Y	Z	x	y	z	dx	dy	dz
1	0.00	0.00	170.00						
A	0.25	−97.37	25.69						
B	−0.81	94.86	27.78						
2	90.00	2.90	169.95	89.998	2.878	169.927	−0.002	−0.022	−0.023
C	87.69	−81.38	25.54	87.699	−81.351	25.572	0.009	0.029	0.032
D	83.66	75.10	29.62	83.653	75.092	29.613	−0.007	−0.008	−0.007
3				172.116	5.757	171.019			
E				165.548	−70.839	25.653			
F				161.659	91.025	30.124			

Comment

Only one iteration cycle was used, and yet the discrepancies are no larger than 0.032 mm. The reason for such a rapid convergence is the fact that the rotations are very small, the largest being 0.021403 radians (1.2°).

Example 10.11

The results of Example 10.9 are used as data for an example on absolute orientation. The coordinates of four points are used, together with the corresponding terrain values, to derive the parameters of absolute orientation. The independent model coordinates (mm) and the ground coordinates (m) are listed below.

	Model coordinates (mm)			Ground coordinates (m)		
Point	x	y	z	E	N	h
A	29.791	−33.564	−145.923	503.30	6077.64	0.67
B	1.690	92.850	−142.345	3942.55	6638.43	52.64
C	80.629	80.660	−146.171	3491.81	4534.98	42.44
D	85.675	−33.135	−148.164	427.51	4574.83	4.77

Eqns (10.47) are used to form observation equations for the solution of the four parameters (r, s, X_T, Y_T). Thus if eqns (10.47) are expressed in the matrix form given by eqn (10.5), viz.

$$\boldsymbol{v} = \boldsymbol{Ap} - \boldsymbol{f}$$

the following matrices may be identified.

$$\boldsymbol{A} = \begin{bmatrix} 29.791 & 33.564 & 1 & 0 \\ -33.564 & 29.791 & 0 & 1 \\ 1.690 & -92.850 & 1 & 0 \\ 92.850 & 1.690 & 0 & 1 \\ 80.629 & -80.660 & 1 & 0 \\ 80.660 & 80.629 & 0 & 1 \\ 85.675 & 33.135 & 1 & 0 \\ -33.135 & 85.675 & 0 & 1 \end{bmatrix}, \quad \boldsymbol{p} = \begin{bmatrix} r \\ s \\ X_T \\ Y_T \end{bmatrix}, \quad \boldsymbol{f} = \begin{bmatrix} 503.30 \\ 6077.64 \\ 3942.55 \\ 6638.43 \\ 3491.81 \\ 4534.98 \\ 427.51 \\ 4574.83 \end{bmatrix}$$

The least squares solution of the parameters is obtained from

$$p = N^{-1}F, \qquad \text{where} \quad N = A^{\mathrm{T}}A \quad \text{and} \quad F = A^{\mathrm{T}}f$$

Thus

$$p = \begin{bmatrix} r \\ s \\ X_T \\ Y_T \end{bmatrix} = \begin{bmatrix} -1.54332203 \\ -26.8677599 \\ 1450.16091 \\ 6826.19091 \end{bmatrix}$$

and, from eqn (10.46), λ and K are found as follows:

$\lambda = (r^2 + s^2)^{1/2}$
$= 26.912049$

$K = \tan^{-1}(s/r) = 1.513418$ rad or $(\pi + 1.513418)$ rad and, since r and s are both negative, it is the latter, which may be written as -1.628175 rad.

λ is applied to the x, y, z coordinates of the four control points and the new values are

Point	x	y	z
A	801.737	−903.276	−3927.087
B	45.481	2498.784	−3830.796
C	2169.892	2170.726	−3933.761
D	2305.690	−891.731	−3987.397

Eqn (10.49) is used with these coordinates to form observation equations for the solution of the remaining three parameters (Ω, Φ, Z_T).

The following matrices may be identified for this solution.

$$A = \begin{bmatrix} -903.276 & -801.737 & 1 \\ 2498.784 & -45.481 & 1 \\ 2170.726 & -2169.892 & 1 \\ -891.731 & -2305.690 & 1 \end{bmatrix}, \quad p = \begin{bmatrix} \Omega \\ \Phi \\ Z_{\mathrm{T}} \end{bmatrix}, \quad f = \begin{bmatrix} 3927.757 \\ 3883.436 \\ 3976.201 \\ 3992.167 \end{bmatrix}$$

The values for the parameters are

$$p = \begin{bmatrix} \Omega \\ \Phi \\ Z_T \end{bmatrix} = \begin{bmatrix} -0.003389 \text{ rad} \\ -0.043045 \text{ rad} \\ 3890.05 \text{ m} \end{bmatrix}$$

The rotation matrix R is formed using eqn (5.20); however, eqn (5.20) is derived from a rotation of axes, therefore the signs of the parameters are reversed in order to achieve the equivalent transformation of coordinates.

Thus

$$\Omega = 0.003389 \qquad \Phi = 0.043045 \qquad K = 1.628175 \qquad \text{and} \qquad R = \begin{bmatrix} -0.057294 & 0.998340 & 0.005852 \\ -0.997430 & -0.057492 & 0.042766 \\ 0.043031 & -0.003386 & 0.999068 \end{bmatrix}$$

Eqn (10.34) is used to transform the model coordinates and the following coordinates are obtained:

Point	x	y	z
A	497.470	5910.501	4.178
B	3919.775	6473.338	56.317
C	3469.944	4368.846	45.974
D	404.476	4407.170	8.603

These values form the starting point for the next iteration, and the iterations continue until the required accuracy is achieved. Successive values for the increments to the seven parameters are listed below, together with the final values. The discrepancies between the transformed coordinates and the given ground coordinates are also listed for each iteration. The results may be compared with those obtained from the same data in Example 7.3.

	Iteration number 1	2	3	Parameter	Final value
Parameter Increment					
$d\lambda$	26.912049	0.999622	0.999999	λ	26.901856
dK	1.628175	0.000520	0.000001	K	1.628696
$d\Phi$	0.043045	-0.000023	-0.000006	Φ	0.043016
$d\Omega$	0.003389	-0.000020	0.000001	Ω	0.003370
dX_T	1450.16	20.90	2.92	X_T	1473.98
dY_T	6826.19	169.58	-3.41	Y_T	6992.36
dZ_T	3890.05	-3.61	-1.46	Z_T	3884.98

Coordinate Discrepancies (m)

		1	2	3
A	dX	-23.83	-2.96	$-.04$
	dY	-167.14	3.37	$-.01$
	dZ	3.51	1.33	$-.13$
B	dX	-22.78	-2.91	0.01
	dY	-165.09	3.42	0.03
	dZ	3.68	1.57	0.10
C	dX	-21.87	-2.92	-0.01
	dY	-166.14	3.41	0.02
	dZ	3.54	1.37	-0.10
D	dX	-23.03	-2.91	0.00
	dY	-167.66	3.46	0.08
	dZ	3.84	1.62	0.15

Comment

(i) Although the coordinate discrepancies after the first two iterations are larger than those obtained in Example 7.3, the final distribution after the third iteration is seen to be better.

(ii) As mentioned earlier, the number of unknown parameters in each group could be reduced to two by adopting the common centroid as a temporary origin. However, this is not possible if the algorithm is used for block aerial triangulation, where the requirement is a simultaneous determination of parameters of several adjoining models.

10.4 General remarks on least-squares estimation

Least-squares estimation—or 'adjustment', as it has been termed in the past—provides the most probable values for the observed quantities. With the introduction of matrix notation and the widespread use of computers, the most difficult stage in the process is the establishment of the mathematical equations (the functional model) which relate the observations to the required quantities (parameters). Because the relationships are generally non-linear, a process of linearization is usually required to produce the linearized functional model. With a few exceptions, it is invariably easier to establish observation equations than condition equations, simply because the equations are usually easier to form, and the number must equal the number of observations. Although the basic functional model might be easy to form, the refinements to it that are sometimes necessary are more difficult to identify. The least-squares residuals provide an indicator as to whether or not the model is suspect. The precision of the measurements may be estimated after a least squares computation by calculating a value for σ_0^2—the unit variance—from

$$\sigma_0^2 = \frac{\boldsymbol{v}^{\mathrm{T}}\boldsymbol{G}^{-1}\boldsymbol{v}}{r} \tag{10.50}$$

and if this is significantly different from unity, there would be cause to suspect the presence of a gross error, or systematic error (i.e. incorrect functional model), or incorrect stochastic model (*a priori* covariance or cofactor matrix).

The other stage in the process which may or may not be demanding is to describe the quality of the observations adequately—in other words, to establish an appropriate covariance matrix for the observations. An *a priori* assessment must be made to facilitate a least-squares procedure, and often there is insufficient data in an observation set to provide reliable information. Therefore past experience, or information supplied by the manufacturer of the equipment, must be resorted to. Even if the number of repeated observations is sufficient to establish certain characteristics internally, it is seldom the case that these reflect the true characteristics. Reference [12] discusses some of the problems involved and the possible solutions.

Once the functional and stochastic models have been formed the remaining processes are nothing more than a series of matrix operations, which are standard operations on most computers. Therefore, provided the matrices have been formed and entered into the computer correctly, the same unique result should be achieved, and this is a distinct advantage over approximate (e.g. semigraphic) methods. The disadvantages are associated with the fact that the sheer volume of numbers makes it difficult on occasions to detect gross errors in the observations, or mistakes in forming the equations. A classic example is block aerial triangulation in photogrammetry, where a gross error is dispersed so efficiently that it often goes undetected. Thus checking the reliability of data has become a problem, particularly in these days of automatic data acquisition, processing and plotting, when many more observations are recorded than in the past.

Least-squares estimates of quantities are unbiased and have minimum variance, and this is so irrespective of whether the observations are normally distributed or not. However, it should be made clear that least squares does not provide a 'true' solution, and in fact some other method (e.g. semigraphic) might, quite fortuitously, produce a result which is closer to the truth in an individual computational problem. On the other hand, the property that least-squares estimates have minimum variance is very useful from the point of view of the design of an observational programme. Provided the functional and stochastic models are known, the algorithm is such that the stochastic properties of the result may be ascertained without observations being required at all. Thus, by computer simulation, the most efficient geometry and observing procedure may be established to solve a particular problem before the observing commences. For more details on this topic the reader is referred to [13] for applications to geodetic networks, and [2] for photogrammetric applications.

It is often convenient to carry out least-squares estimation in phases. Examples arise if the computing facilities are limited, and the number of parameters in a problem is such that they cannot all be solved simultaneously. Other examples are when not all of the observed data is available at a particular time, and it is necessary to compute all the parameters, which will be upgraded by further computation at some future date when the additional data is available. The same results may be achieved in a phased computation provided the stochastic properties of the results of the initial phase are adopted as *a priori* characteristics in the latter phase. The method using observation equations is well suited to this type of problem, and very simple illustrations of this principle have been demonstrated in Examples 8.6 and 10.8. More complicated applications arise in the updating of control coordinates in a national survey organization as new observations become available, or in the design by computer simulation of observing programmes, when certain observations may be added (or subtracted) in order to optimize the programme. Further details on this topic are given in [13], [14] and [27].

Appendix A: Linearization using Taylor's Theorem

The Taylor series expansion for a non-linear function, $y = f(x)$, is

$$y = f(x_0) + f'(x_0)\Delta x + \frac{1}{2!}f''(x_0)\Delta x^2 + \cdots \text{etc.} \tag{A1}$$

where

x_0 is an approximate value for the variable x;
$f(x_0)$ is the zero-order term, evaluated by letting $x = x_0$;
$f'(x_0), f''(x_0), \ldots$ are first- and higher-order derivatives, evaluated by letting $x = x_0$; therefore they become constant coefficients in the first- and higher-order terms; and

$$\Delta x = x - x_0.$$

The linearized form of the function, at the point $x = x_0$, is obtained by adopting only the zero-order and first-order terms in eqn. A1; therefore it becomes

$$y = f(x_0) + f'(x_0)\Delta x \tag{A2}$$

For example, let the function be

$$y = 1 + 2x + x^2$$

and let $x_0 = -2$; therefore

$$y = (1 + 2x_0 + x_0^2) + (2 + 2x_0)\Delta x$$

and, substituting for x_0, the linearized form is

$$y = 1 - 2\Delta x.$$

If the function contains more than one independent variable—such as x_1 and x_2—the linearized form of $y = f(x_1, x_2)$ is

$$y = f(x_{1_0}, x_{2_0}) + \left(\frac{\partial f}{\partial x_1}\right)_0 \cdot \Delta x_1 + \left(\frac{\partial f}{\partial x_2}\right)_0 \cdot \Delta x_2 \tag{A3}$$

where

x_{1_0}, x_{2_0} are approximate values for the variables x_1 and x_2 respectively;

$\left(\frac{\partial f}{\partial x_1}\right)_0, \left(\frac{\partial f}{\partial x_2}\right)_0$ are partial derivatives of $f(x_1, x_2)$ with respect to x_1 and x_2, evaluated by letting $x_1 = x_{1_0}$, and $x_2 = x_{2_0}$; and $\Delta x_1 = x_1 - x_{1_0}$; $\Delta x_2 = x_2 - x_{2_0}$.

Eqn (A3) may be written

$$y = y_0 + j_1 \Delta x_1 + j_2 \Delta x_2 \tag{A4}$$

or, in matrix notation,

$$y = y_0 + \boldsymbol{J} \cdot \boldsymbol{\Delta} \tag{A5}$$

where

$\boldsymbol{J}$ is a row vector containing the partial derivatives; and
$\boldsymbol{\Delta}$ is a column vector containing the variables Δx_1 and Δx_2.

For example, if

$$y = (x_1^2 + x_2^2)^{1/2}$$

and

$$x_{1_0} = -1, \quad x_{2_0} = 2,$$

the linearized form becomes

$$y = (x_{1_0}^2 + x_{2_0}^2)^{1/2} + \left(\frac{x_{1_0}}{(x_{1_0}^2 + x_{2_0}^2)^{1/2}}\right) \cdot \Delta x_1 + \left(\frac{x_{2_0}}{x_{1_0}^2 + x_{2_0}^2)^{1/2}}\right) \cdot \Delta x_2$$

which may be written

$$y = y_0 + \frac{x_{1_0}}{y_0} \cdot \Delta x_1 + \frac{x_{2_0}}{y_0} \cdot \Delta x_2$$

and when values for x_{1_0} and x_{2_0} are substituted,

$$y = (5)^{1/2} - \frac{1}{(5)^{1/2}} \cdot \Delta x_1 + \frac{2}{(5)^{1/2}} \cdot \Delta x_2$$

Thus

$$\boldsymbol{J} = \left(\left(\frac{\partial f}{\partial x_1}\right)_0 \left(\frac{\partial f}{\partial x_2}\right)_0\right) = \left(\left(\frac{-1}{(5)^{1/2}}\right)\left(\frac{2}{(5)^{1/2}}\right)\right)$$

and

$$\boldsymbol{\Delta} = \begin{pmatrix} \Delta x_1 \\ \Delta x_2 \end{pmatrix}$$

Extending the concept further, if there are n independent variables and m

functions containing some or all of these variables, eqn (A5) may be written

$$\boldsymbol{y} = \boldsymbol{y}_0 + \boldsymbol{J}\cdot\boldsymbol{\Delta} \tag{A6}$$

where

$\boldsymbol{y}$ is the vector $(m \times 1)$ which represents the m functions;
$\boldsymbol{y}_0$ is the vector $(m \times 1)$ which represents the functions with approximate values assigned to the variables;
$\boldsymbol{J}$ is the matrix $(m \times n)$ of partial derivatives, known as the *Jacobian matrix*; the elements are evaluated by substituting approximate values for the variables; and
$\boldsymbol{\Delta}$ is the vector $(n \times 1)$ of variables.

Thus

$$\boldsymbol{y} = \begin{pmatrix} y_1 \\ y_2 \\ \vdots \\ y_m \end{pmatrix}, \quad \boldsymbol{y}_0 = \begin{pmatrix} y_{1_0} \\ y_{2_0} \\ \vdots \\ y_{m_0} \end{pmatrix} = \begin{pmatrix} f_1(x_{1_0}, & x_{2_0}, & \cdots x_{n_0}) \\ f_2(x_{1_0}, & x_{2_0}, & \cdots x_{n_0}) \\ \vdots & & \\ f_m(x_{1_0}, & x_{2_0}, & \cdots x_{n_0}) \end{pmatrix},$$

$$\boldsymbol{J} = \begin{pmatrix} \frac{\partial y_1}{\partial x_1} & \frac{\partial y_1}{\partial x_2} & \cdots & \frac{\partial y_1}{\partial x_n} \\ \frac{\partial y_2}{\partial x_1} & \frac{\partial y_2}{\partial x_2} & \cdots & \frac{\partial y_2}{\partial x_n} \\ \vdots & & & \\ \frac{\partial y_m}{\partial x_1} & \frac{\partial y_m}{\partial x_2} & \cdots & \frac{\partial y_m}{\partial x_n} \end{pmatrix}, \quad \text{and} \quad \boldsymbol{\Delta} = \begin{pmatrix} \Delta x_1 \\ \Delta x_2 \\ \vdots \\ \vdots \\ \Delta x_n \end{pmatrix} \tag{A7}$$

For example, let the functions be

$$E' = E + D\sin\alpha$$
$$N' = N + D\cos\alpha$$

where E, N, D, α are the independent variables, with approximate values

$$E_0 = 12.00; \quad N_0 = 15.00; \quad D_0 = 10.00; \quad \alpha_0 = \pi/6\,\text{rad}$$

Therefore approximate values for E' and N' are:

$$E'_0 = 17.00; \quad N'_0 = 23.66$$

The linearized equation is

$$\boldsymbol{y} = \boldsymbol{y}_0 + \boldsymbol{J}\cdot\boldsymbol{\Delta}$$

where

$$\boldsymbol{y} = \begin{pmatrix} E' \\ N' \end{pmatrix}, \quad \boldsymbol{y}_0 = \begin{pmatrix} E'_0 \\ N'_0 \end{pmatrix} = \begin{pmatrix} 17.00 \\ 23.66 \end{pmatrix}$$

$$\boldsymbol{J}=\begin{pmatrix} \dfrac{\partial E'}{\partial E} & \dfrac{\partial E'}{\partial N} & \dfrac{\partial E'}{\partial D} & \dfrac{\partial E'}{\partial \alpha} \\[2ex] \dfrac{\partial N'}{\partial E} & \dfrac{\partial N'}{\partial N} & \dfrac{\partial N'}{\partial D} & \dfrac{\partial N'}{\partial \alpha} \end{pmatrix} = \begin{pmatrix} 1 & 0 & \sin\alpha & D\cos\alpha \\[2ex] 0 & 1 & \cos\alpha & -D\sin\alpha \end{pmatrix}$$

$$\boldsymbol{\Delta}=\begin{pmatrix} \Delta E \\ \Delta N \\ \Delta D \\ \Delta\alpha \end{pmatrix}$$

and, when approximate values are introduced,

$$J=\begin{pmatrix} 1 & 0 & 0.500 & 8.66 \\ 0 & 1 & 0.866 & -5.00 \end{pmatrix}$$

Appendix B: Linearized collinearity equations

The collinearity equations are developed in section 5.7 and are expressed in a compact matrix form for a point A by eqn (5.36). For an arbitrary point the equations are

$$\left.\begin{aligned} x &= -f\cdot\frac{\boldsymbol{M}_1\boldsymbol{X}}{\boldsymbol{M}_3\boldsymbol{X}} \\ y &= -f\cdot\frac{\boldsymbol{M}_2\boldsymbol{X}}{\boldsymbol{M}_3\boldsymbol{X}} \end{aligned}\right\} \tag{B1}$$

where

- x, y are the image coordinates with respect to principal point; f is the nominal focal length;
- $\boldsymbol{M}_1, \boldsymbol{M}_2, \boldsymbol{M}_3$ are row vectors containing the appropriate elements of the rotation matrix $\boldsymbol{M}$—i.e. $\boldsymbol{M}_1 = (m_{i1}\, m_{i2}\, m_{i3})$, where the m_{iJ}'s are functions of the three rotation parameters (w, ϕ, k);
- $\boldsymbol{X}$ is the column vector of coordinate differences between the object coordinates of the point and the projection centre C, and expressed in the object system—i.e. $\boldsymbol{X} = ((X - X_C)(Y - Y_C) \times (Z - Z_C))^T$.

Let F and G be the functions in eqn (B_1).

Then
$$\frac{\partial F}{\partial w} = \frac{-f}{(\boldsymbol{M}_3\boldsymbol{X})^2}\left(\boldsymbol{M}_3\boldsymbol{X}\cdot\frac{\partial(\boldsymbol{M}_1\boldsymbol{X})}{\partial w} - \boldsymbol{M}_1\boldsymbol{X}\cdot\frac{\partial(\boldsymbol{M}_3\boldsymbol{X})}{\partial w}\right) \tag{B2}$$

or, written as a determinant,

$$\frac{\partial F}{\partial w} = \frac{-f}{(\boldsymbol{M}_3\boldsymbol{X})^2}\begin{vmatrix} \boldsymbol{M}_3\boldsymbol{X} & \boldsymbol{M}_1\boldsymbol{X} \\ \dfrac{\partial(\boldsymbol{M}_3\boldsymbol{X})}{\partial w} & \dfrac{\partial(\boldsymbol{M}_1\boldsymbol{X})}{\partial w} \end{vmatrix} \tag{B3}$$

Also, let the partial derivatives of $\boldsymbol{M}$ with respect to w, ϕ, k be written as

$$\frac{\partial \boldsymbol{M}}{\partial w} = \begin{pmatrix} m_{w1} \\ m_{w2} \\ m_{w3} \end{pmatrix}, \quad \frac{\partial \boldsymbol{M}}{\partial \phi} = \begin{pmatrix} m_{\phi 1} \\ m_{\phi 3} \\ m_{\phi 2} \end{pmatrix}, \quad \frac{\partial \boldsymbol{M}}{\partial k} = \begin{pmatrix} m_{k1} \\ m_{k2} \\ m_{k2} \end{pmatrix} \tag{B4}$$

Therefore eqn (B3) may be written

$$\frac{\partial F}{\partial w}=\frac{-f}{(\boldsymbol{M}_3\boldsymbol{X})^2}\begin{vmatrix}\boldsymbol{M}_3\boldsymbol{X} & \boldsymbol{M}_1\boldsymbol{X}\\ \boldsymbol{M}_{w3}\boldsymbol{X} & \boldsymbol{M}_{w1}\boldsymbol{X}\end{vmatrix} \tag{B5}$$

ie.

$$\frac{\partial F}{\partial w}=\frac{-f}{(\boldsymbol{M}_3\boldsymbol{X})^2}(\boldsymbol{M}_3\boldsymbol{X}\cdot\boldsymbol{M}_{w1}\boldsymbol{X}-\boldsymbol{M}_1\boldsymbol{X}\cdot\boldsymbol{M}_{w3}\boldsymbol{X})=\frac{-f}{\boldsymbol{M}_3\boldsymbol{X}}\left(\boldsymbol{M}_{w1}\boldsymbol{X}+\frac{x}{f}\cdot\boldsymbol{M}_{w3}\boldsymbol{X}\right) \tag{B6}$$

Similarly

$$\frac{\partial F}{\partial \phi}=\frac{-f}{(\boldsymbol{M}_3\boldsymbol{X})^2}\begin{vmatrix}\boldsymbol{M}_3\boldsymbol{X} & \boldsymbol{M}_1\boldsymbol{X}\\ \boldsymbol{M}_{\phi3}\boldsymbol{X} & \boldsymbol{M}_{\phi1}\boldsymbol{X}\end{vmatrix}=\frac{-f}{\boldsymbol{M}_3\boldsymbol{X}}\left(\boldsymbol{M}_{\phi1}\boldsymbol{X}+\frac{x}{f}\cdot\boldsymbol{M}_{\phi3}\boldsymbol{X}\right) \tag{B7}$$

$$\frac{\partial F}{\partial k}=\frac{-f}{\boldsymbol{M}_3\boldsymbol{X}}\left(\boldsymbol{M}_{k1}\boldsymbol{X}+\frac{x}{f}\cdot\boldsymbol{M}_{k3}\boldsymbol{X}\right) \tag{B8}$$

Also

$$\frac{\partial F}{\partial X}=\frac{-f}{(\boldsymbol{M}_3\boldsymbol{X})^2}\begin{vmatrix}\boldsymbol{M}_3\boldsymbol{X} & \boldsymbol{M}_1\boldsymbol{X}\\ \dfrac{\partial(\boldsymbol{M}_3\boldsymbol{X})}{\partial X} & \dfrac{\partial(\boldsymbol{M}_1\boldsymbol{X})}{\partial X}\end{vmatrix}=\frac{-f}{\boldsymbol{M}_3\boldsymbol{X}}\left(m_{11}+\frac{x}{f}\cdot m_{31}\right) \tag{B9}$$

$$\frac{\partial F}{\partial Y}=\frac{-f}{\boldsymbol{M}_3\boldsymbol{X}}\left(m_{12}+\frac{x}{f}\cdot m_{32}\right) \tag{B10}$$

$$\frac{\partial F}{\partial Z}=\frac{-f}{\boldsymbol{M}_3\boldsymbol{X}}\left(m_{13}+\frac{x}{f}\cdot m_{33}\right) \tag{B11}$$

and

$$\frac{\partial F}{\partial X_C}=\frac{f}{\boldsymbol{M}_3\boldsymbol{X}}\left(m_{11}+\frac{x}{f}\cdot m_{31}\right) \tag{B12}$$

$$\frac{\partial F}{\partial Y_C}=\frac{f}{\boldsymbol{M}_3\boldsymbol{X}}\left(m_{12}+\frac{x}{f}\cdot m_{32}\right) \tag{B13}$$

$$\frac{\partial F}{\partial Z_C}=\frac{f}{\boldsymbol{M}_3\boldsymbol{X}}\left(m_{13}+\frac{x}{f}\cdot m_{33}\right) \tag{B14}$$

If $\boldsymbol{M}$ is given by eqn (5.20), and is expressed as $\boldsymbol{M}=\begin{pmatrix}m_{11} & m_{12} & m_{13}\\ m_{21} & m_{22} & m_{23}\\ m_{31} & m_{32} & m_{33}\end{pmatrix}$,

then

$$\frac{\partial M}{\partial w}=\begin{pmatrix}\boldsymbol{M}_{w1}\\ \boldsymbol{M}_{w2}\\ \boldsymbol{M}_{w3}\end{pmatrix}$$

$$= \begin{pmatrix} 0 & (\cos w \sin\phi \cos k - \sin w \sin k) & (\sin w \sin\phi\cos k + \cos w \sin k) \\ 0 & (-\cos w \sin\phi \sin k - \sin w \cos k) & (-\sin w \sin\phi \sin k + \cos w \cos k) \\ 0 & -\cos w \cos\phi & -\sin w \cos\phi \end{pmatrix}$$

$$= \begin{pmatrix} 0 & -m_{13} & m_{12} \\ 0 & -m_{23} & m_{22} \\ 0 & -m_{33} & m_{32} \end{pmatrix} \tag{B15}$$

$$\frac{\partial M}{\partial \phi} = \begin{pmatrix} \boldsymbol{M}_{\phi 1} \\ \boldsymbol{M}_{\phi 2} \\ \boldsymbol{M}_{\phi 3} \end{pmatrix} = \begin{pmatrix} -\sin\phi\cos k & \sin w \cos\phi \cos k & -\cos w \cos\phi \cos k \\ \sin\phi \sin k & -\sin w \cos\phi \sin k & \cos w \cos\phi \sin k \\ \cos\phi & \sin w \sin\phi & -\cos w \sin\phi \end{pmatrix} \tag{B16}$$

$$\frac{\partial M}{\partial k} = \begin{pmatrix} \boldsymbol{M}_{k1} \\ \boldsymbol{M}_{k2} \\ \boldsymbol{M}_{k3} \end{pmatrix}$$

$$= \begin{pmatrix} -\cos\phi \sin k & (-\sin w \sin\phi \sin k + \cos w \cos k) & (\cos w \sin\phi \sin k + \sin w \cos k) \\ -\cos\phi\cos k & (-\sin w \sin\phi\cos k - \cos w \sin k) & (\cos w \sin\phi \cos k - \sin w \sin k) \\ 0 & 0 & 0 \end{pmatrix}$$

$$= \begin{pmatrix} m_{21} & m_{22} & m_{23} \\ -m_{11} & -m_{12} & -m_{13} \\ 0 & 0 & 0 \end{pmatrix} \tag{B17}$$

Therefore, substituting eqns (B15), (B16), (B17) in (B6), (B7), (B8), the partial derivatives are given finally as

$$\left.\begin{aligned}
\frac{\partial F}{\partial w} &= \frac{-f}{\boldsymbol{M}_3\boldsymbol{X}}\Big(-m_{13}(Y-Y_C) + m_{12}(Z-Z_C) \\
&\quad + \frac{x}{f}\cdot(-m_{33}(Y-Y_C) + m_{32}(Z-Z_C))\Big) \\
\frac{\partial F}{\partial \phi} &= \frac{-f}{\boldsymbol{M}_3\boldsymbol{X}}\Big(-\sin\phi\cos k(X-X_C) + \sin w \cos\phi \cos k(Y-Y_C) \\
&\quad - \cos w \cos\phi\cos k(Z-Z_c) + \frac{x}{f}\cdot(\cos\phi\,(X-X_C) \\
&\quad + \sin w \sin\phi(Y-Y_C) - \cos w \sin\phi(Z-Z_C))\Big) \\
\frac{\partial F}{\partial k} &= \frac{-f}{\boldsymbol{M}_3\boldsymbol{X}}(m_{21}(X-X_C) + m_{22}(Y-Y_C) + m_{23}(Z-Z_C)) \\
\frac{\partial F}{\partial X} &= \frac{-f}{\boldsymbol{M}_3\boldsymbol{X}}\left(m_{11} + \frac{x}{f}\cdot m_{31}\right) \\
\frac{\partial F}{\partial Y} &= \frac{-f}{\boldsymbol{M}_3\boldsymbol{X}}\left(m_{12} + \frac{x}{f}\cdot m_{32}\right)
\end{aligned}\right\} \tag{B18}$$

$$\frac{\partial F}{\partial Z} = \frac{-f}{\boldsymbol{M}_3\boldsymbol{X}}\left(m_{13} + \frac{x}{f}\cdot m_{33}\right)$$

$$\frac{\partial F}{\partial X_C} = \frac{f}{\boldsymbol{M}_3\boldsymbol{X}}\left(m_{11} + \frac{x}{f}\cdot m_{31}\right)$$

$$\frac{\partial F}{\partial Y_C} = \frac{f}{\boldsymbol{M}_3\boldsymbol{X}}\left(m_{12} + \frac{x}{f}\cdot m_{32}\right)$$

$$\frac{\partial F}{\partial Z_C} = \frac{f}{\boldsymbol{M}_3\boldsymbol{X}}\left(m_{13} + \frac{x}{f}\cdot m_{33}\right)$$

A similar set of equations may be derived for the function *G*, and numerical values for all partial derivatives may be obtained by substituting approximate values for the unknowns.

Appendix C: Useful mathematical formulae

Signs of trigonometric functions in individual quadrants

Function	Quadrant 1	2	3	4
$\sin A$	+	+	−	−
$\cos A$	+	−	−	+
$\tan A$	+	−	+	−
$\cot A$	+	−	+	−

Values of the trigonometric functions

Function	Angle in degrees 0°	90°	180°	270°
sin	0	1	0	−1
cos	1	0	−1	0
tan	0	−	0	−
cot	−	0	−	0

Reduction of trigonometric functions to the first quadrant

Function	$B = 90° \pm A$	$B = 180° \pm A$	$B = 270° \pm A$	$B = 360° \pm A$
$\sin B$	$\pm \cos A$	$\mp \sin A$	$-\cos A$	$\pm \sin A$
$\cos B$	$\mp \sin A$	$-\cos A$	$\pm \sin A$	$+\cos A$
$\tan B$	$\mp \cot A$	$\pm \tan A$	$\mp \cot A$	$\pm \tan A$
$\cot B$	$\mp \tan A$	$\pm \cot A$	$\mp \tan A$	$\pm \cot A$

Trigonometric formulae

$\tan A = \sin A/\cos A$ $\quad \cot A = 1/\tan A$ $\quad \operatorname{cosec} A = 1/\sin A$ $\quad \sec A = 1/\cos A$

$\sin^2 A + \cos^2 A = 1$ $\quad \sec^2 A = 1 + \tan^2 A$ $\quad \operatorname{cosec}^2 A = 1 + \cot^2 A$

$\sin^2 A = (1 - \cos 2A)/2$ $\quad \cos^2 A = (1 + \cos 2A)/2$ $\quad \tan^2 A = (1 - \cos 2A)/(1 + \cos 2A)$

$$\sin A = \frac{2 \tan A/2}{1 + \tan^2 A/2} \qquad \cos A = \frac{1 - \tan^2 A/2}{1 + \tan^2 A/2} \qquad \tan A = \frac{2 \tan A/2}{1 - \tan^2 A/2}$$

$$\sin 2A = 2 \sin A \cos A \qquad \cos 2A = \cos^2 A - \sin^2 A = 2\cos^2 A - 1 = 1 - 2\sin^2 A$$

$$\tan 2A = 2 \tan A/(1 - \tan^2 A)$$

$$\sin(A + B) = \sin A \cos B + \cos A \sin B \qquad \cos(A + B) = \cos A \cos B - \sin A \sin B$$

$$\sin(A + B) = \sin A \cos B - \cos A \sin B \qquad \cos(A - B) = \cos A \cos B + \sin A \sin B$$

$$\tan(A + B) = \frac{\tan A + \tan B}{1 - \tan A \tan B} \qquad \cot(A + B) = \frac{\cot A \cot B - 1}{\cot B + \cot A}$$

$$\tan(A - B) = \frac{\tan A - \tan B}{1 + \tan A \tan B} \qquad \cot(A - B) = \frac{\cot A \cot B + 1}{\cot B - \cot A}$$

$$\sin A + \sin B = 2\sin\frac{A+B}{2}\cos\frac{A-B}{2} \qquad \cos A + \cos B = 2\cos\frac{A+B}{2}\cos\frac{A-B}{2}$$

$$\sin A - \sin B = 2\cos\frac{A+B}{2}\sin\frac{A-B}{2} \qquad \cos A - \cos B = -2\sin\frac{A+B}{2}\sin\frac{A-B}{2}$$

$$\tan A + \tan B = \frac{\sin(A+B)}{\cos A\cos B} \qquad \cot A + \cot B = \frac{\sin(B+A)}{\sin A\sin B}$$

$$\tan A - \tan B = \frac{\sin(A-B)}{\cos A\cos B} \qquad \cot A - \cot B = \frac{\sin(B-A)}{\sin A\sin B}$$

$$\tan A + \cot B = \frac{\cos(A-B)}{\cos A\sin B} \qquad \tan A - \cot B = \frac{-\cos(A+B)}{\cos A\sin B}$$

$$2\sin A\cos B = \sin(A+B) + \sin(A-B) \qquad 2\sin A\sin B = \cos(A-B) - \cos(A+B)$$
$$2\cos A\sin B = \sin(A+B) - \sin(A-B) \qquad 2\cos A\cos B = \cos(A+B) + \cos(A-B)$$

$$\tan A\tan B = \frac{\tan A + \tan B}{\cot A + \cot B} \qquad \cot A\cot B = \frac{\cot A + \cot B}{\tan A + \tan B} \qquad \tan A\cot B = \frac{\tan A + \cot B}{\tan B + \cot A}$$

Plane triangle formulae

Consider a plane triangle with sides a, b, c opposite the angles A, B, C. Thus $A + B + C = 180°$. Let Δ be the area of the triangle; also

$$s = \frac{a+b+c}{2}, \quad \text{and} \quad r^2 = \frac{(s-a)(s-b)(s-c)}{s}$$

Then

$$a/\sin A = b/\sin B = c/\sin C = abc/2\Delta \qquad a^2 = b^2 + c^2 - 2bc\cos A$$

$$\frac{a+b}{a-b} = \frac{\tan\frac{A+B}{2}}{\tan\frac{A-B}{2}} \qquad a = b\cos C + c\cos B \qquad \tan\frac{A}{2} = \frac{(b-c)}{(b+c)}\cot\frac{B-C}{2} = \frac{r}{s-a}$$

$$\sin^2\frac{A}{2} = \frac{(s-b)(s-c)}{bc} \qquad \cos^2\frac{A}{2} = \frac{s(s-a)}{bc} \qquad \tan^2\frac{A}{2} = \frac{(s-b)(s-c)}{s(s-a)}$$

$$\Delta = \tfrac{1}{2}bc\sin A = rs = (s(s-a)(s-b)(s-c))^{1/2}$$

Spherical triangle formulae

Consider a spherical triangle with sides a, b, c opposite the angles A, B, C. Thus $A + B + C = 180° + E$, where E is the spherical excess. Let R be the radius of the sphere, and let Δ be the area of the triangle.

Then $E° = \dfrac{\Delta.180°}{\pi R^2}$

If R is unity, the following relationships are valid:

$\sin a/\sin A = \sin b/\sin B = \sin c/\sin C$

$\cos a = \sin b \sin c \cos A + \cos b \cos c$ $\cos A = \sin B \sin C \cos a - \cos B \cos C$

$\tan a = \sin b/(\cot A \sin C + \cos b \cos C)$ $\tan A = \sin B/(\cot a \sin c - \cos c \cos B)$

$\cos a \sin b = \sin a \cos b \cos C + \sin c \cos A$ $\cos A \sin B = \sin C \cos a - \sin A \cos B \cos c$

$\cot a \sin b = \sin C \cot A + \cos C \cos b$ $\cot A \sin B = \sin c \cot a - \cos c \cos B$

Napier's rules for a right-angled triangle:
Let $C = 90°$.

$\sin a = \sin c \sin A = \tan b \cot B$ $\cos A = \cos a \sin B = \tan b \cot c$

$\sin b = \sin c \sin B = \tan a \cot A$ $\cos B = \cos b \sin A = \tan a \cot c$

$\cos c = \cos a \cos b = \tan A \cot B$

Differential calculus

In the following formulae *u*, *v*, and *w* are variables which are functions of *x*, while *a*, *c*, and *r* are constants. Also, ln and log signify logarithms to the base *e* and 10 respectively.

$$\frac{d}{dx}(u + v - w) = \frac{du}{dx} + \frac{dv}{dx} - \frac{dw}{dx} \qquad \frac{d}{dx}(cu) = c.\frac{du}{dx} \qquad \frac{d}{dx}(u/c) = \frac{1}{c}.\frac{du}{dx}$$

$$\frac{d}{dx}(uv) = u.\frac{dv}{dx} + v.\frac{du}{dx} \qquad \frac{d}{dx}(u/v) = v.\frac{\frac{du}{dx} - u.\frac{dv}{dx}}{v^2} \qquad \frac{d}{dx}(u^r) = ru^{r-1}.\frac{du}{dx}$$

$$\frac{d}{dx}(\log u) = \log u.\frac{1}{u}.\frac{du}{dx} \qquad \frac{d}{dx}(\ln u) = \frac{1}{u}.\frac{du}{dx}$$

$$\frac{d}{dx}(a^u) = a^u.\ln a.\frac{du}{dx} \qquad \frac{d}{dx}(e^u) = e^u.\frac{du}{dx}$$

Standard derivatives:

$f(x)$	$f'(x)$	$f(x)$	$f'(x)$
x^r	$r.x^{r-1}$	x^x	$x^x.(1 + \ln x)$
e^x	e^x	$\ln x$	$1/x$
a^x	$a^x.\ln a$	$\log x$	$\log e.1/x$
$\sin x$	$\cos x$	$\text{arc}\sin x$	$1/(1 - x^2)^{1/2}$
$\cos x$	$-\sin x$	$\text{arc}\cos x$	$-1/(1 - x^2)^{1/2}$
$\tan x$	$\sec^2 x$	$\text{arc}\tan x$	$1/(1 + x^2)$
$\cot x$	$\text{cosec}^2 x$	$\text{arc}\cot x$	$-1/(1 + x^2)$
$\sec x$	$\sec x \tan x$	$\text{arc}\sec x$	$\frac{1}{x(x^2 - 1)^{1/2}}$
$\text{cosec}\, x$	$-\text{cosec}\, x \cot x$	$\text{arc cosec}\, x$	$\frac{-1}{x(x^2 - 1)^{1/2}}$

References

1. *Admiralty Manual of Hydrographic Surveying* (Vol. 1, 1965). HMSO, London.
2. Amer, F.A.A.F. (1981) Theoretical reliability of elementary photogrammetric procedures. *ITC Journal* **3**, **4**.
3. American Society of Photogrammetry (1980) *Manual of Photogrammetry* (4th edn). American Society of Photogrammetry, Falls Church, Virginia.
4. Ashkenazi, V., Cross, P.A., Davis, M.J.K. and Procter, D.W. (1972) The readjustment of the retriangulation of Great Britain. *Ordinance Survey Professional Papers*, New Series No. **24**.
5. Atkinson, A.P. (1958–59) Triangulation by the Direction Method. *Empire Survey Review* **14**(110), 364–377; **15**(111), 6–15.
6. Atkinson, K.B. (ed.) (1980) *Developments in Close-Range Photogrammetry*. Applied Science Publishers, London.
7. Barrell, H. and Sears, J.E. (1939) The refraction and dispersion of air for the visible spectrum. *Phil. Trans. Roy. Soc. Lond.* **A238**, 1–64.
8. Bomford, A.G. (1962) Transverse Mercator arc-to-chord and finite distance scale factor formulae. *Empire Survey Review* **16** (125), 318–327.
9. Bomford, A.G. (1967) The geodetic adjustment of Australia. *Survey Review* **19**(144), 52–71.
10. Burnside, C.D. (1982) *Electromagnetic Distance Measurement*. Granada, London.
11. Cooper, M.A.R. (1974) *Fundamentals of Survey Measurement and Analysis*. Granada, London.
12. Cooper, M.A.R. (1981) A priori and a posteriori analysis, or game and guess. *Survey Review* **26**(201), 137–140.
13. Cross, P.A. (1981) Computer aided design of geodetic networks. *Proc. VI Int. Symp. on Geodetic Computations* (III), 13–21.
14. Cross, P.A. (1983) *Advanced Least Squares Applied to Position-fixing*, Working Paper No. 6, Department of Land Surveying, North-East London Polytechnic.
15. Davis, R.E., Foote, F.S., Anderson, J.M. and Mikhail, E.M. (1981) *Surveying: Theory and Practice*. McGraw-Hill, New York.
16. Dixon, J.L. (1959) Triangulation by the Direction Method. *Empire Survey Review* **15**(114), 155–161.
17. Essen, L. and Froome, K.D. (1951) The refractive indices and dielectric constants of air and its principal constituents at 24000 Mc/s. *Proc. Phys. Soc. Lond.* **B64**, 862–875.
18. Froome, K.D. and Essen, L. (1969) *The Velocity of Light and Radio Waves*. Academic Press, London.
19. Granshaw, S.I. (1980) Bundle adjustment methods in engineering photogrammetry. *Photogrammetric Record* **10**(56), 181–207.
20. Halliday, G.H. (1962) Triangulation by the Direction (Graphical) Method. *Empire Survey Review* **16** (125), 310–318.
21. Hawkins, D.M. (1980) *Identification of Outliers*. Chapman and Hall, London.
22. Hodges, D.J. and Greenwood, J.B. (1971) *Optical Distance Measurement*. Butterworths, London.
23. International Union of Geodesy and Geophysics (1975) Resolutions, XVIth General Assembly, 1975, Grenoble, France. *Bulletin Géodesique*, NS118, p. 365.
24. Karara, H.M. (1979) *Handbook of Non-Topographic Photogrammetry*. American Society of Photogrammetry, Falls Church, Virginia.
25. Laurila, S.H. (1976) *Electronic Surveying and Navigation*. John Wiley, New York.
26. List, R.J. (1958) *Smithsonian Meteorological Tables*, 6th edn. (Revised), Publication 4014, Smithsonian Miscellaneous Collection, Washington.
27. Mikhail, E.M. (1976) *Observations and Least Squares*. Dun-Donnelley, New York.
28. Mikhail, E.M. and Gracie, G. (1981) *Analysis and Adjustment of Survey Measurements*. Van Nostrand Reinhold, New York.

29. *Military Engineering*, Vol. XIII, Part VI—Survey Computations, 1978.
30. Oliver, J.G. and Clendinning, J. (1978) *Principles of Surveying*, Vol. 1, Plane Surveying. Van Nostrand Reinhold, London.
31. Ordnance Survey (1983) *Transverse Mercator Projection—Constants, Formulae and Methods*. Information Leaflet, HMSO, London.
32. Rainsford, H.F. (1958) *Survey Adjustments and Least Squares*. Constable, London.
33. Richardus, P. (1984) *Project Surveying*. North Holland, Amsterdam.
34. Rueger, J.M. (1982) *Introduction to Electronic Distance Measurement*. School of Surveying, University of New South Wales, Sydney.
35. Schut, G.H. (1957) *Analytical Aerial Triangulation at the National Research Council*. National Research Council of Canada, Ottawa.
36. Schut, G.H. (1969) Photogrammetric Refraction. *Photogrammetric Engineering* **35**(1), 79–86.
37. Schwendener, H.R. (1972) Electronic Distancers for Short Ranges: Accuracy and Checking Procedures. *Survey Review* **21**(164), 273–281.
38. Thompson, E.H. (1969) *Introduction to the Algebra of Matrices with Some Applications*. Adam Hilger, London.
39. US Army (1958) *Universal Transverse Mercator Grid*, Technical Manual TM 5–241–8, US Government Printing Office, Washington.
40. Veress, S.A. (1974) *Adjustment by Least Squares*. American Congress on Surveying and Mapping, Washington.
41. Williams, H.S. (1961) Triangulation by the Direction Method—a further note. *Empire Survey Review* **16**(119), 18–27.

Index